Advanced Inorganic Semiconductor Materials, 2nd Edition

Advanced Inorganic Semiconductor Materials, 2nd Edition

Guest Editors

Sake Wang
Minglei Sun
Nguyen Tuan Hung

MDPI

Basel • Beijing • Wuhan • Barcelona • Belgrade • Novi Sad • Cluj • Manchester

Guest Editors
Sake Wang
College of Science
Jinling Institute of Technology
Nanjing
China

Minglei Sun
Department of Physics
University of Antwerp
Antwerpen
Belgium

Nguyen Tuan Hung
Frontier Research Institute for
Interdisciplinary Sciences
Tohoku University
Sendai
Japan

Editorial Office
MDPI AG
Grosspeteranlage 5
4052 Basel, Switzerland

This is a reprint of the Special Issue, published open access by the journal *Inorganics* (ISSN 2304-6740), freely accessible at: https://www.mdpi.com/journal/inorganics/special_issues/4L19I7955Z.

For citation purposes, cite each article independently as indicated on the article page online and as indicated below:

Lastname, A.A.; Lastname, B.B. Article Title. *Journal Name* **Year**, *Volume Number*, Page Range.

ISBN 978-3-7258-4298-8 (Hbk)
ISBN 978-3-7258-4297-1 (PDF)
https://doi.org/10.3390/books978-3-7258-4297-1

Contents

About the Editors

Sake Wang

Sake Wang earned his Ph.D. in physics at the Southeast University, China, in 2016. During this period, he was awarded the national scholarship for doctoral students. Since 2021, he has been an Associate Professor at the Jinling Institute of Technology, China. He was a Visiting Scientist at Tohoku University, Japan, from 2019 to 2021. His current interests focus on theoretical studies of spin and valley transport, as well as valley-optoelectronic devices in two-dimensional materials. He is PIs of the National Science Foundation for Young Scientists of China and the Natural Science Foundation of Jiangsu Province. He has published 1 book (first author, IOP Publishing) and 76 papers with more than 4,100 citations; in 31 of these papers he was the first author or the corresponding author. Four of his first-authored and corresponding-authored papers are in the top 1% of ESI highly cited papers. He was ranked in the World's Top 2% most-cited scientists in 2023 and 2024 by Stanford University. In addition, he is an associate member of the IOP. He also served as an Associate Editor of the *Journal of Superconductivity and Novel Magnetism* (Springer Publishing) since 2020 and as a Guest Editor of the *Journal of Physics D: Applied Physics* (IOP Publishing) since 2023, as well as having been an outstanding reviewer of five *SCI* journals.

Minglei Sun

Dr. Minglei Sun is a research scientist at the Department of Materials Science and Engineering at the University of Texas in Dallas. His research focuses on first-principles studies of nanomaterials and their applications, particularly in energy conversion and storage. He is actively engaged in exploring the theoretical design and optimization of advanced materials for sustainable energy solutions and electronic devices. Currently, he is working on Bayesian optimization and nonlinear physics.

Nguyen Tuan Hung

Nguyen Tuan Hung received his Ph.D. in Physics from Tohoku University in March 2019. Since April 2019, he has been an Assistant Professor at the Frontier Research Institute for Interdisciplinary Sciences (FRIS), Tohoku University. He was a visiting scholar at the Chinese Academy of Sciences (2017). At the Massachusetts Institute of Technology (2023–2024), he was a researcher as a part of the Young Leaders Overseas Program. In addition, he received the Aoba Society Prize for the Promotion of Science from Tohoku University (2017), Research Fellowships for Young Scientists from the Japan Society for the Promotion of Science (2018), and Prominent Research Fellow from Tohoku University (2021–2025).

Preface

We are pleased to present this Special Issue, titled "Advanced Inorganic Semiconductor Materials, 2nd Edition", a continuation and expansion of the foundational work laid out in our previous edition. This reprint gathers a diverse and impactful collection of original research papers and topical reviews that reflect the rapid progress and growing sophistication in the field of inorganic semiconductors.

The scope of this Special Issue spans a broad range of cutting-edge topics, including two-dimensional materials, metal–halide perovskites, photocatalysts, transition metal dichalcogenides, nanostructured heterojunctions, dielectric and magnetic semiconductors, and advanced thin-film technologies. From fundamental studies on electronic and magnetic tuning via atomic intercalation, to application-oriented work on photocatalysis and optoelectronics, this Special Issue offers an integrated view of the field's current frontiers. Topical reviews further enrich the volume by addressing key trends and challenges in halide perovskite nanocrystals, single-crystal growth techniques, aluminum nitride film development, and correlated phase transitions in nitride systems.

The motivation behind this second edition is to showcase novel strategies for designing, synthesizing, and utilizing inorganic semiconductor materials across emerging applications, including energy conversion, light emission, environmental remediation, biomedical sensing, and spintronic devices. We aimed to assemble contributions that not only report significant experimental and theoretical advances, but also contextualize them within the broader framework of functional materials science.

This Special Issue is intended for a wide audience that includes materials scientists, condensed matter physicists, chemists, and engineers working in academia, research institutions, and the industry. We believe the contents will be equally valuable to graduate students seeking comprehensive updates and to senior researchers looking for new perspectives or collaborations.

We express our deep gratitude to all contributing authors from China, Mexico, Norway, Portugal, Romania, Saudi Arabia, and Vietnam for choosing this Special Issue as a venue for disseminating their outstanding research. Their work represents the global and interdisciplinary nature of modern materials science, and we are proud to feature their insights and innovations.

We also acknowledge the invaluable support of peer reviewers, editorial staff, and technical teams who made this publication possible through their thoughtful engagement and dedication.

We sincerely hope that you find the articles in this Special Issue as enlightening and inspiring as we did. We invite all readers and contributors to join us again for the forthcoming "Advanced Inorganic Semiconductor Materials, 3rd Edition", and continue advancing the frontier of this exciting research area.

Sake Wang, Minglei Sun, and Nguyen Tuan Hung
Guest Editors

inorganics

MDPI

Editorial

Advanced Inorganic Semiconductor Materials, 2nd Edition

Sake Wang [1,*], Minglei Sun [2] and Nguyen Tuan Hung [3]

1 College of Science, Jinling Institute of Technology, 99 Hongjing Avenue, Nanjing 211169, China
2 NANOlab Center of Excellence, Department of Physics, University of Antwerp, Groenenborgerlaan 171, 2020 Antwerp, Belgium; minglei.sun@uantwerpen.be
3 Frontier Research Institute for Interdisciplinary Sciences, Tohoku University, Sendai 980-8578, Japan; nguyen.tuan.hung.e4@tohoku.ac.jp
* Correspondence: isaacwang@jit.edu.cn

1. Introduction

Building upon our previous edition [1], this second edition of "Advanced Inorganic Semiconductor Materials" continues to explore the evolution of materials chemistry and physics in the realm of semiconductors [2,3]. This edition reflects a vibrant cross-section of contemporary research and review contributions focused on the synthesis, functional design, and applications of inorganic semiconductors. With emerging priorities such as energy sustainability, biocompatibility, spintronics, and optoelectronics [4], the 11 featured articles reflect the extraordinary depth and breadth of this field.

2. Semiconductor Architectures and Optical Phenomena

Two-dimensional metal-halide perovskites [5] are now redefining RTP. In a standout study (Contribution 1 [6]), the strategic tuning of alkyl chain lengths in organic amines resulted in perovskites (4-POMACC and 4-POEACC) exhibiting highly efficient RTP, particularly 4-POMACC with an extended lifetime (254 ms) and a 9.5% quantum yield. By highlighting intersystem crossing mechanisms via computational methods, the work provides a blueprint for next-generation photonic and encryption applications.

The work finds synergy with a broader review on halide perovskites (Contribution 2 [7]), which explores biomedical prospects through coated nanocrystals with enhanced biocompatibility and reduced toxicity. As both structural tunability and stability issues are addressed through innovative encapsulation strategies, the biomedical field is poised to benefit from these optoelectronic breakthroughs.

3. Semiconductors in Environmental Remediation and Energy

In the realm of environmental catalysis, photocatalysts incorporating biogenic materials and nanostructures stand out. A TiO_2-cuttlebone composite (Contribution 3 [8]) exhibited high solar-driven activity for tetracycline hydrochloride degradation, aided by the formation of carbonate radicals and a recyclability-focused 3D-printed photocatalytic device. Furthermore, the broccoli-like $Ag/Cu_2O/ZnO$ nanowire heterostructure [9] (Contribution 4 [10]) effectively degrades methyl orange under visible light, driven by plasmon-enhanced charge separation and p–n heterojunctions [11].

These studies underscore a clear trend: the hybridization of inorganic semiconductors with morphological or biological motifs can significantly enhance photocatalytic activity, opening doors to green chemistry and energy applications.

4. Tailoring Magnetic and Electronic Properties for Spintronics

The study by Liu et al. (Contribution 5 [12]) exemplifies the theoretical exploration of spintronic materials. By intercalating *3d* transition metals [13] into bilayer TMDs, a wide spectrum of magnetic phases—ranging from half-metals to ferromagnetic semiconductors—was uncovered via density functional theory. These materials, particularly V- and Cr-doped MoS_2 and VS_2, offer promising platforms for spin-filtering and magnetic logic devices.

CrN nanoparticles (Contribution 6 [14]) are also revisited in a thorough review focusing on correlated electronic and structural phase transitions [15]. By dissecting synthesis-dependent variability, the review establishes CrN as an excellent system for probing fundamental spin–lattice interactions and phase coupling phenomena. These insights can inform the design of robust multifunctional devices.

5. Functional Oxides and Thin Film Technologies

The use of NiO_x as a p-type semiconductor [16] in heterojunction photodiodes is elegantly explored in a practical device study (Contribution 7 [17]). The work highlights how deposition temperature and post-annealing influence responsivity and power efficiency. Notably, rapid thermal annealing substantially enhances photoresponse but introduces voltage memory effects, offering critical insight for sensor design.

Complementing this is the review on AlN thin films (Contribution 8 [18]), which chronicles advances in magnetron sputtering techniques. By discussing the interplay between sputtering parameters and film quality, the review underscores AlN's emerging importance in radio frequency and thermal management applications. Both papers reinforce the growing relevance of transition metal oxides [19] and nitrides in thin-film electronics.

6. Composite and Hybrid Semiconductor Systems

Hybrid systems often combine the best of inorganic and organic properties. A study on polysilane–barium titanate composites (Contribution 9 [20]) demonstrates how ultrasonication [21] can yield a ultraviolet-stable, semiconducting material with enhanced dielectric properties. The resulting composite exhibits promising electromagnetic shielding characteristics and improved conductivity, suggesting applications in electronics exposed to high-radiation or harsh environments.

The coupling of dielectric and semiconducting properties in such hybrid systems is increasingly vital, especially for applications in smart electronics and multifunctional materials. These approaches illustrate how a careful matrix-filler design can push materials performance beyond classical boundaries.

7. Crystallization and Dimensionality Control

The fabrication of single crystals is critical to achieving predictable and reproducible properties. The Bridgman method, comprehensively reviewed in Contribution 10 [22], has re-emerged as a powerful tool for growing high-quality metal halide single crystals. This review outlines the evolution of dimensional control (zero-dimensional to 3D) in halide structures, emphasizing crystal purity and application-specific tailoring. The method's resurgence stems from its capacity to overcome the limitations of solvent-based techniques, ensuring scalability and performance uniformity.

Alongside the 2D phosphorescent perovskites [23] in Contribution 1 [6], this review further validates the significance of dimensionality in tuning optical and electronic responses.

8. Perovskites at the Forefront

Three of the eleven papers (Contributions 1 [6], 2 [7], and 11 [24]) focus on perovskites [25], affirming their central role in this field. The comprehensive review by Tayari et al. (Contribution 11 [24]) discusses perovskite structure–property relationships and the roles of polarization, charge transport, and magnetic interactions. It ties together fundamental insights with device-level challenges such as scalability and environmental resilience. These reviews, in conjunction with focused studies on optical RTP performance and biocompatibility, position perovskites as true multifunctional materials with strong potential across sectors.

9. Conclusion: Toward a Unified Future in Inorganic Semiconductors

This second edition of "Advanced Inorganic Semiconductor Materials" reveals a dynamic and rapidly diversifying field. We see the merging of computational design, environmentally inspired synthesis, device engineering, and high-throughput fabrication. The 11 papers collected here capture a thematic convergence around tunability, dimensional control, hybrid functionality, and application-focused optimization.

From RTP-enabled cryptography (Contribution 1 [6]), to solar-assisted water purification (Contribution 3 [8]), and spintronic platforms via intercalated TMDs (Contribution 5 [12]), these contributions provide a glimpse into the future of electronics, energy, and biomedical technologies [26]. While challenges remain—particularly in stability, integration, and scalability—this volume affirms that the pursuit of tailored functionality in inorganic semiconductors is not only fruitful but essential.

This edition will stimulate continued interdisciplinary collaboration [27], inspiring both curiosity and application-driven exploration in materials science, which we sincerely hope will contribute to the next edition of this SI "Advanced Inorganic Semiconductor Materials: 3rd Edition" [28].

Funding: S.W. was funded by the China Scholarship Council (No. 201908320001), the Natural Science Foundation of Jiangsu Province (No. BK20211002), and Qinglan Project of Jiangsu Province of China. N.T.H. was funded by financial support from the Frontier Research Institute for Interdisciplinary Sciences, Tohoku University, Japan. M.S. was supported by funding from Research Foundation-Flanders (FWO; no. 12A9923N).

Acknowledgments: The authors thank all the staff in MDPI Publishing and the editors of *Inorganics* for establishing and running this SI, as well as reviewers around the globe who spent their valuable time thoroughly reviewing and improving the articles published in this SI. We also feel grateful to all the authors from China, Mexico, Norway, Portugal, Romania, Saudi Arabia, and Vietnam for choosing this SI to publish their excellent science. During the preparation of this manuscript, the authors used ChatGPT (OpenAI, GPT-4.5, May 2025 version) for language editing and grammar suggestions. The authors have thoroughly reviewed and edited the AI-generated output and take full responsibility for the content of this publication.

Conflicts of Interest: The authors declare no conflicts of interest.

Abbreviations

The following abbreviations are used in this manuscript:

2D	two-dimensional
3D	three-dimensional
AlN	aluminum nitride
RTP	room-temperature phosphorescence
SI	Special Issue
TMD	transition metal dichalcogenides

References

1. Wang, S.; Sun, M.; Hung, N.T. (Eds.) *Advanced Inorganic Semiconductor Materials*, 1st ed.; MDPI: Basel, Switzerland, 2024.
2. Reedijk, J.; Poeppelmeier, K.R. (Eds.) *Comprehensive Inorganic Chemistry II: From Elements to Applications*, 2nd ed.; Elsevier Ltd.: Amsterdam, The Netherlands, 2013.
3. Tian, H.; Ren, C.; Wang, S. Valleytronics in two-dimensional materials with line defect. *Nanotechnology* **2022**, *33*, 212001. [CrossRef]
4. Batabyal, S.K.; Pradhan, B.; Mohanta, K.; Bhattacharjee, R.R.; Banerjee, A. *Carbon Quantum Dots for Sustainable Energy and Optoelectronics*; Woodhead Publishing Series in Electronic and Optical Materials; Elsevier: Cambridge, UK, 2023.
5. Liu, Z.; Qin, X.; Chen, Q.; Jiang, T.; Chen, Q.; Liu, X. Metal–Halide Perovskite Nanocrystal Superlattice: Self-Assembly and Optical Fingerprints. *Adv. Mater.* **2023**, *35*, 2209279. [CrossRef] [PubMed]
6. Wang, S.; Zhu, H.; Sheng, M.; Shao, B.; He, Y.; Liu, Z.; Li, M.; Zhou, G. Achieving Ultralong Room-Temperature Phosphorescence in Two-Dimensional Metal-Halide Perovskites by Tuning Alkyl Chain Length. *Inorganics* **2025**, *13*, 108. [CrossRef]
7. Wang, G.; Qi, Y.; Zhou, Z.; Liu, Z.; Wang, R. Research Progress of Halide Perovskite Nanocrystals in Biomedical Applications: A Review. *Inorganics* **2025**, *13*, 55. [CrossRef]
8. Li, Q.; Liu, P.; Lin, H.; Xue, H.; Mao, J. A Novel TiO_2-Cuttlebone Photocatalyst for Highly Efficient Catalytic Degradation of Tetracycline Hydrochloride. *Inorganics* **2024**, *12*, 319. [CrossRef]
9. Manh, D.H.; Thanh, T.D.; Kim, D.H.; Phan, T.L. Increased magnetocaloric response of $La_{2/3}Ca_{1/3}MnO_3$/Gd nanocomposites in a large temperature range. *Curr. Appl. Phys.* **2024**, *60*, 70–78. [CrossRef]
10. Thu, P.T.; Bach, T.N.; Le Phong, T.; Tung, D.H.; Ky, V.H.; Tung, D.K.; Lam, V.D.; Manh, D.H.; Dan, N.H.; Anh, T.X.; et al. Synthesis and Characterization of Broccoli-like Ag/Cu_2O Nanostructures on ZnO Nanowires Using the Plasma–Liquid Interaction Method. *Inorganics* **2024**, *12*, 80. [CrossRef]
11. Do, H.T.; Nguyen, D.B.; Nguyen, T.H.P.; Le, T.T.H.; Nguyen, T.T.T. Increasing the flame resistance of cellulose by nonthermal plasma coupled with chemical treatments. *Appl. Phys. Express* **2025**, *18*, 026001. [CrossRef]
12. Liu, Y.; Yang, G.; He, Z.; Wang, Y.; Niu, X.; Wang, S.; Liu, Y.; Zhang, X. Tunable Electronic and Magnetic Properties of 3*d* Transition Metal Atom-Intercalated Transition Metal Dichalcogenides: A Density Functional Theory Study. *Inorganics* **2024**, *12*, 237. [CrossRef]
13. Wang, S.; Yu, J. Magnetic Behaviors of 3d Transition Metal-Doped Silicane: a First-Principle Study. *J. Supercond. Nov. Magn.* **2018**, *31*, 2789–2795. [CrossRef]
14. Alam, K. Synthesis and Study of Correlated Phase Transitions of CrN Nanoparticles. *Inorganics* **2024**, *12*, 247. [CrossRef]
15. Alam, K.; Ponce-Pérez, R.; Sun, K.; Foley, A.; Takeuchi, N.; Smith, A.R. Study of the structure, structural transition, interface model, and magnetic moments of CrN grown on MgO(001) by molecular beam epitaxy. *J. Vac. Sci. Technol. A* **2023**, *41*, 053411. [CrossRef]
16. Mateos-Anzaldo, D.; Nedev, R.; Perez-Landeros, O.; Curiel-Alvarez, M.; Castillo-Saenz, J.; Arias-Leon, A.; Valdez-Salas, B.; Silva-Vidaurri, L.; Martinez-Guerra, E.; Osorio-Urquizo, E.; et al. High-performance broadband photodetectors based on sputtered NiO_x/n-Si heterojunction diodes. *Opt. Mater.* **2023**, *145*, 114422. [CrossRef]
17. Nedev, R.; Mateos-Anzaldo, D.; Osuna-Escalante, E.; Perez-Landeros, O.; Curiel-Alvarez, M.; Osorio-Urquizo, E.; Castillo-Saenz, J.; Lopez-Medina, J.; Valdez-Salas, B.; Nedev, N. Effect of Deposition Temperature and Thermal Annealing on the Properties of Sputtered NiO_x/Si Heterojunction Photodiodes. *Inorganics* **2025**, *13*, 11. [CrossRef]
18. Jadoon, N.A.; Puvanenthiram, V.; Mosa, M.A.; Sharma, A.; Wang, K. Recent Advances in Aluminum Nitride (AlN) Growth by Magnetron Sputtering Techniques and Its Applications. *Inorganics* **2024**, *12*, 264. [CrossRef]
19. Feng, T.; Luo, X.; Liu, Z.; Liu, X.; Yan, X.; Li, G.; Zhang, W.; Wang, K. Nanoarchitectonics of highly flexible iron-oxide nanoporous electrodes on stainless steel substrate for wearable supercapacitors. *Appl. Phys. Rev.* **2024**, *11*, 041414. [CrossRef]
20. Rotaru, R.; Fortună, M.E.; Ungureanu, E.; Sacarescu, L. Polysilane–Barium Titanate Polymeric Composite Obtained through Ultrasonication. *Inorganics* **2024**, *12*, 213. [CrossRef]
21. Rotaru, R.; Fortună, M.E.; Ungureanu, E.; Brezuleanu, C.O. Effects of Ultrasonication in Water and Isopropyl Alcohol on High-Crystalline Cellulose: A Fourier Transform Infrared Spectrometry and X-ray Diffraction Investigation. *Polymers* **2024**, *16*, 2363. [CrossRef]
22. Zhu, H.; Wang, S.; Sheng, M.; Shao, B.; He, Y.; Liu, Z.; Zhou, G. Bridgman Method for Growing Metal Halide Single Crystals: A Review. *Inorganics* **2025**, *13*, 53. [CrossRef]
23. Liu, Z.; Qin, X.; Chen, Q.; Chen, Q.; Jing, Y.; Zhou, Z.; Zhao, Y.S.; Chen, J.; Liu, X. Highly Stable Lead-Free Perovskite Single Crystals with NIR Emission Beyond 1100 nm. *Adv. Opt. Mater.* **2022**, *10*, 2201254. [CrossRef]
24. Tayari, F.; Teixeira, S.S.; Graca, M.P.F.; Nassar, K.I. A Comprehensive Review of Recent Advances in Perovskite Materials: Electrical, Dielectric, and Magnetic Properties. *Inorganics* **2025**, *13*, 67. [CrossRef]
25. Tayari, F.; Iben Nassar, K.; Benamara, M.; Essid, M.; Soreto Teixeira, S.; Graça, M.P.F. Sol–gel synthesized $(Bi_{0.5}Ba_{0.5}Ag)_{0.5}$ $(NiMn)_{0.5}O_3$ perovskite ceramic: An exploration of its structural characteristics, dielectric properties and electrical conductivity. *Ceram. Int.* **2024**, *50*, 11207–11215. [CrossRef]

26. Roque, A.C.A.; Gracanin, D.; Lorenz, R.; Tsanas, A.; Bier, N.; Fred, A.; Gamboa, H. (Eds.) Biomedical Engineering Systems and Technologies. In *Communications in Computer and Information Science, Proceedings of the 15th International Joint Conference, BIOSTEC 2022, Virtual Event, 9–11 February 2022*; Revised Selected Papers; Springer: Cham, Switzerland, 2023; Volume 1814.
27. Wang, S.; Tian, H. *Two-Dimensional Valleytronic Materials: From Principles to Device Applications*; IOP Publishing: Bristol, UK, 2025.
28. Inorganics | Special Issue: Advanced Inorganic Semiconductor Materials: 3rd Edition. Available online: https://www.mdpi.com/journal/inorganics/special_issues/XU563P063S (accessed on 13 May 2025).

inorganics

MDPI

Article

Achieving Ultralong Room-Temperature Phosphorescence in Two-Dimensional Metal-Halide Perovskites by Tuning Alkyl Chain Length

Suqin Wang [1,†], Hui Zhu [1,†], Ming Sheng [1], Bo Shao [1], Yu He [1], Zhuang Liu [2,*], Min Li [3] and Guangtao Zhou [1,*]

1 College of Engineering, Shandong Xiehe University, Jinan 250109, China
2 School of Materials Science and Engineering, Shandong University, Jinan 250061, China
3 School of Materials Science and Engineering, Nanjing University of Posts & Telecommunications, Nanjing 210023, China
* Correspondence: liu.zhuang@sdu.edu.cn (Z.L.); charlie@sdxiehe.edu.cn (G.Z.)
† These authors contributed equally to this work.

Abstract: Two-dimensional (2D) metal-halide perovskites with highly efficient room-temperature phosphorescence (RTP) are rare due to their complex structures and intricate intermolecular interactions. In this study, by varying the alkyl chain length in organic amines, we synthesized two 2D metal-halide perovskites, namely 4-POMACC and 4-POEACC, both of which exhibit significant RTP emission. Notably, 4-POMACC demonstrates a stronger green RTP emission with a significantly longer lifetime (254 ms) and a higher photoluminescence quantum yield (9.5%) compared to 4-POEACC. A thorough investigation of structural and optical properties reveals that shorter alkyl chains can enhance the optical performance due to reduced molecular vibrations and more effective exciton recombination. Computational calculations further show that the smaller energy gap between S_1 and T_n in 4-POMA facilitates intersystem crossing, thereby improving RTP performance. Based on their remarkable phosphorescence properties, we demonstrated their applications in information encryption. This work offers a novel design strategy that could inspire the development of next-generation RTP materials.

Keywords: room-temperature phosphorescence; two-dimensional; metal-halide perovskites; chain length; information encryption

1. Introduction

Room-temperature phosphorescence (RTP) arising from the triplet excitons of organic molecules has garnered increased attention due to its substantial potential in a wide range of research fields, including display and lighting, secure encryption, chemical sensing, and information storage [1–5]. Among the various luminescent material systems, metal-halide perovskites offer numerous opportunities for the development of RTP materials, due to their wide-ranging structural and compositional versatility [6–10]. In particular, two dimensional (2D) metal-halide perovskites can be readily synthesized through a straightforward and low-temperature process, presenting an obvious advantage over organic materials [11–16]. Moreover, the rigid structure of 2D metal-halide perovskites helps reduce molecular vibrations, thereby suppressing non-radiative recombination and enhancing exciton emission [17–20]. Usually, the choice of metal cation largely determines the electronic structure of the 2D metal-halide perovskite and, hence, its emissive properties. In recent years, a number of Cd-based 2D metal-halide perovskites have been explored

https://doi.org/10.3390/inorganics13040108

mainly because of their interesting structural characteristics and the large band gap, which suggest that they may be good candidates for application as a luminescence matrix [21–24].

Organic spacer cations play a key role in the physicochemical properties of 2D metal-halide perovskites, that their size and chain length is crucial in altering the inorganic octahedra framework and determining the band gap [25–27]. In the past, extensive research has been conducted on the impact of the organic component chain length on the fluorescent properties of 2D metal-halide perovskite materials. Ye et al. investigated the photoluminescence properties of 2D metal-halide perovskite crystals synthesized with organic amines with different alkyl chain lengths [28]. The results prove that the chain length of organic amines can affect the arrangement of the inorganic layer octahedra, and the spectrum exhibits an obvious blue shift with extending alkyl chain. This study indicates that the alkyl chain length of the organic amines can be used to tune the structure and luminescent properties of 2D metal-halide perovskites. In addition, Liu et al. used two organic amines with different alkyl chain lengths as the organic components, and lead bromide as the inorganic component to synthesize two 2D metal-halide perovskites [29]. The photoluminescence from the as-prepared samples is attributed to the recombination of free excitons and self-trapped excitons. Moreover, the deeper exploration of the emission mechanisms demonstrates that by varying the alkyl chain length of the organic amines, the energy levels of the free and self-trapped excitons can be tuned accordingly. Furthermore, Palstra et al. found that the different chain lengths of alkyl in the organic amines influenced the optical bandgap of materials by varying the framework of the inorganic layer [30]. For example, short-chain organic amines resulted in corner-shared octahedra, while long-chain amines lead to both corner-sharing and face-sharing octahedra. Therefore, adjusting the alkyl chain length of organic amines is a powerful tool for tuning the optical band gap and luminescent properties of 2D metal-halide perovskites. However, investigations on the modulation of RTP in 2D metal-halide perovskites using organic amines with varying alkyl chain lengths are still relatively scarce.

In this work, two organic amines with different alkyl chain lengths, namely 4-phenoxybenzylamine (4-POMA) and 4-phenoxyphenethylamine (4-POEA), are selected as the organic components, with $CdCl_2$ as the inorganic component. Using a simple synthesis method, two 2D metal-halide perovskite materials (4-POMACC and 4-POEACC) with distinct RTP properties are synthesized. Photophysical characterizations reveal that both materials exhibit dual-peak emissions, while there are notable differences in their RTP lifetimes. Specifically, 4-POMACC has lifetimes of 254 ms, whereas 4-POEACC has a lifetime of only 68 ms. Additionally, the quantum efficiencies of the two materials are 9.50% and 8.06%, respectively. The interlayer spacing calculations are employed to further investigate the origins of their RTP characteristics. Based on their RTP properties, the materials are explored for anti-counterfeiting applications. The research findings of this work demonstrate that the alkyl chain length of organic amines has a significant impact on the RTP properties of 2D metal-halide perovskites.

2. Results and Discussion

By performing powder X-ray diffraction (PXRD) analysis on 4-POMACC and 4-POEACC, the layered structure characteristic of 2D metal-halide perovskites can be confirmed. As shown in Figure 1, PXRD patterns indicate that the synthesized materials exhibit structural features similar to those reported for 2D metal-halide perovskites, that both 4-POMACC and 4-POEACC display strong periodic diffraction peaks corresponding to the (00l) crystal planes, with no significant diffraction peaks observed from other crystal planes. This result suggests that both materials preferentially grow along the c-axis, favoring the formation of a layered structure with alternating inorganic and organic layers [31,32]. The diffraction peaks of the (00l) planes are sharp and intense, indicating that both materials possess good crystallinity. In Figure 1a, when comparing the PXRD pattern of 4-POMACC with its corresponding organic amine salt (4-POMACl) and inorganic component ($CdCl_2$), it is observed that the diffraction peaks of 4-POMACC are slightly shifted relative to those of 4-POMACl and do not correspond to the peak positions of $CdCl_2$. This result confirms the successful synthesis of a well-crystallized 2D metal-halide perovskite material. Similarly, in Figure 1b, 4-POEACC exhibits a clear layered structural feature with improved crystallinity and orientation compared to 4-POEACl, and it does not display the characteristic diffraction peaks of $CdCl_2$.

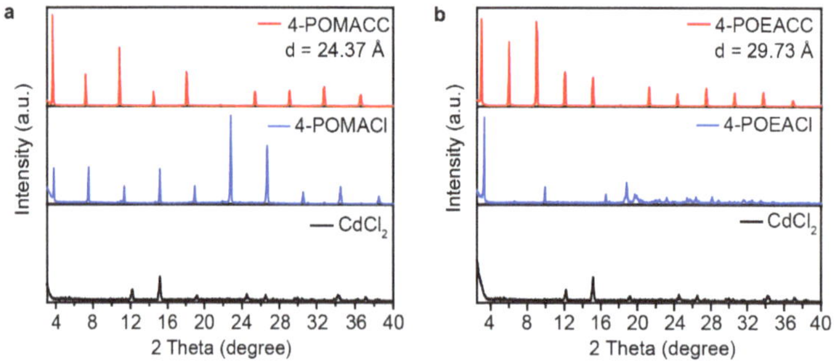

Figure 1. PXRD patterns of (**a**) 4-POMACC and (**b**) 4-POEACC, as well as their corresponding organic amine salts (4-POMACl and 4-POEACl) and inorganic salts ($CdCl_2$).

Two-dimensional metal-halide perovskites possess a layered structure, and one of their key characteristics is described by interlayer spacing (d), which refers to the distance between adjacent inorganic layers within the material's structure. This spacing can be calculated using PXRD data and Bragg's equation ($2d\sin\theta = n\lambda$, where θ denotes the diffraction angle, λ stands for the X-ray wavelength, and n refers to the diffraction order). The calculated results in this work are 24.37 Å for 4-POMACC and 29.73 Å for 4-POEACC. This difference arises because 4-POEACl contains an additional methyl group compared to 4-POMACl. Therefore, the larger organic molecule leads to greater interlayer spacing and looser packing of organic molecules. As a result, the molecular vibration becomes stronger, which is detrimental to luminescence [33].

The vibrational peak positions obtained from Fourier transform infrared (FTIR) spectroscopy can be used to identify specific functional groups and chemical bonds within a material, allowing for further inference about the internal structure and vibrational modes of organic components. As shown in Figure S3, all four species (4-POMACl, 4-POEACl, 4-POMACC, and 4-POEACC) exhibit a hydroxyl (-OH) vibrational peak between 3400 and 3500 cm^{-1}, likely due to water absorption from the air. Comparing the FTIR patterns of 4-POMACC and 4-POEACC, one can find that both of them show the presence of amino ($-NH_2$) vibrational peaks, corresponding to the vibrational modes of the functional groups in the reactants. In Figure S3a, the N-H vibrational peak of 4-POMACC (3176 cm^{-1}) is shifted to a higher frequency compared to that in 4-POMACl (2995 cm^{-1}), which is because the amino group ($-NH_3^+$) in 4-POMACC forms hydrogen bonds (N-H$\cdots$Cl) with the inorganic layer. The formation of hydrogen bonds generally reduces the vibrational frequency (or peak value) of N-H. In contrast, in 4-POMACl, the $-NH_3^+$ group forms ionic bonds with free Cl^- ions, resulting in a stronger ionic interaction compared to the hydrogen bonding in 4-POMACC. This difference in bond types causes a noticeable shift in the N-H vibrational peaks. These observations further confirm that in 4-POMACC, the organic and inorganic layers are connected through N-H$\cdots$Cl hydrogen bonds, which facilitate the formation of the 2D perovskite structure. Similarly, Figure S3b demonstrates that 4-POEACC exhibits the same structural characteristic, indicating the presence of N-H···Cl hydrogen bonds as well.

Scanning electron microscopy (SEM) is used to observe the microstructure of materials, providing insight into their morphological characteristics. As shown in Figure 2a,b, both 4-POMACC and 4-POEACC exhibit a layered structure formed by the stacking of sheet-like growths, indicating that the synthesized materials possess the layered characteristic [34]. Notably, the sheets of 4-POEACC appear to be thicker than those of 4-POMACC, likely due to the increased carbon chain length in the organic component, which may lead to poorer sheet formation and a tendency for agglomeration. Elemental mapping, as shown in Figure 2c,d, reveals that the elements Cd, Cl, and C are evenly distributed throughout in both 2D metal-halide perovskites. To further confirm the 2D structure of the synthesized materials, energy-dispersive X-ray spectroscopy (EDS) analysis was performed. As seen in Figure 2e,f, the actual molar ratios of Cd to Cl are 1:4.03 and 1:3.51, respectively, which are close to the theoretical molar ratio of 1:4 for 2D metal-halide perovskites. This result confirms that both materials exhibit the characteristic features of 2D metal-halide perovskites, and their chemical formulas are $(4\text{-POMA})_2CdCl_4$ and $(4\text{-POEA})_2CdCl_4$, respectively.

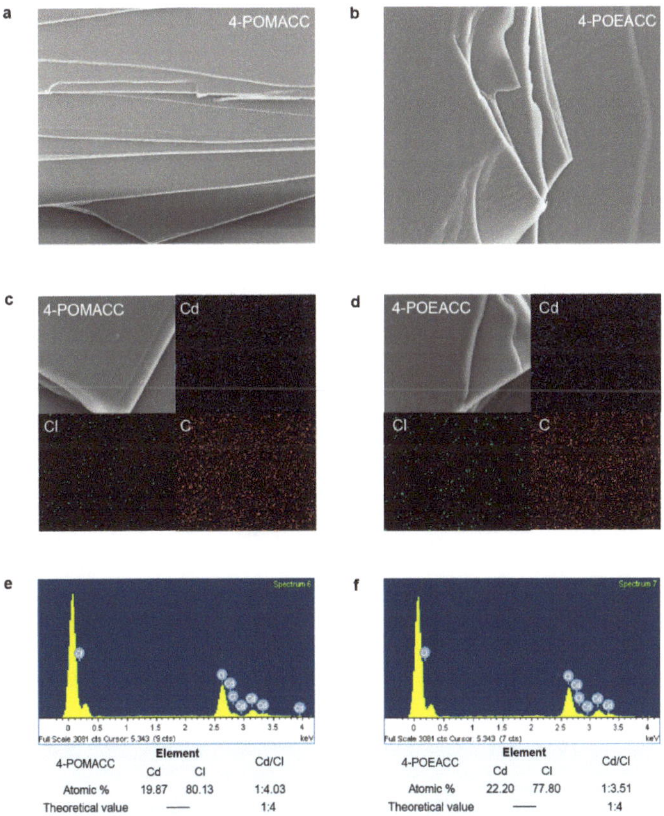

Figure 2. SEM images of (**a**) 4-POMACC and (**b**) 4-POEACC. Element mapping results of (**c**) 4-POMACC and (**d**) 4-POEACC. Energy-dispersive X-ray spectroscopy results of (**e**) 4-POMACC and (**f**) 4-POEACC. The picture size in (**a**,**b**) and (**c**,**d**) are 10 × 10 um and 2 × 2 um, respectively.

Thermal stability is a key property of metal-halide perovskites, in that the greater the thermal stability, the more capable the metal-halide perovskite is of maintaining long-term performance in high-temperature environments [35]. From the thermogravimetric analysis (TGA) shown in Figure S4a,c, it is evident that the initial thermal decomposition temperature of 4-POMACC is 263 °C, which is 41 °C higher than that of 4-POMACl. This improvement in thermal stability can be attributed to the formation of the 2D metal-halide structure in 4-POMACC compared to 4-POMACl, which increases the rigidity of the materials. The derivative thermogravimetry (DTG) curve of 4-POMACl reveals two distinct weight loss peaks, which are due to the two-step decomposition of the organic components. The absence of a clear weight loss plateau suggests that these two decomposition processes overlap. In contrast, the DTG curve of 4-POMACC shows an additional weight loss peak, which is attributed to the decomposition of the inorganic components. This further confirms that 4-POMACC is a 2D metal-halide perovskite material, formed by the self-assembly of both organic and inorganic components. In detail, 4-POMACC exhibits a weight loss of 5.96% in the temperature range of 263 °C to 297 °C, and this loss reaches 44.43% at 589 °C. It is speculated that the lost groups transform from HCl and NH_2 to $C_{13}H_{11}O$. Beyond 589 °C, reactions involving the inorganic components occur, including the volatilization and decomposition of $CdCl_2$, followed by the oxidation of elemental cadmium in the presence of air. For 4-POMACl, nearly all organic components decompose before 603 °C, which is close to the decomposition temperature (589 °C) of organic components in 4-POMACC. Similarly, 4-POEACC decomposes at 253 °C, a lower temperature compared to 4-POMACC, indicating that 4-POMACC has good thermal stability. Furthermore, as seen in Figure S4b,d, the decomposition temperature of 4-POEACC is 38 °C higher than that of 4-POEACl, and like 4-POMACC, it undergoes decomposition of both organic and inorganic components. This suggests that 4-POEACC is also a 2D metal-halide perovskite material formed through the self-assembly of organic and inorganic constituents.

We further study the optical properties of these two 2D metal-halide perovskite materials. Figure 3a,b display the solid-state UV diffuse reflectance spectra of 4-POMACC and 4-POEACC. In the 200 nm to 400 nm range, 4-POMACC exhibits strong absorption, while in the visible light region from 400 nm to 800 nm, it shows weaker absorption, with the most intense absorption peak at 275 nm. Similarly, 4-POEACC shows absorption in the 200 nm to 400 nm range, comparable to that of 4-POMACC, but its absorption in the visible region is limited to the 400 nm to 600 nm range, with the strongest absorption peak at 280 nm. When comparing the absorption peak positions of the two materials, it is observed that the peak position of 4-POEACC is red-shifted relative to that of 4-POMACC. This red shift can be attributed to the longer alkyl chain of the organic amine in 4-POEACC, which enhances the intermolecular interactions among the organic amine chains [36]. As a result, the coordination field of the halide ion is weakened, leading to a reduction in the splitting energy of the inorganic cadmium ions, thereby causing the absorption peak to shift to a longer wavelength. In addition, Tauc plots were further conducted to determine the optical band gaps of these two materials. As shown in Figure 3c,d, their indirect optical band gaps are 4.25 eV for 4-POMACC and 4.19 eV for 4-POEACC, respectively. The close similarity in the indirect optical band gaps of these two materials can be attributed to their analogous coordination environments and symmetry, especially evident in the inorganic layers, where both materials are composed of $CdCl_2$. Additionally, the organic components of the two materials are nearly identical, differing only by a single $-CH_2$ group.

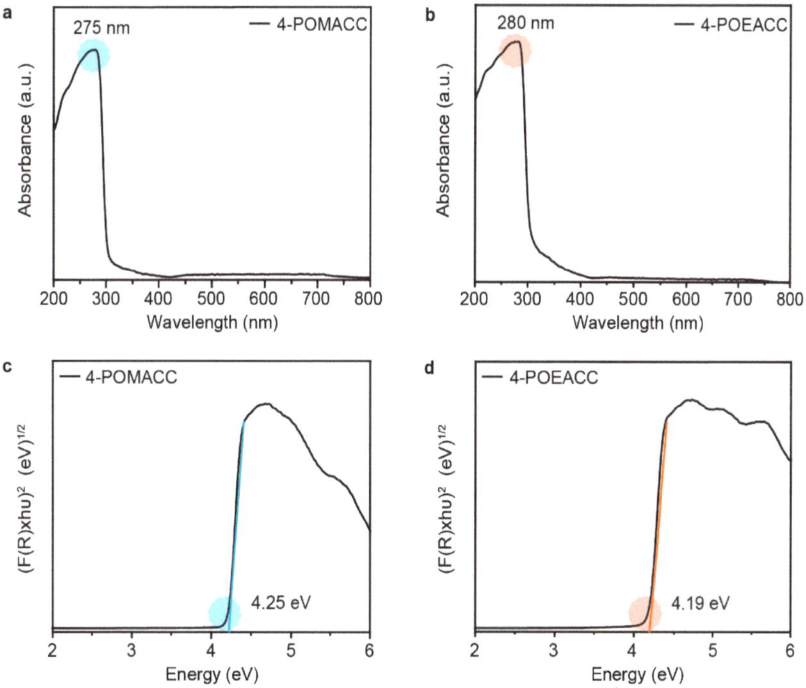

Figure 3. Diffuse reflectance spectra of (**a**) 4-POMACC and (**b**) 4-POEACC. Tauc plots of (**c**) 4-POMACC and (**d**) 4-POEACC according to their absorption spectra.

To investigate the effect of different organic amines on the luminescent properties of the materials, we then conducted photophysical characterizations. Phosphorescence spectra were obtained using an Edinburgh FLS980 fluorescence spectrophotometer with a 10 ms delay time after excitation using a microsecond flash lamp. Figure 4a displays the steady-state photoluminescence spectrum (black curve) and phosphorescence spectrum (cyan curve) of 4-POMACC. The steady-state photoluminescence spectrum was measured under excitation at a wavelength of 270 nm, while the phosphorescence spectrum was recorded under excitation at 286 nm. The steady-state spectrum of 4-POMACC shows a sharp emission peak at 305 nm, along with a weaker emission band in the 450–700 nm range, which may be attributed to the phosphorescence. Figure 4b shows the steady-state photoluminescence spectrum (brown curve) and phosphorescence spectrum (green curve) of 4-POEACC. The version 4-POEACC exhibits a sharp emission peak at 310 nm but lacks the long-wavelength emission observed in 4-POMACC. After turning off the UV lamp, 4-POMACC exhibits a green RTP. The phosphorescence spectrum reveals a dual-peak emission at 495 nm and 595 nm, with the intensity of the first peak being stronger. Although 4-POEACC's RTP is barely visible to the naked eye, it also displays dual-peak emissions at 505 nm and 615 nm, but with an intensity pattern opposite to that of 4-POMACC. Comparing the emission peaks of 4-POMACC and 4-POEACC, it is evident that both the fluorescence and phosphorescence emissions exhibit red shifts as the alkyl chain length of the organic amine increases, consistent with the red shift observed in their respective absorption spectra.

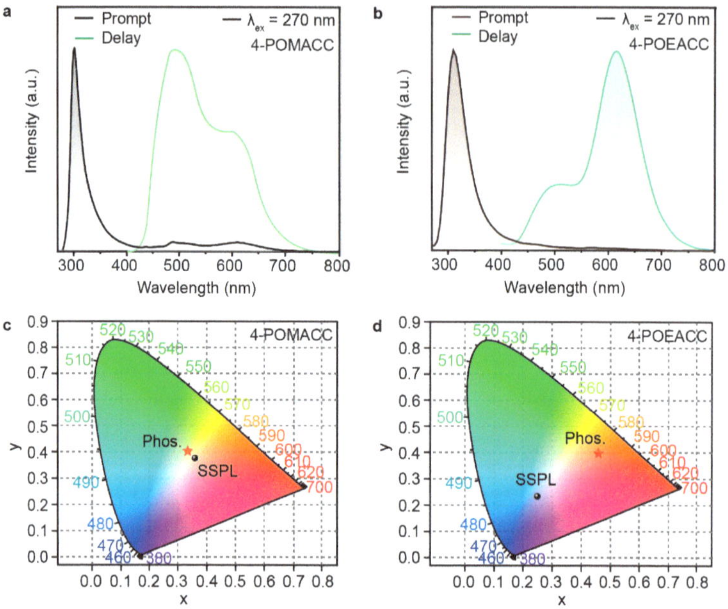

Figure 4. Steady-state photoluminescence and phosphorescence spectra of (**a**) 4-POMACC and (**b**) 4-POEACC excited at 270 nm. CIE coordinate diagrams of (**c**) 4-POMACC and (**d**) 4-POEACC.

To gain a clearer understanding of the colors exhibited by the materials, we used the Commission Internationale de l'Eclairage (CIE) standard color measurement system to process and calculate the steady-state and phosphorescence spectra data. As shown in Figure 4c,d, the fluorescence color coordinates of 4-POMACC are (0.36, 0.38), corresponding to a hue that leans toward the orange–yellow region. The phosphorescence color coordinates are (0.33, 0.41), indicating a hue that leans toward the yellow–green region, with both fluorescence and phosphorescence occupying similar color spaces. In contrast, 4-POEACC exhibits fluorescence color coordinates of (0.25, 0.23), placing its hue in the ultraviolet region. The phosphorescence color coordinates are (0.46, 0.40), indicating a shift toward the orange–red region. The significant difference between the fluorescence and phosphorescence color regions for 4-POEACC contrasts sharply with the close color proximity observed in 4-POMACC.

Time-resolved spectroscopy was then employed to measure the lifetimes of these two materials, providing a metric for comparing the RTP characteristics of the two samples. Figure 5a,c display the fluorescence and phosphorescence decay curves of 4-POMACC. As illustrated, the fluorescence emission peak at 305 nm exhibits a lifetime of only 1.98 ns, characteristic of singlet exciton radiative recombination. In contrast, the phosphorescence emissions at 495 nm and 595 nm have lifetimes of 322 ms and 240 ms, respectively, indicative of triplet exciton radiative recombination. Similarly, Figure 5b,d show the fluorescence and phosphorescence decay curves of 4-POEACC. The fluorescence emission peak at 310 nm has a lifetime of 1.12 ns, also representative of singlet exciton emission. However, the phosphorescence emissions at 505 nm and 615 nm exhibit much shorter lifetimes of 68 ms and 33 ms, respectively, reflecting triplet exciton emission. As a result, for both fluorescence and phosphorescence, 4-POMACC demonstrates significantly longer lifetimes than 4-POEACC.

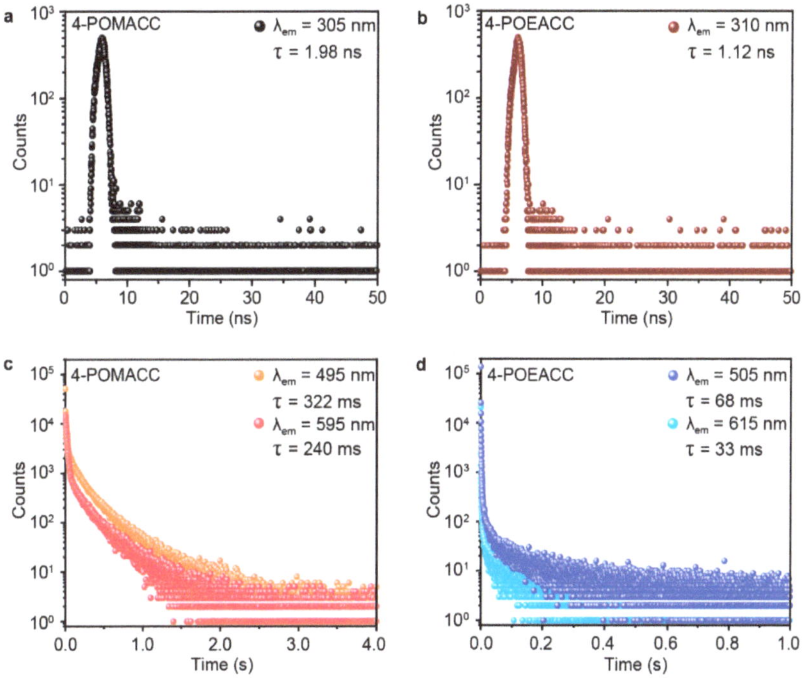

Figure 5. Time-resolved spectroscopy of (**a**,**b**) fluorescence and (**c**,**d**) phosphorescence for 4-POMACC (**left**) and 4-POEACC (**right**).

To provide a more comprehensive comparison of phosphorescence performance across the entire spectral range, time-resolved emission spectroscopy (TRES) was conducted. Figure S5a,b present the TRES patterns for 4-POMACC and 4-POEACC under 286 nm and 284 nm excitation, respectively. The results reveal that both 4-POMACC and 4-POEACC exhibit a gradual decrease in emission intensity over time. Notably, 4-POMACC maintains an emission intensity at level 3 even after a delay of 2 s. In contrast, 4-POEACC reaches the same intensity level only after a delay of 0.5 s. This comparison demonstrates that 4-POMACC has a longer phosphorescence lifetime across the entire spectral range.

The photoluminescence quantum yield (PLQY) is defined as the ratio of the number of emitted photons to the number of absorbed photons, providing a direct measure of the material's luminescence intensity [37]. As illustrated in Figure S6a,b, the PLQY of 4-POMACC increased from 6.00% for 4-POMACl to 9.50%. This enhancement is attributed to the rigid structure of 4-POMACC. The PLQY of 4-POEACC improved compared to 4-POEACl, as shown in Figure S6c,d. It is speculated that the introduced inorganic halide framework can effectively suppress the vibrations of organic molecules and reduce non-radiative recombination [38].

To elucidate the origin of the luminescence in these two 2D metal-halide perovskites, we performed spectral comparisons between the materials and their corresponding organic amine salts. Figure 6a,b show the fluorescence and phosphorescence spectra for 4-POMACC and 4-POMACl. It is observed that the fluorescence emission peak of 4-POMACC is centered around 305 nm, which is similar to the fluorescence emission peak of 4-POMACl. Additionally, 4-POMACC exhibits dual peaks of phosphorescence around 495 nm and 595 nm, whereas 4-POMACl displays a single peak of phosphorescence centered at 495 nm. This indicates that the fluorescence at 305 nm and the phosphorescence at 495 nm in 4-POMACC originate from the organic amine salt 4-POMACl. Based on previous reports on perovskite defect luminescence, the phosphorescence emission peak at 595 nm is likely due to defect emissions caused by distortions in the inorganic lattice [39]. Similarly, as shown in Figure 6c,d, the fluorescence emission at 310 nm in 4-POEACC is attributed to the organic amine salt 4-POEACl, while the phosphorescence emissions at 505 nm and 615 nm are sourced from the organic amine salt 4-POEACl, with the 615 nm emission also being ascribed to defect emission due to lattice distortions in the inorganic layer. Note that, the fluorescence spectra of the two compounds in this work appear similar because their emission originates from A-site organic parts, but not from the bandgap emission, which is similar to previously reported Cd-based organic–inorganic hybrid halide perovskites, such as ABA_2CdCl_4 and $(F_2CHCH_2NH_3)_2CdBr_4$ [40,41].

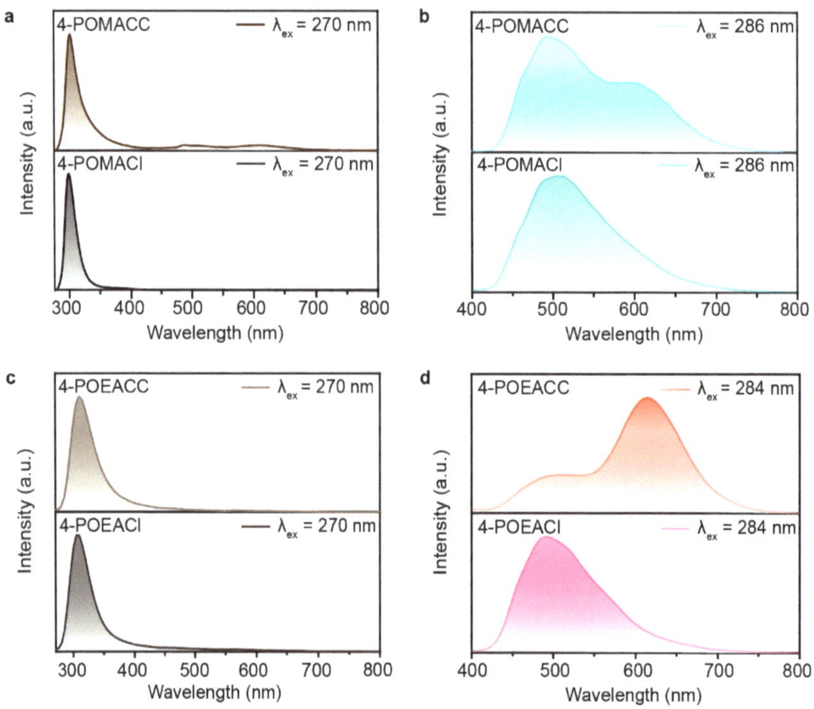

Figure 6. (**a**) Steady-state spectra of 4-POMACC and 4-POMACl. (**b**) Phosphorescence spectra of 4-POMACC and 4-POMACl. (**c**) Steady-state spectra of 4-POEACC and 4-POEACl. (**d**) Phosphorescence spectra of 4-POEACC and 4-POEAC.

To further verify that the 595 nm emission peak of 4-POMACC and the 615 nm emission peak of 4-POEACC are due to defect emission from lattice distortions in the inorganic layers, we conducted spectral tests under varying excitation powers. The relationship between emission intensity and excitation power was analyzed using a power law equation ($y = y_0 + x^n$), where n is a parameter that indicates the type of recombination mechanism [42]. When n is less than 1, it typically corresponds to defect state recombination; when n ranges from 1 to 2, it indicates free exciton recombination; and when n is equal to 2, it is associated with band-to-band recombination. From Figure S7, the fitting results show that the value of n for 4-POMACC is 0.84 and for 4-POEACC is 0.92. Since both values of n are less than 1, this suggests that the 595 nm emission in 4-POMACC and the 615 nm emission in 4-POEACC can be attributed to defect state luminescence.

To investigate the effects of excitation intensity and duration on the optical properties of materials, we conducted corresponding kinetic tests. As shown in Figure 7a,b, the emission intensity of both 4-POMACC and 4-POEACC increases with rising radiation flux. At 10% of the radiation flux (with an excitation intensity of 30 μW/cm^2), both materials can be effectively excited. Additionally, as shown in Figure 7c,d, both 4-POMACC and 4-POEACC exhibit stable luminescence during the 10 s excitation period; however, 4-POMACC shows greater stability compared to 4-POEACC during excitation. After 10 s of excitation, 4-POMACC displays a distinct exponential decay, whereas 4-POEACC shows a less pronounced exponential decay. Furthermore, as the excitation duration is extended, the emission intensity of both 4-POMACC and 4-POEACC also increases. This outcome suggests that the optical properties of these materials are influenced by both radiation flux and excitation duration.

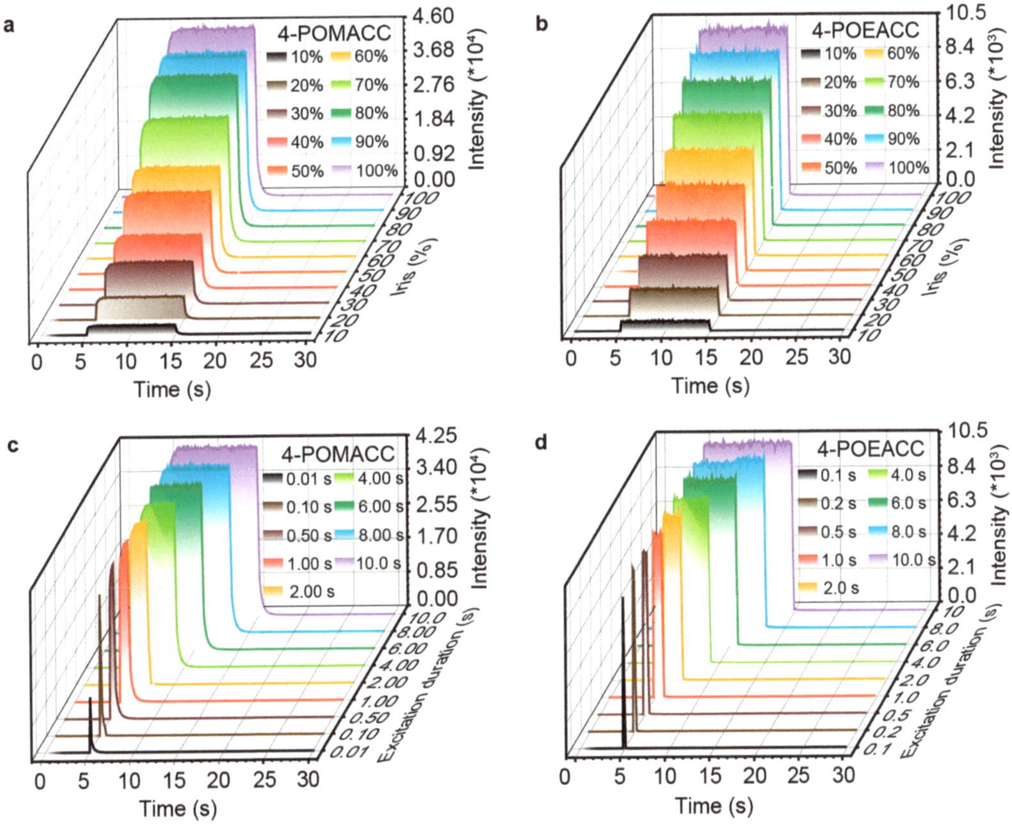

Figure 7. Emission intensity of 4-POMACC as functions of (**a**) radiation intensity and (**c**) excitation duration (excitation at 286 nm, emission at 495 nm). Emission intensity of 4-POEACC as functions of (**b**) radiation intensity and (**d**) excitation duration (excitation at 284 nm, emission at 505 nm).

To quantitatively analyze the impact of radiation flux and excitation duration on the optical properties of the two materials, we further compared their fluorescence and phosphorescence intensities under identical radiation flux and excitation durations. Figure S8a,b display their fluorescence and phosphorescence intensities as a function of varying radiation fluxes. At the same radiation flux, 4-POMACC exhibits significantly higher fluorescence and phosphorescence intensities than 4-POEACC. Furthermore, for the same range of radiation flux variations, 4-POMACC shows a greater increase in intensity compared to 4-POEACC. These results indicate that the fluorescence and phosphorescence properties of 4-POMACC surpass those of 4-POEACC. Similarly, Figure S8c,d present the fluorescence and phosphorescence intensities of both materials under different excitation durations. However, as the excitation duration is extended, the increase in fluorescence and phosphorescence intensities for 4-POMACC is less pronounced, whereas the intensities for 4-POEACC remain relatively unchanged. This phenomenon, known as light activation, reflects an enhancement in luminescence intensity with prolonged excitation duration. The results suggest that 4-POMACC exhibits a notable light activation effect, while 4-POEACC shows minimal or no light activation.

To investigate the effect of different excitation wavelengths on the luminescent properties of these two metal-halide perovskite materials, we performed excitation-delay emission spectroscopy. The testing conditions involved collecting the excitation emission spectra 25 milliseconds after the excitation light was turned off. As shown in Figure S9, both 4-POMACC and 4-POEACC exhibit two luminescent centers, with their positions closely matching the phosphorescence emission observed in previous tests. Both materials display a single dominant excitation range, with 4-POMACC showing the strongest excitation from 260 nm to 290 nm, while 4-POEACC's peak excitation range is slightly different, focusing from 270 nm to 290 nm. Additionally, the luminescent centers of both materials do not shift with changes in excitation wavelength, indicating that the materials exhibit no excitation dependency and possess stable luminescent characteristics.

To gain a deeper understanding of the phosphorescence emission mechanism of 4-POMACC, we conducted a temperature test for both samples. Figure 8a shows the temperature-dependent phosphorescence spectra of 4-POMACC. It is evident that the phosphorescence intensity decreases as the temperature increases, which is attributed to the enhancement of molecular vibrations at high temperatures, leading to increased non-radiative transitions and reduced luminescence efficiency [43]. Figure 8b presents the CIE diagrams at different temperatures, illustrating that the color of the phosphorescence shifts with temperature changes, demonstrating a temperature-dependent color characteristic. Figure 8c plots the relationship between the emission peak position (peak 1, from organic molecules) and its full-width at half-maximum (FWHM) when temperature changes. As the temperature rises, the peak position exhibits a red shift, likely due to increased molecular vibrations at higher temperatures, leading to characteristics of aggregated luminescence [44]. The FWHM also increases, indicating enhanced coupling between phonons and electrons [45]. Figure 8d shows the temperature dependence of emission peak position (peak 2, from defect state) and its FWHM. With increasing temperature, the peak position shifts to the blue, possibly due to lattice expansion of the inorganic layers, which enlarges the bandgap. The FWHM also gradually increases, reflecting increased coupling between phonons and electrons. Similarly, the phosphorescence emission mechanism of 4-POEACC was also conducted as shown in Figure S10. It is evident that, as the temperature increases, the phosphorescence intensity gradually decreases, owing to the increased non-radiative transitions. Although 4-POEACC also shows a color dependence on temperature, the extent of color change with temperature is less pronounced compared to 4-POMACC. In addition, as temperature rises, the position of peak 1 remains unchanged, while the FWHM increases, indicating the enhanced coupling between phonons and electrons. The blue-shifted peak 2 with increasing temperature again confirms the lattice expansion of inorganic layers.

To fully understand the differences in luminescent performance between the two metal-halide perovskites, density functional theory (DFT) calculations were performed. Since the luminescence of both materials primarily originates from the organic molecules, we calculated the orbital distributions and intersystem crossing (ISC) rates for the organic components of each material. Figure 9a,b present the orbital distribution diagrams for 4-POMA and 4-POEA, respectively. These diagrams reveal that the HOMO and LUMO energy levels of the two organic components have very similar band gaps, and the orbital distributions of both HOMO and LUMO are concentrated on the benzene rings. DFT calculations of the spin orbital coupling constants (ξ) for the S_1 and T_n (n = 1, 2, 3) states of the two organic components indicate that both 4-POMA and 4-POEA have excellent intersystem crossing. But, according to energy level diagrams of 4-POMA and 4-POEA (Figure 9c,d), the energy gap between T_3 and T_2 in 4-POMA is significantly greater than that of 4-POEA. Therefore, the internal conversion rate between T_3 and T_2 of 4-POMA is smaller than that of 4-POEA, which leads to a longer phosphorescence lifetime of 4-POME.

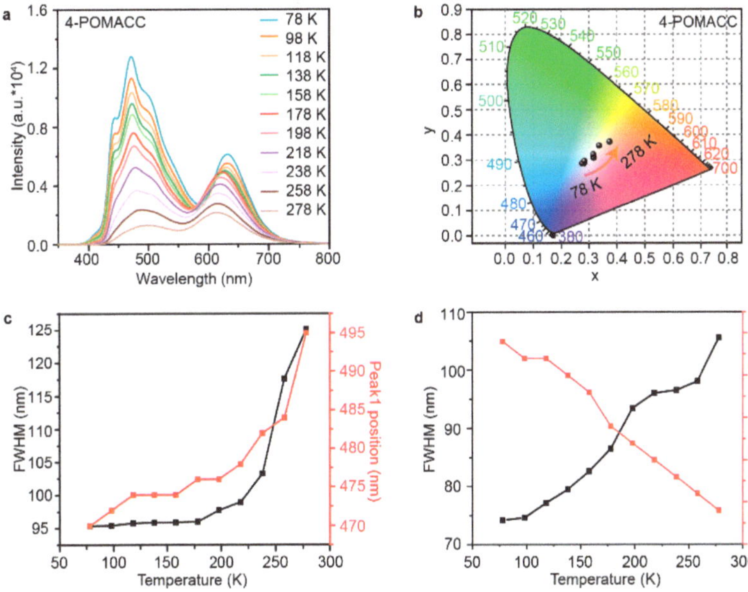

Figure 8. (**a**) Temperature-dependent emission spectra of 4-POMACC excited at 286 nm and (**b**) the corresponding CIE coordinate diagram. The intensity and FWHM variations in the two emission peaks from (**c**) organic molecule (peak 1) and (**d**) defect states (peak 2).

Encryption is a crucial technology in the information society nowadays [46–48]. Leveraging the distinct RTP properties of these two materials, we applied them in the field of information encryption. We coated molds marked with the number "9" with 4-POMACC and 4-POEACC materials. As illustrated in Figure 10, the area within the red dashed box is filled with 4-POEACC, and the area within the white dashed box is filled with 4-POMACC. Under UV light, the two materials exhibit different colors, so that the 4-POEACC region displays the UV light color, while the 4-POMACC region emits green light. At this point, the displayed information in the shape of a "9" appears to be false. Given the distinct RTP characteristics of these two materials, after 1 s of UV light being turned off, the pattern of the "9" will reveal only partial information (Figure 9). The area coated with 4-POMACC will continue to emit green light, revealing the true information. This approach effectively provides an encryption method based on the differing luminescent properties of the materials.

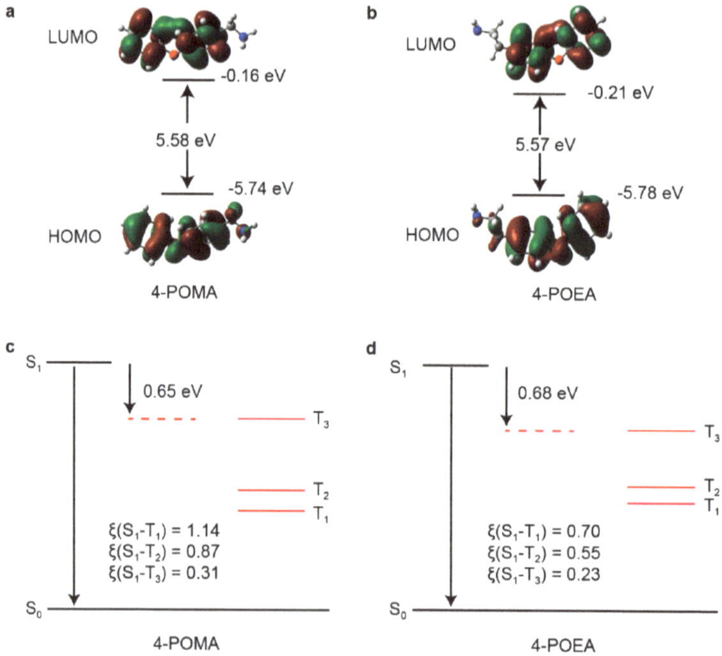

Figure 9. Orbital distribution diagrams for (**a**) 4-POMA and (**b**) 4-POEA. Energy level diagrams for (**c**) 4-POMA and (**d**) 4-POEA.

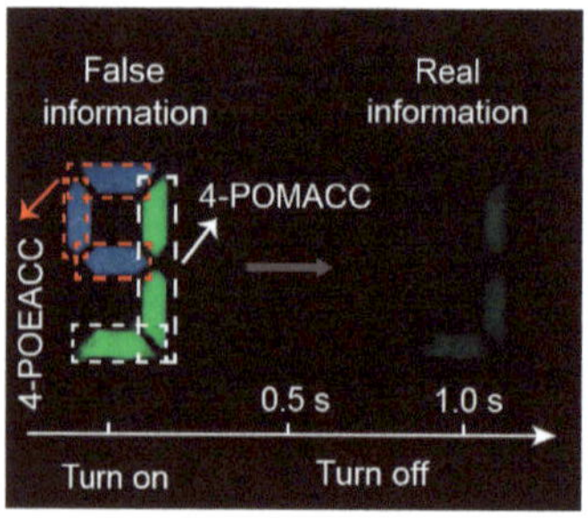

Figure 10. Demonstration of information encryption using 4-POMACC and 4-POEACC.

3. Materials and Methods

3.1. Materials

4-Phenoxybenzylamine (4-POMA, 98%), 4-phenoxyphenethylamine (4-POEA, 97%) and cadmium chloride ($CdCl_2$, 99%) were purchased from Aladdin (Shanghai, China) or Energy Chemical (Shanghai, China). HCl (37 wt.%, in water), ethanol and acetone were purchased from Nanjing Chemical Reagent Co., Ltd. (Nanjing, China). All reagents and solvents were used without further purifications.

3.2. Synthesis of Organic Amines Salts

Substance 4-POMACl. As shown in Figure S1, over an ice-water bath, 10 mL of ethanol was placed in a 50 mL round-bottom flask. To this system, 1.6 mmol (286 μL) of 4-POMA was added, followed by the dropwise addition of 4 mL of HCl with continuous stirring. As the solution was stirred, the product 4-POMACl began to precipitate out. After allowing the reaction to proceed for 1 h, the product was collected. The product was then subjected to vacuum filtration using a filtration apparatus. During the filtration process, the products were washed six times with acetone. The washed products were stored in a centrifuge tube, sealed with parafilm, and placed in a vacuum drying oven for further characterization. The final yield product in this synthesis procedure was 0.3380 g (approximately 90%).

Substance 4-POEACl. The synthesis of 4-POEACl salts were carried out in a similar way to that of 4-POMACl by replacing 4-POMA with 4-POEA (1.6 mmol, 314 μL). The final yield product in this synthesis procedure was 0.2901 g (approximately 85%).

3.3. Synthesis of 2D Metal-Halide Perovskites

Substance 4-POMACC. As shown in Figure S2, under ambient conditions, 4-POMACl (0.8 mmol, 0.1886 g) and $CdCl_2$ (0.4 mmol, 0.0733 g) were placed in a 25 mL round-bottom flask. A measure of 5 mL of ethanol was added to this mixture, and the system was thoroughly shaken. The mixture was then heated to 85 °C and stirred continuously at this temperature for 1 h until a clear solution was obtained. The clear solution was then transferred to a clean sample vial. After allowing the vial to stand at room temperature for a period of time, the product 4-POMACC began to precipitate out. The product was then collected through vacuum filtration, washed 6 to 7 times with acetone, and stored in a centrifuge tube, sealed with parafilm. Finally, the product was placed in a vacuum drying oven for further characterizations. The final yield product in this synthesis procedure was 0.2279 g (approximately 87%).

Substance 4-POEACC. The synthesis of 4-POEACC was carried out in a similar way to that of 4-POMACC by replacing 4-POMAC with 4-POEAC (0.8 mmol, 0.1998 g), and the heating temperature was set to 80 °C. The final yield product was 0.2294 g (approximately 84%).

3.4. Characterization

Crystalline structures of the samples were measured by powder X-ray diffraction (PXRD) using a Bruker D8 Advance X diffractometer (Cu Kα: λ = 1.5418 Å, Bruker AXS GmbH, Karlsruhe, Germany). Thermogravimetric analyses (TGA) were conducted on a DTG-60 Shimadzu thermal analyst system (Shimadzu, Kyoto, Japan) with a heating rate of 10 °C/min and a nitrogen flow rate of 50 cm^3/min. Fourier transform infrared (FTIR) spectra were obtained on a Bruker VERTEX 70 (Bruker AXS GmbH, Karlsruhe, Germany). Scanning electron microscope (SEM) images were acquired using a Hitachi SU4800-II cFEG SEM (Hitachi, Tokyo, Japan) at an accelerating voltage of 2.0 kV. Ultraviolet/visible (UV/Vis) and fluorescence spectra were recorded on a Jasco V-750 spectrophotometer (Jasco, Tokyo, Japan) and Edinburgh FLS980 (EI, Edinburgh, Germany), respectively. The absolute photoluminescence quantum yield (PLQY) was obtained using an Edinburgh FLS980 fluorescence spectrophotometer equipped with an integrating sphere. For fluorescence decay measurements, picosecond pulsed light-emitting diode (EPLED-380, Yida, Shenzhen, China, wavelength: 377 nm, pulse width: 947.7 ps; EPLED-295, Yida, Shenzhen, China, wavelength: 300 nm, pulse width: 833.7 ps) were used. Phosphorescence spectra were obtained using an Edinburgh FLS980 fluorescence spectrophotometer with a 10 ms delay time after excitation using a microsecond flash lamp. The microsecond flash lamp produces short, typically a few μs, and high irradiance optical pulses for phosphorescence decay measurements ranging from microseconds to seconds. The kinetic measurements, afterglow spectra and lifetimes were measured using an Edinburgh FLS980 fluorescence spectrophotometer and a microsecond flash-lamp (μF900). The lifetime and time-resolved emission spectra were obtained on Edinburgh FLSP 920 fluorescence spectrophotometer (EI, Edinburgh, Germany) equipped with a xenon arc lamp (Xe 900), a nanosecond hydrogen flash-lamp (nF 920), or a microsecond flash-lamp (μF900). The lifetimes (τ) of the luminescence were obtained by fitting the decay curve with a multi-exponential decay function.

Excitation–phosphorescence mapping was measured using Hitachi F-4700 (Hitachi, Tokyo, Japan) with a 25 ms delay time under ambient conditions.

3.5. Computational Methods

DFT calculations were performed to investigate the singlet-to-triplet exciton transformation using the Gaussian 09 package. The ground-state structures were optimized at the B3LYP/6-31G(d) level of theory. The first triplet excited state (T_1) was subsequently optimized at the same level based on the optimized ground-state geometry. Using time-dependent DFT (TD-DFT), the vertical excitation energies of the designed materials were calculated at the B3LYP/6-31G(d) level. Based on the optimized T_1 geometric structure, the spin-orbit coupling (SOC) between the S_1 state and the T_1, T_2, and T_3 states of the two materials was computed using the Dalton 2020 software package at the B3LYP/6-31G* level of theory. The Hirshfeld surfaces and 2D fingerprint plots were calculated by using Crystal Explorer 17. And d_i is the distance between the Hirshfeld surface and the nearest atom inside the Hirshfeld surface, while de is the distance between the Hirshfeld surface and the nearest atom outside the Hirshfeld surface. The intense of the color in the fingerprint plots represents the contribution of a pair (d_i, d_e) on the Hirshfeld surface.

4. Conclusions

In this work, two organic amines with different alkyl chain lengths were selected as organic components and $CdCl_2$ was selected as the inorganic framework due to its heavy-atom effect, which can enhance the intersystem crossing rate and strengthen triplet-state luminescence to synthesize 2D metal-halide perovskites 4-POMACC and 4-POEACC. Layer-spacing calculations suggest that 4-POMACC is more densely packed compared to 4-POEACC, which is favorable for RTP emission. Comprehensive studies of the optical properties confirm that these two materials exhibit significant differences in their RTP characteristics. Specifically, 4-POMACC has a much longer RTP lifetime than 4-POEACC. DFT calculations reveal a greater likelihood of intersystem crossing in 4-POMA and thus more effective triplet-state emission. Based on the RTP properties of these 2D metal-halide perovskites, the materials were applied in anti-counterfeiting applications. This study provides a new approach for regulating RTP in 2D metal-halide perovskites and enriches the variety of RTP materials.

Supplementary Materials: The following supporting information can be downloaded at: https://www.mdpi.com/article/10.3390/inorganics13040108/s1, Figure S1: The synthesis process of preparing organic amines salts in this work; Figure S2: The synthesis process of preparing 2D metal-halide perovskites in this work; Figure S3: Fourier transform infrared patterns of (a) 4-POMACC and (b) 4-POEACC, as well as their corresponding organic amine salts (4-POMACl and 4-POEACl); Figure S4: Thermogravimetry and derivative thermogravimetry curves of (a) 4-POMACC, (b) 4-POEACC, (c) 4-POMACl, and (d) 4-PAEACl; Figure S5: Time-resolved emission spectroscopy results of (a) 4-POMACC and (b) 4-POEACC; Figure S6: Photoluminescence quantum efficiency measurement results of (a) 4-POMACl, (b) 4-POMACC, (c) 4-POEACl, and (d) 4-POEACC; Figure S7: Double-exponential linear fitting of excitation power vs. phosphorescence intensity for (a) 4-POMACC and (b) 4-POEACC; Figure S8: (a) Fluorescence and (b) phosphorescence intensities of 4-POMACC and 4-POEACC at differing radiation flux. (c) Fluorescence and (d) phosphorescence intensities of 4-POMACC and 4-POEACC under different excitation duration; Figure S9: Time-resolved phosphorescence emission spectroscopy of (a) 4-POMACC and (b) 4-POEACC; Figure S10: (a) Temperature-dependent emission spectra of 4-POEACC excited at 284 nm and (b) the corresponding CIE coordinate diagram. The intensity and FWHM variations of two emission peaks from (c) organic molecule (peak 1) and (d) defect states (peak 2).

Author Contributions: Conceptualization, Z.L. and G.Z.; methodology, H.Z.; software, S.W., B.S. and Y.H.; validation, Z.L., B.S. and Y.H.; formal analysis, M.S. and M.L.; investigation, G.Z.; resources, G.Z.; data curation, H.Z.; writing—original draft preparation, H.Z. and S.W.; writing—review and editing, B.S. and Y.H.; visualization, S.W., M.S. and M.L.; supervision, Z.L. and G.Z.; project administration, B.S., Y.H. and G.Z.; funding acquisition, Z.L. and G.Z. All authors have read and agreed to the published version of the manuscript.

Funding: This research was funded by Shandong Provincial Natural Science Foundation, grant number ZR2024QF165.

Institutional Review Board Statement: Not applicable.

Informed Consent Statement: Not applicable.

Data Availability Statement: The original contributions presented in the study are included in the article/Supplementary Materials; further inquiries can be directed to the corresponding authors.

Acknowledgments: We thank Zhiyan Zhou from School of Stomatology, Shandong University, for her help with language polishing.

Conflicts of Interest: The authors declare no conflicts of interest.

References

1. Wang, B.; Mu, Y.; Zhang, H.; Shi, H.; Chen, G.; Yu, Y.; Yang, Z.; Li, J.; Yu, J. Red Room-Temperature Phosphorescence of CDs@Zeolite Composites Triggered by Heteroatoms in Zeolite Frameworks. *ACS Cent. Sci.* **2019**, *5*, 349–356. [CrossRef] [PubMed]
2. Ma, X.-K.; Liu, Y. Supramolecular Purely Organic Room-Temperature Phosphorescence. *Acc. Chem. Res.* **2021**, *54*, 3403–3414. [CrossRef]
3. Zhi, J.; Zhou, Q.; Shi, H.; An, Z.; Huang, W. Organic Room Temperature Phosphorescence Materials for Biomedical Applications. *Chem.—Asian J.* **2020**, *15*, 947–957. [CrossRef]
4. Yang, J.; Zhen, X.; Wang, B.; Gao, X.; Ren, Z.; Wang, J.; Xie, Y.; Li, J.; Peng, Q.; Pu, K.; et al. The Influence of the Molecular Packing on the Room Temperature Phosphorescence of Purely Organic Luminogens. *Nat. Commun.* **2018**, *9*, 840. [CrossRef] [PubMed]
5. Peng, Q.; Ma, H.; Shuai, Z. Theory of Long-Lived Room-Temperature Phosphorescence in Organic Aggregates. *Acc. Chem. Res.* **2021**, *54*, 940–949. [CrossRef]
6. Manser, J.S.; Saidaminov, M.I.; Christians, J.A.; Bakr, O.M.; Kamat, P.V. Making and Breaking of Lead Halide Perovskites. *Acc. Chem. Res.* **2016**, *49*, 330–338. [CrossRef]
7. Zhang, W.; Eperon, G.E.; Snaith, H.J. Metal Halide Perovskites for Energy Applications. *Nat. Energy* **2016**, *1*, 16048. [CrossRef]
8. Jena, A.K.; Kulkarni, A.; Miyasaka, T. Halide Perovskite Photovoltaics: Background, Status, and Future Prospects. *Chem. Rev.* **2019**, *119*, 3036–3103. [CrossRef] [PubMed]
9. Stranks, S.D.; Snaith, H.J. Metal-Halide Perovskites for Photovoltaic and Light-Emitting Devices. *Nat. Nanotechnol.* **2015**, *10*, 391–402. [CrossRef]
10. Stoumpos, C.C.; Kanatzidis, M.G. The Renaissance of Halide Perovskites and Their Evolution as Emerging Semiconductors. *Acc. Chem. Res.* **2015**, *48*, 2791–2802. [CrossRef]
11. Ortiz-Cervantes, C.; Carmona-Monroy, P.; Solis-Ibarra, D. Two-Dimensional Halide Perovskites in Solar Cells: 2D or Not 2D? *ChemSusChem* **2019**, *12*, 1560–1575. [CrossRef]
12. Ghimire, S.; Klinke, C. Two-Dimensional Halide Perovskites: Synthesis, Optoelectronic Properties, Stability, and Applications. *Nanoscale* **2021**, *13*, 12394–12422. [CrossRef]
13. Shi, E.; Gao, Y.; Finkenauer, B.P.; Akriti, A.; Coffey, A.H.; Dou, L. Two-Dimensional Halide Perovskite Nanomaterials and Heterostructures. *Chem. Soc. Rev.* **2018**, *47*, 6046–6072. [CrossRef] [PubMed]
14. Liu, Y.; Xiao, H.; Goddard, W.A. Two-Dimensional Halide Perovskites: Tuning Electronic Activities of Defects. *Nano Lett.* **2016**, *16*, 3335–3340. [CrossRef] [PubMed]
15. Mao, L.; Stoumpos, C.C.; Kanatzidis, M.G. Two-Dimensional Hybrid Halide Perovskites: Principles and Promises. *J. Am. Chem. Soc.* **2019**, *141*, 1171–1190. [CrossRef] [PubMed]
16. Huo, C.; Cai, B.; Yuan, Z.; Ma, B.; Zeng, H. Two-Dimensional Metal Halide Perovskites: Theory, Synthesis, and Optoelectronics. *Small Methods* **2017**, *1*, 1600018. [CrossRef]
17. Ba, Q.; Jana, A.; Wang, L.; Kim, K.S. Dual Emission of Water-Stable 2D Organic–Inorganic Halide Perovskites with Mn(II) Dopant. *Adv. Funct. Mater.* **2019**, *29*, 1904768. [CrossRef]

18. Quan, L.N.; Zhao, Y.; García De Arquer, F.P.; Sabatini, R.; Walters, G.; Voznyy, O.; Comin, R.; Li, Y.; Fan, J.Z.; Tan, H.; et al. Tailoring the Energy Landscape in Quasi-2D Halide Perovskites Enables Efficient Green-Light Emission. *Nano Lett.* **2017**, *17*, 3701–3709. [CrossRef]
19. Yu, J.; Kong, J.; Hao, W.; Guo, X.; He, H.; Leow, W.R.; Liu, Z.; Cai, P.; Qian, G.; Li, S.; et al. Broadband Extrinsic Self-Trapped Exciton Emission in Sn-Doped 2D Lead-Halide Perovskites. *Adv. Mater.* **2019**, *31*, 1806385. [CrossRef]
20. Smith, M.D.; Karunadasa, H.I. White-Light Emission from Layered Halide Perovskites. *Acc. Chem. Res.* **2018**, *51*, 619–627. [CrossRef]
21. Nawab, G.; Rahman, A.U.; Haq, I.U.; Ali, A.; Abdelkader, A.; Ismail, A.H.; Alomar, M.; Khan, I. Structural and Optoelectronic Properties of 2D Halide Perovskites Cs2MBr4 (M = Zn, Cd, Hg): A First Principle Study. *Opt. Quantum Electron.* **2024**, *56*, 871. [CrossRef]
22. Yangui, A.; Pillet, S.; Bendeif, E.-E.; Lusson, A.; Triki, S.; Abid, Y.; Boukheddaden, K. Broadband Emission in a New Two-Dimensional Cd-Based Hybrid Perovskite. *ACS Photonics* **2018**, *5*, 1599–1611. [CrossRef]
23. Cai, T.; Yang, H.; Hills-Kimball, K.; Song, J.-P.; Zhu, H.; Hofman, E.; Zheng, W.; Rubenstein, B.M.; Chen, O. Synthesis of All-Inorganic Cd-Doped $CsPbCl_3$ Perovskite Nanocrystals with Dual-Wavelength Emission. *J. Phys. Chem. Lett.* **2018**, *9*, 7079–7084. [CrossRef]
24. Holder, C.F.; Fanghanel, J.; Xiong, Y.; Dabo, I.; Schaak, R.E. Phase-Selective Solution Synthesis of Perovskite-Related Cesium Cadmium Chloride Nanoparticles. *Inorg. Chem.* **2020**, *59*, 11688–11694. [CrossRef] [PubMed]
25. Hoffman, J.M.; Malliakas, C.D.; Sidhik, S.; Hadar, I.; McClain, R.; Mohite, A.D.; Kanatzidis, M.G. Long Periodic Ripple in a 2D Hybrid Halide Perovskite Structure Using Branched Organic Spacers. *Chem. Sci.* **2020**, *11*, 12139–12148. [CrossRef]
26. Zhou, N.; Zhou, H. Spacer Organic Cation Engineering for Quasi-2D Metal Halide Perovskites and the Optoelectronic Application. *Small Struct.* **2022**, *3*, 2100232. [CrossRef]
27. Li, X.; Hoffman, J.M.; Kanatzidis, M.G. The 2D Halide Perovskite Rulebook: How the Spacer Influences Everything from the Structure to Optoelectronic Device Efficiency. *Chem. Rev.* **2021**, *121*, 2230–2291. [CrossRef]
28. Gan, L.; Li, J.; Fang, Z.; He, H.; Ye, Z. Effects of Organic Cation Length on Exciton Recombination in Two-Dimensional Layered Lead Iodide Hybrid Perovskite Crystals. *J. Phys. Chem. Lett.* **2017**, *8*, 5177–5183. [CrossRef]
29. Deng, C.; Zhou, G.; Chen, D.; Zhao, J.; Wang, Y.; Liu, Q. Broadband Photoluminescence in 2D Organic–Inorganic Hybrid Perovskites: $(C_7H_{18}N_2)PbBr_4$ and $(C_9H_{22}N_2)PbBr_4$. *J. Phys. Chem. Lett.* **2020**, *11*, 2934–2940. [CrossRef]
30. Kamminga, M.E.; Fang, H.-H.; Filip, M.R.; Giustino, F.; Baas, J.; Blake, G.R.; Loi, M.A.; Palstra, T.T.M. Confinement Effects in Low-Dimensional Lead Iodide Perovskite Hybrids. *Chem. Mater.* **2016**, *28*, 4554–4562. [CrossRef]
31. Smith, M.D.; Connor, B.A.; Karunadasa, H.I. Tuning the Luminescence of Layered Halide Perovskites. *Chem. Rev.* **2019**, *119*, 3104–3139. [CrossRef]
32. Gong, J.; Hao, M.; Zhang, Y.; Liu, M.; Zhou, Y. Layered 2D Halide Perovskites beyond the Ruddlesden–Popper Phase: Tailored Interlayer Chemistries for High-Performance Solar Cells. *Angew. Chem.* **2022**, *134*, e202112022. [CrossRef]
33. Gao, S.; Wang, S.; Wu, J.; Lin, Z. Regulation and Application of Organic Luminescence from Low-Dimensional Organic–Inorganic Hybrid Metal Halides. *J. Mater. Chem. C* **2023**, *11*, 16890–16911. [CrossRef]
34. Dhanabalan, B.; Castelli, A.; Palei, M.; Spirito, D.; Manna, L.; Krahne, R.; Arciniegas, M. Simple Fabrication of Layered Halide Perovskite Platelets and Enhanced Photoluminescence from Mechanically Exfoliated Flakes. *Nanoscale* **2019**, *11*, 8334–8342. [CrossRef] [PubMed]
35. Liu, Z.; Qin, X.; Chen, Q.; Jiang, T.; Chen, Q.; Liu, X. Metal–Halide Perovskite Nanocrystal Superlattice: Self-Assembly and Optical Fingerprints. *Adv. Mater.* **2023**, *35*, 2209279. [CrossRef]
36. Bright, D.W.; Dias, F.B.; Galbrecht, F.; Scherf, U.; Monkman, A.P. The Influence of Alkyl-Chain Length on Beta-Phase Formation in Polyfluorenes. *Adv. Funct. Mater.* **2009**, *19*, 67–73. [CrossRef]
37. Crosby, G.A.; Demas, J.N. Measurement of Photoluminescence Quantum Yields. Review. *J. Phys. Chem.* **1971**, *75*, 991–1024. [CrossRef]
38. Duan, X.; Song, W.; Qiao, J.; Li, X.; Cai, Y.; Wu, H.; Zhang, J.; Hao, X.; Tang, Z.; Ge, Z.; et al. Ternary Strategy Enabling High-Efficiency Rigid and Flexible Organic Solar Cells with Reduced Non-Radiative Voltage Loss. *Energy Environ. Sci.* **2022**, *15*, 1563–1572. [CrossRef]
39. Liu, Z.; Qin, X.; Chen, Q.; Chen, Q.; Jing, Y.; Zhou, Z.; Zhao, Y.S.; Chen, J.; Liu, X. Highly Stable Lead-Free Perovskite Single Crystals with NIR Emission Beyond 1100 Nm. *Adv. Opt. Mater.* **2022**, *10*, 2201254. [CrossRef]
40. Peng, Y.; Ma, J.; Zhao, Y.; You, D.; Yao, Y.; Deng, Z.; Liao, J.; Chang, Y.; Shen, W.; Li, M.; et al. Multilevel Stimulus-Responsive Room Temperature Phosphorescence Achieved by Efficient Energy Transfer from Triplet Excitons to Mn^{2+} Pairs in 2D Hybrid Metal Halide. *Adv. Funct. Mater.* **2024**, 2420311. [CrossRef]
41. Luo, B.; Liang, D.; Sun, S.; Xiao, Y.; Lian, X.; Li, X.; Li, M.-D.; Huang, X.-C.; Zhang, J.Z. Breaking Forbidden Transitions for Emission of Self-Trapped Excitons in Two Dimensional $(F_2CHCH_2NH_3\ )_2CdBr_4$ Perovskite through Pb Alloying. *J. Phys. Chem. Lett.* **2020**, *11*, 199–205. [CrossRef] [PubMed]

42. Goushi, K.; Yamada, T.; Otomo, A. Excitation Intensity Dependence of Power-Law Blinking Statistics in Nanocrystal Quantum Dots. *J. Phys. Chem. C* **2009**, *113*, 20161–20168. [CrossRef]
43. Kim, Y.H.; Arunkumar, P.; Kim, B.Y.; Unithrattil, S.; Kim, E.; Moon, S.-H.; Hyun, J.Y.; Kim, K.H.; Lee, D.; Lee, J.-S.; et al. A Zero-Thermal-Quenching Phosphor. *Nat. Mater.* **2017**, *16*, 543–550. [CrossRef]
44. Yang, J.; Fang, M.; Li, Z. Organic Luminescent Materials: The Concentration on Aggregates from Aggregation-induced Emission. *Aggregate* **2020**, *1*, 6–18. [CrossRef]
45. Gong, Z.; Li, Z.; Zhong, Y. Circularly Polarized Luminescence of Coordination Aggregates. *Aggregate* **2022**, *3*, e177. [CrossRef]
46. Su, Y.; Phua, S.Z.F.; Li, Y.; Zhou, X.; Jana, D.; Liu, G.; Lim, W.Q.; Ong, W.K.; Yang, C.; Zhao, Y. Ultralong Room Temperature Phosphorescence from Amorphous Organic Materials toward Confidential Information Encryption and Decryption. *Sci. Adv.* **2018**, *4*, eaas9732. [CrossRef]
47. Liu, S.; Liu, X.; Yuan, J.; Bao, J. Multidimensional Information Encryption and Storage: When the Input Is Light. *Research* **2021**, *2021*, 7897849. [CrossRef]
48. Tan, J.; Li, Q.; Meng, S.; Li, Y.; Yang, J.; Ye, Y.; Tang, Z.; Qu, S.; Ren, X. Time-Dependent Phosphorescence Colors from Carbon Dots for Advanced Dynamic Information Encryption. *Adv. Mater.* **2021**, *33*, 2006781. [CrossRef]

inorganics

MDPI

Review

Research Progress of Halide Perovskite Nanocrystals in Biomedical Applications: A Review

Guiyun Wang [1,†], Yanxia Qi [1,†], Zhiyan Zhou [2,*], Zhuang Liu [3] and Ruowei Wang [1,*]

1 College of Nursing, Shandong Xiehe University, Jinan 250109, China
2 School of Stomatology, Cheeloo College of Medicine, Shandong University & Shandong Key Laboratory of Oral Tissue Regeneration & Shandong Engineering Research Center of Dental Materials and Oral Tissue Regeneration & Shandong Provincial Clinical Research Center for Oral Diseases, Jinan 250012, China
3 School of Materials Science and Engineering, Shandong University, Jinan 250061, China; liu.zhuang@sdu.edu.cn
* Correspondence: zhouzhiyan@sdu.edu.cn (Z.Z.); wangruowei@sdxiehe.edu.cn (R.W.)
† These authors contributed equally to this work.

Abstract: Halide perovskite nanocrystals have rapidly emerged as a prominent research topic in materials science over the past decade owing to their exceptional optoelectronic properties and tunability. Their distinctive characteristics, including high light absorption coefficients, high quantum yields, narrow-band emissions, low defect densities, and adjustable chemical compositions and sizes, position them as highly promising candidates for applications in optoelectronic devices, energy conversion units, and other related systems. However, due to the toxicity and instability of halide perovskite nanocrystals, their widespread application in the biomedical field has been limited in the past. In recent years, numerous innovative coating strategies have been reported to effectively enhance the stability of halide perovskite nanocrystals while confining their toxic metal ions within the coating layers, thereby significantly improving their biocompatibility. This review provides a comprehensive summary of the recent progress of halide perovskite nanocrystals in the field of biomedicine. It covers coating strategies to enhance stability and biocompatibility, as well as the applications of coated halide perovskite nanocrystals in biomedicine, with a particular focus on their unique advantages in bioimaging and chemical sensing. Finally, we address unresolved issues and challenges, such as the metabolic pathways and final products of halide perovskite nanocrystals in vivo. We hope to inspire researchers in the field and provide direction for future studies.

Keywords: halide perovskite nanocrystal; coating; stability; biocompatibility; biomedical application; bioimaging; chemical sensing

1. Introduction

Halide perovskite nanocrystals, characterized by the ABX_3 crystal structure, consist of an organic or inorganic cation at the A site (e.g., methylamine MA^+ or Cs^+), a divalent metal cation at the B site (e.g., Pb^{2+} or Sn^{2+}), and halide anions (e.g., Cl^-, Br^-, or I^-) occupying the X site [1–3]. Over the past decade, these nanomaterials have emerged as a cornerstone in materials science, attributed to their outstanding optoelectronic properties and exceptional versatility [4,5]. Their distinctive features—including superior light absorption coefficients, impressive quantum efficiencies, sharp emission spectra, minimal defect states, and customizable chemical and morphological attributes—position them as promising candidates for a wide array of applications, ranging from advanced optoelectronic devices and energy harvesting systems to scintillator technologies and beyond [6–10].

1.1. Advantages of Halide Perovskite Nanocrystals and Their Applications

The remarkable light absorption capacity and luminescence efficiency of halide perovskite nanocrystals make them highly advantageous for optoelectronic applications [11,12]. With photoluminescence quantum yields (PLQY) approaching nearly 100% and exceptionally narrow emission bandwidths (commonly exhibiting a full width at half maximum, FWHM, of less than 25 nm), they deliver unparalleled color purity [13,14]. Moreover, the emission wavelength can be precisely controlled across the visible spectrum and extended into the near-infrared (NIR) region by fine-tuning the nanocrystals' chemical composition and size [15–17]. The synthesis of halide perovskite nanocrystals is relatively simple and encompasses scalable techniques, such as solution-phase methods and hot-injection protocols [18–21]. These methods typically enable large-scale production under mild conditions, ensuring cost-effectiveness and adaptability for diverse application scenarios. Additionally, the morphologies of halide perovskite nanocrystals can be engineered with great flexibility, spanning zero-dimensional (0D) quantum dots, one-dimensional (1D) nanowires, two-dimensional (2D) nanosheets, and three-dimensional (3D) bulk crystals [22–26]. This structural versatility significantly expands their applicability across various domains. Beyond their exceptional optical attributes, halide perovskite nanocrystals demonstrate excellent charge carrier mobility (20–100 $cm^2 V^{-1} s^{-1}$) and extended carrier diffusion lengths (1–10 μm) [27–29]. These properties further enhance their performance in optoelectronic devices and energy-related technologies, cementing their status as transformative materials in modern science and engineering.

Halide perovskite nanocrystals have gained widespread recognition for their diverse applications across multiple fields. Notably, they have shown tremendous promise in solar cells, light-emitting diodes (LEDs), photodetectors, and lasers [30–32]. For instance, perovskite quantum dot-based LEDs, celebrated for their exceptional color purity and tunable spectral output, are emerging as transformative components in display technologies [33–35]. Similarly, perovskite photodetectors, characterized by broad spectral sensitivity and high detection efficiency, have proven invaluable for advancements in communication and imaging systems [36–38]. The utilization of halide perovskite nanocrystals is also expanding into photocatalysis [39,40]. Recent studies highlight their superior catalytic performance in critical reactions, including water splitting for hydrogen generation, carbon dioxide reduction, and the degradation of organic pollutants [41–43]. Beyond optoelectronic and catalytic domains, halide perovskite nanocrystals are increasingly recognized for their potential in energy storage technologies, paving the way for novel solutions in renewable energy systems [44–46]. However, compared to other semiconductor materials, such as small-molecule organic dyes, which have already been widely utilized in the biomedical field, the application of halide perovskite nanocrystals in this domain appears to be in its nascent stages.

1.2. Biotoxicity and Instability of Halide Perovskite Nanocrystals

Although halide perovskite nanocrystals exhibit remarkable optical and electronic properties that position them as promising candidates for medical applications, their biotoxicity and inherent instability remain significant barriers to their widespread adoption in biomedical fields and clinical practice [47]. The biotoxicity of perovskite nanocrystals is largely attributed to their chemical composition, particularly the presence of Pb^{2+} ions [48]. While Pb^{2+} contributes to the superior optoelectronic performance of certain perovskite materials, its accumulation in biological systems poses severe risks, including cellular and organ damage (Figure 1) [49,50]. Extensive studies have demonstrated that Pb^{2+} ions exhibit toxic effects, potentially causing neurotoxicity, liver and kidney impairment, immune suppression, and reproductive dysfunction [51]. During biomedical applications, if the lead content in perovskite nanocrystals is not adequately controlled or encapsulated, Pb^{2+} may be released into the body, triggering adverse biological responses and increasing long-term health risks [52]. Moreover, the nanoscale dimensions of perovskite nanocrystals allow them to penetrate cell membranes with ease, interacting with intracellular biomolecules [53]. This interaction can result in oxidative stress, deoxyribonucleic acid (DNA) damage, and apoptosis. Additionally, the biocompatibility of perovskite nanocrystals can be influenced by surface defects and surface passivation strategies, potentially leading to cytotoxicity and immune responses [54]. These challenges underscore the critical need for innovative approaches to mitigate the biotoxicity and enhance the stability of halide perovskite nanocrystals for their safe application in biomedical fields.

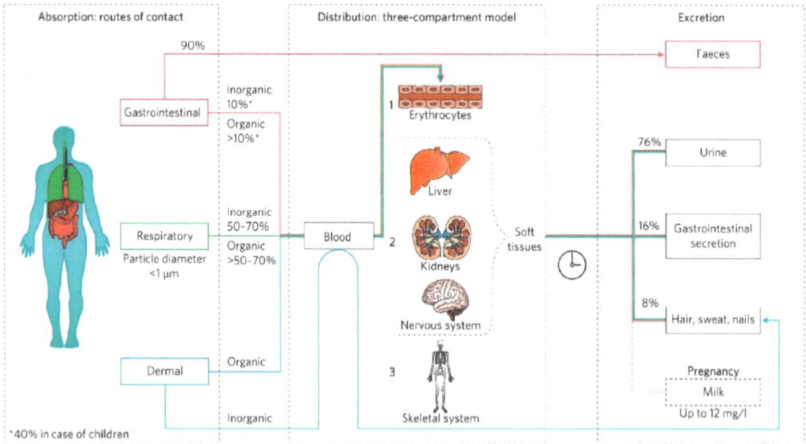

Figure 1. A schematic illustration depicting the absorption, distribution, and excretion pathways of lead compounds within the human body. For each exposure route, the percentage indicates the proportion of lead absorbed. Lead not absorbed through gastrointestinal ingestion is expelled via feces, while unabsorbed lead from respiratory or dermal exposure is presumed to be exhaled or removed through natural bodily processes, respectively. Once absorbed, lead accumulates in three primary compartments of the body, where it is retained for durations specific to each organ before eventual excretion. Reprinted with permission from ref. [49]. Copyright 2016, Springer Nature.

Furthermore, lead leakage from perovskite materials into the soil can be absorbed by plants and subsequently enter the food chain at rates up to ten times higher than other lead contaminants introduced through human activities, thereby posing severe biological toxicity risks (Figure 2) [55]. Experimental findings underscore the necessity of the meticulous handling of lead from halide perovskites. Notably, it has been emphasized that the permissible lead content in perovskite-based materials must be significantly lower than that in other lead-containing electronic devices. Consequently, researchers strongly advocate for the complete substitution of lead with more inert and environmentally benign metals to facilitate the safe advancement of perovskite technologies.

Figure 2. The image illustrates mint plants cultivated in (**a**) uncontaminated control soil and (**b**) soil contaminated with Pb^{2+} from perovskite materials. The lead concentrations detected in the leaves, stems, and roots are displayed adjacent to each corresponding section of the plants. Reprinted with permission from ref. [55]. Copyright 2020, Springer Nature.

A major limitation of perovskite materials is their poor chemical and environmental stability [56–58]. Halide perovskite nanocrystals are highly susceptible to degradation when exposed to humidity, oxygen, ultraviolet (UV) light, and heat, challenges that are particularly pronounced in the complex environment of the human body. This instability leads to significant performance deterioration over time, restricting their potential for biomedical applications. Moisture sensitivity is one of the primary factors compromising the stability of halide perovskites [59]. Exposure to moisture can degrade their crystal structure, causing optical performance to decline or even leading to complete crystal dissociation. Within biological environments, the presence of both moisture and oxygen exacerbates this degradation, further limiting their durability for applications like biological imaging and therapeutic interventions. Additionally, halide perovskite nanocrystals are prone to structural and functional degradation under UV light and elevated temperatures [60,61]. This is particularly problematic for biomedical uses, such as in vivo or in vitro photothermal therapy (PTT), where exposure to UV light and heat is often unavoidable [62–64].

Surface defects further compromise the stability of halide perovskites [65,66]. The high surface energy of these materials makes them susceptible to degradation, especially in the presence of water or oxygen. Such defects not only degrade the optical performance of the material but may also induce cytotoxic responses in biological systems. Addressing these stability challenges is crucial for advancing the safe and effective use of halide perovskite nanocrystals in biomedical fields.

1.3. Current Solutions to Address Biotoxicity and Instability

To address the challenges outlined above, current research focuses on several key strategies. To mitigate the toxicity associated with lead, significant efforts have been directed toward developing lead-free perovskite materials, such as substituting tin (Sn^{2+}) for lead [67,68]. These lead-free alternatives effectively reduce toxicity risks. However, studies have shown that exposure to SnI_2 can be more acutely toxic to zebrafish than PbI_2, with toxicity levels closely correlated to the pH of the growth medium (Figure 3) [49]. Additionally, despite their reduced toxicity, lead-free halide perovskites face persistent challenges, including limited stability and lower luminescence efficiency, highlighting the need for continued optimization to enhance their practical viability [16,69].

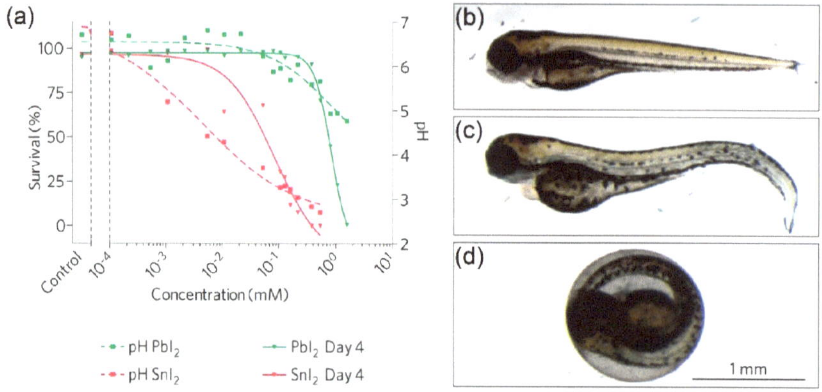

Figure 3. (**a**) Survival rate of zebrafish, 4 days post-fertilization, plotted against toxicant concentration and pH. (**b**) Healthy zebrafish (control group). (**c**) Zebrafish exposed to PbI_2, displaying brain hemovascular defects, heart edema, and dorsal curvature. (**d**) Zebrafish embryo exposed to SnI_2, exhibiting failure to hatch at 96 h post-fertilization. Reprinted with permission from ref. [49]. Copyright 2016, Springer Nature.

Surface modification and encapsulation techniques offer effective solutions to enhance the biocompatibility and chemical stability of perovskite nanocrystals [8,70,71]. As early as around thirty years ago, studies had already reported that halide perovskite $(C_6H_5C_2H_4NH_3)_2PbI_4$ encapsulated in polymethyl methacrylate (PMMA) polymers could be fabricated on SiO_2 substrates by performing spin-coating and subsequent annealing processes [72,73]. The as-prepared halide perovskite compounds were dispersed as nanometer-sized crystals, and, more importantly, their stability was considerably improved. Similar strategies have been in use up to the present day. For example, coating halide perovskite nanocrystals with organic or inorganic materials—such as polymers or silica—can isolate toxic components and reduce moisture and oxygen infiltration, thereby significantly improving their stability [74,75]. Furthermore, surface modifications can further bolster the in vivo stability of halide perovskite nanocrystals, ensuring they retain exceptional performance over extended periods of use.

Simultaneously, to guarantee the long-term stability of perovskite nanocrystals in biomedical applications, it is crucial to establish a robust stability assessment system. This system should encompass a thorough evaluation of their degradation rate, long-term stability, and biotoxicity within the in vivo environment. Such an assessment framework will provide the scientific evidence necessary to support the clinical translation of these materials.

This review provides a comprehensive overview of the research advancements in halide perovskite nanocrystals within the biomedical field, building upon the background discussed. It covers strategies for enhancing the stability of halide perovskite nanocrystals, application cases in biological imaging, and their use in environmental monitoring and biosensing. We believe that this review not only highlights the significant progress made in these areas but also serves as a source of inspiration for further research in the biological sciences, thereby fostering the continued exploration of halide perovskite nanocrystals in the biomedical domain.

2. Halide Perovskite Nanocrystals for Bioimaging Applications

Halide perovskite materials inherently suffer from reduced stability in aqueous environments and exhibit even greater instability in complex biological settings. To overcome this limitation, halide perovskite nanocrystals must be encapsulated within protective matrices [76]. Currently, two primary strategies are employed to enhance the water stability of halide perovskite nanocrystals: (i) surface coating of individual nanocrystals and (ii) encapsulation of multiple nanocrystals within water-resistant matrices [77,78]. The surface coating approach involves applying water-protective shells, such as silica or titanium dioxide, directly onto individual perovskite nanocrystals. Alternatively, encapsulation techniques embed multiple perovskite nanocrystals within robust matrices made from polymers or phospholipids, effectively shielding them from water exposure.

For bioimaging applications, the challenge becomes even more complex, as halide perovskite nanocrystals often interact directly with biological environments, including blood, cells, and tissues—highly intricate aqueous systems composed of ions, salts, proteins, and other biomolecules. Consequently, the protective coatings for halide perovskite nanocrystals must not only be water-resistant but also demonstrate exceptional stability in biological fluids to ensure their functionality and reliability in biomedical applications. Despite recent progress in using halide perovskite nanocrystals as imaging tags for living cells, the in vivo stability remains largely unproven, representing a significant challenge for clinical translation [79].

A recent study has introduced a universal approach aimed at advancing the potential bioimaging applications of halide perovskites [80]. Specifically, researchers have proposed an effective method for encapsulating $CsPbBr_3$ halide perovskite nanocrystals within organized microscale polymer carriers, rendering them water-resistant (Figure 4a,b). Following the comprehensive structural and optical characterization of the resulting materials, their stabilities in various biological fluids, including phosphate-buffered saline and human serum, and within a cellular microenvironment (CT26 cells) were evaluated. For the first time, the in vivo stability of encapsulated halide perovskite nanocrystals was assessed using Dilute Brown Non-Agouti (DBA) mice (Figure 4c). Cell–halide perovskite nanocrystal interactions were visualized using confocal laser scanning microscopy, with cell membranes being stained red and cell nuclei being stained blue. The bright green fluorescence emitted by the halide perovskite nanocrystals highlighted the water-resistant properties of the organic polymer coating, demonstrating its stability in an aqueous biological environment (Figure 4d).

To assess the in vivo stability of $CsPbBr_3$ nanocrystals encapsulated in polymer microscale carriers, the carriers were intramuscularly injected into DBA mice. Following injection, the mice were euthanized, and their tissues were collected and prepared for histological analysis using fluorescent staining. The results showed bright fluorescence from the nanocrystals exclusively in the muscle tissue, with no detectable fluorescence in other tissues (Figure 4e). This finding indicates that the microscale carriers containing halide perovskite nanocrystals remained localized at the injection site without dispersing to other tissues. Notably, the encapsulated nanocrystals retained their fluorescence for up to 72 h post-injection, highlighting the durability of the polymer coating even under in vivo conditions. Compared to highly efficient commercial up-conversion nanoparticles doped with rare-earth ions, halide perovskite nanocrystals offer significant advantages in both cost and performance [81–83]. The fabrication of halide perovskite nanocrystals is approximately one tenth the cost of up-conversion nanoparticles, and they exhibit a much broader excitation spectral range, thereby enhancing their versatility for a wide range of applications. Thus, the encapsulation strategy developed in this study presents a promising pathway for utilizing halide perovskite nanocrystals as stable fluorescence tags in bioimaging, ensuring both durability and functionality in complex biological environments.

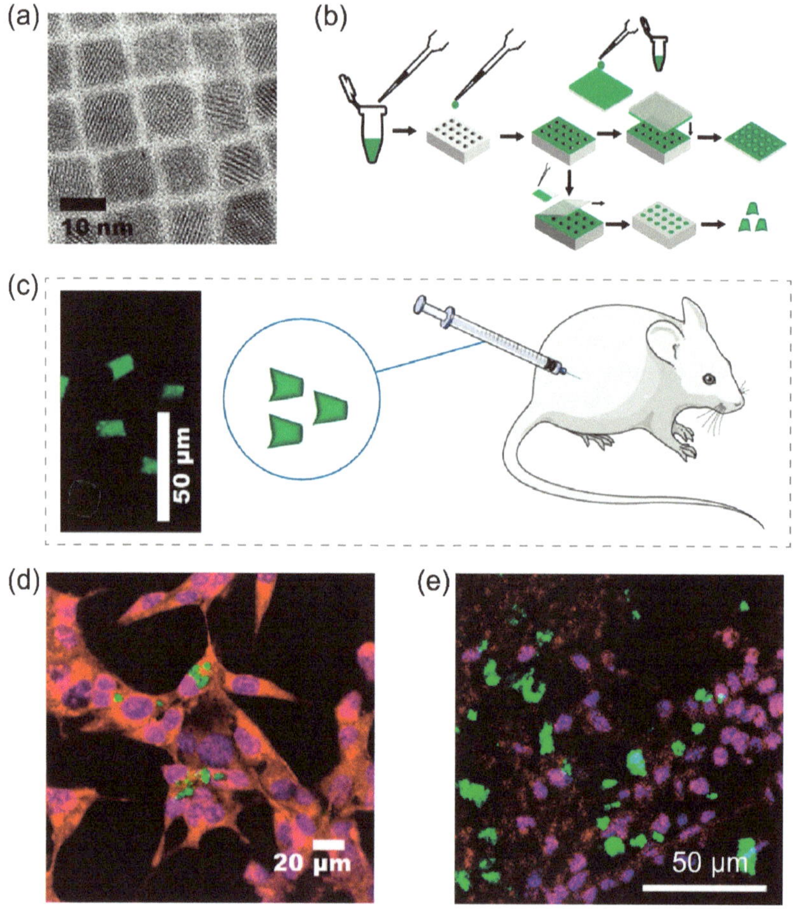

Figure 4. (**a**) High-resolution transmission electron microscopy (HRTEM) image of $CsPbBr_3$ nanocrystals. (**b**) Schematic illustration depicting the fabrication strategies for microscale carriers loaded with $CsPbBr_3$ nanocrystals. (**c**) Confocal laser scanning microscopy image of microscale carriers loaded with $CsPbBr_3$ nanocrystals. (**d**) Confocal laser scanning microscopy image of CT-26 cells incubated with microscale carriers containing $CsPbBr_3$ nanocrystals for 24 h. (**e**) Confocal laser scanning microscopy image of muscle tissue following intramuscular injection of microscale carriers loaded with $CsPbBr_3$ nanocrystals. Reprinted with permission from ref. [80]. Copyright 2022, American Chemical Society.

Composite particles, such as the aforementioned microscale carriers, are too large to be efficiently internalized into the cytoplasm, limiting their suitability for small-cell imaging applications [84,85]. In a recent study, insoluble $CsPbBr_3/CsPb_2Br_5$ composite nanocrystals were successfully synthesized through a simple water-assisted chemical transformation of halide perovskite nanocrystals (Figure 5a) [86]. Water plays two critical roles in this process: (i) it removes CsBr from $CsPbBr_3/Cs_4PbBr_6$, and (ii) it alters the coordination number of Pb^{2+}, changing it from six in $CsPbBr_3$ and Cs_4PbBr_6 to eight in $CsPb_2Br_5$. Figure 5b shows photographs of $CsPbBr_3/CsPb_2Br_5$ composite nanocrystals dispersed in toluene and water under UV light (365 nm), where the composite nanocrystals emit bright green fluorescence in both solvents. Remarkably, the water-dispersed $CsPbBr_3/CsPb_2Br_5$ composite nanocrystals exhibit exceptional photoluminescent (PL) properties, maintaining a high PLQY of 80% and a narrow FWHM of 16 nm. These characteristics closely resemble those of the original $CsPbBr_3/CsPb_2Br_5$ nanocrystals (PLQY of 83%) and pure $CsPbBr_3$ nanocrystals dispersed in toluene (PLQY of approximately 90%). These properties position $CsPbBr_3/CsPb_2Br_5$ composite nanocrystals as highly effective fluorescent probes for live HeLa cell imaging.

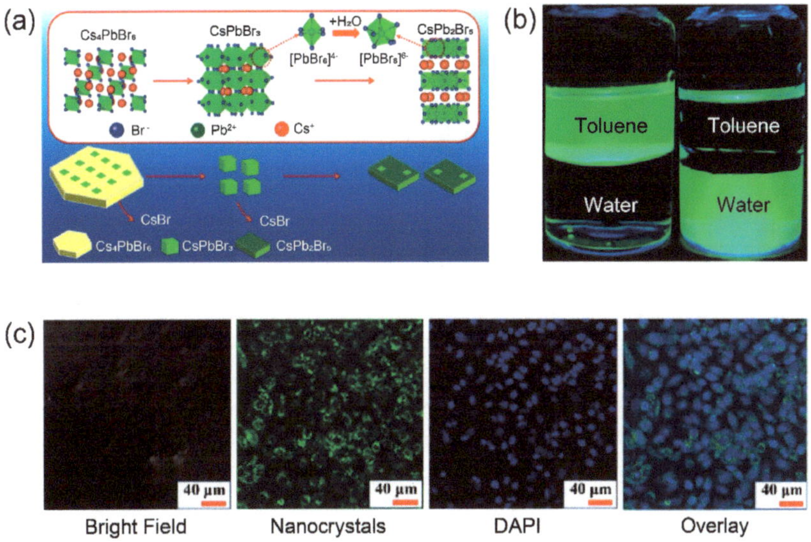

Figure 5. (**a**) Schematic diagram illustrating the water-assisted transformation process, which depicts the conversion of $CsPbBr_3/Cs_4PbBr_6$ composite nanocrystals into $CsPbBr_3/CsPb_2Br_5$ composite nanocrystals. (**b**) Photographs showcasing $CsPbBr_3/CsPb_2Br_5$ composite nanocrystals dispersed in toluene and water taken under UV light (365 nm). (**c**) Bright-field and fluorescence images of HeLa cells incubated with $CsPbBr_3/CsPb_2Br_5$ nanocrystals for 24 h, with nuclei stained in blue using 4,6-diamidino-2-phenylindole (DAPI) dye. The overlay images reveal the cellular uptake of the nanocrystals. Reprinted with permission from ref. [86]. Copyright 2019, American Chemical Society.

High water stability is a critical requirement for the effective application of $CsPbBr_3/CsPb_2Br_5$ composite nanocrystals in cell imaging. The potentials of the as-prepared $CsPbBr_3/CsPb_2Br_5$ composite nanocrystals for cellular imaging were further explored. As shown in Figure 5c, fluorescence images of HeLa cells incubated at 37 °C with $CsPbBr_3/CsPb_2Br_5$ composite nanocrystals for 24 h reveal green fluorescence localized around the nucleus. This observation confirms that the composite nanocrystals successfully penetrate the cell membranes and enter the cells. In contrast, larger $CsPbBr_3$ nanocrystals encapsulated in polystyrene ($CsPbBr_3$@PS) have been previously used as cellular labeling agents [87]. However, the luminescence intensities of the labeled cells decreased with extended incubation times. The differences in performance between $CsPbBr_3$@PS and $CsPbBr_3/CsPb_2Br_5$ composite nanocrystals could be attributed to two factors. First, a larger quantity of $CsPbBr_3/CsPb_2Br_5$ composite nanocrystals was internalized into the cytoplasm due to their nanoscale size, which differed from the processes that the larger $CsPbBr_3$@PS particles underwent. Second, the $CsPbBr_3/CsPb_2Br_5$ nanocrystals exhibited highly stable luminescence in aqueous solutions. Therefore, this study not only introduces a novel method for preparing insoluble, narrow-band, and brightly luminescent composite nanocrystals but also broadens the potential applications of halide perovskite nanocrystals in biomedicine.

Despite the presence of an effective protective coating, the fluorescence intensity of halide perovskite nanocrystals inevitably diminishes over time due to inherent defects, leading to a deterioration in performance [88–90]. For instance, the fluorescence intensity of the encapsulated halide perovskite nanocrystals, as depicted in Figures 4 and 5, significantly declines after just one week, rendering them unsuitable for prolonged bioimaging applications [80,86]. Consequently, researchers are investigating methods to extend the fluorescence longevity of halide perovskite nanocrystals. A recent study introduced an innovative, stepwise approach for encapsulating $CsPbBr_3$ nanocrystals within water-soluble polymer capsules [91]. This adaptable protocol accommodates nanocrystals with diverse surface ligands. Notably, the encapsulated nanocrystals exhibit a remarkable PLQY of approximately 60%, which remains stable for over two years when the capsules are dispersed in water. The capsule fabrication process is illustrated in Figure 6a, while Figure 6b,c showcase images of the capsules in water under UV light. These capsule solutions display a milky dispersion and emit vibrant green fluorescence upon UV exposure. Remarkably, this fluorescence persists for several months. A representative TEM image reveals the formation of well-defined, spherical capsules. Within these capsules, both aggregated and individual nanocrystals are present, with the latter situated at the core and surrounded by a polymer shell approximately 25-nm thick (Figure 6d).

Following encapsulation and redispersion in water, researchers selected a batch of samples exhibiting high PLQY. These samples were stored in sealed vials at room temperature, with their PLQY monitored over a two-year period. Figure 6e depicts the temporal evolution of the sample's PLQY. The protocol employed in this study demonstrated exceptional efficacy in preventing water infiltration through the hydrophobic regions of the polymer. Consequently, the capsules maintained their emission profile and PLQY while dispersed in water, with the Cs–oleate and DDAB-coated $CsPbBr_3$ nanocrystals showing PLQY values of 60% and 55%, respectively, after 12 months. This underscores the robust protective properties of the polymer shell surrounding the nanocrystals.

To fully assess the potential of capsules for cell labeling, a low concentration—1000 times below the toxicity threshold for Pb and Cs ions—was selected, and the cells were incubated for 24 h. The capsules demonstrated excellent biocompatibility, with no statistically significant cytotoxicity observed. The uptake of the capsules was clearly evident as their green fluorescence was detected after all incubation periods when excited at 400 nm. The sustained emission observed within the acidic lysosomal environment of the cells further confirms the suitability of these capsules for cellular and bioimaging applications. This property offers distinct advantages over conventional fluorescent dyes, especially due to the photostability of these inorganic nanocrystals in aqueous environments. Consequently, the methodology presented in this study significantly expands the potential applications of these luminescent materials, particularly in the fields of biology and biomedicine.

The previously discussed cases involve utilizing UV light to excite halide perovskite nanocrystals, resulting in visible light emission for bioimaging applications. However, in practical applications, UV light has limited penetration through the skin, which poses challenges in reaching deeper tissues [92,93]. In contrast to fluorescence-based techniques, X-Ray attenuation-based imaging offers virtually no limitations in penetration depth [94–97]. Conventional X-Ray imaging, such as chest X-Rays, presents several advantages, including lower cost, significantly reduced radiation exposure, and broader accessibility, compared to more advanced X-Ray techniques like computed tomography (CT). However, leveraging conventional X-Ray imaging as a novel modality for detecting deep-seated cancers requires the development of highly X-Ray-attenuating, cancer-specific nanomaterials [98,99].

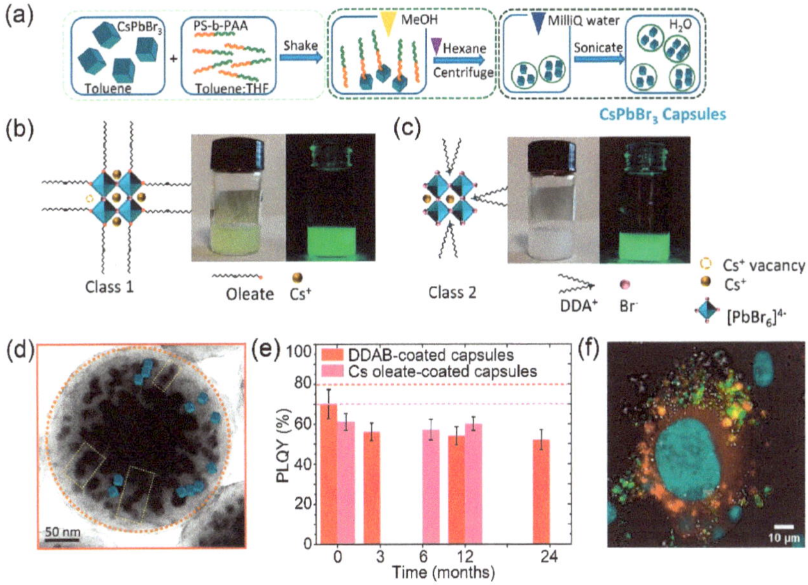

Figure 6. (**a**) A schematic illustration outlines the room-temperature fabrication process for polymer capsules containing $CsPbBr_3$ nanocrystals. This procedure involves encapsulating the nanocrystals within a water-soluble polymer matrix, ensuring both protection and enhanced stability. The photographs in (**b**,**c**) showcase aged samples of Cs–oleate-coated and didodecyl dimethylammonium bromide (DDAB)-coated $CsPbBr_3$ nanocrystals, respectively, encapsulated in polymer capsules and dispersed in deionized water. Images taken under both daylight and UV light demonstrate that both samples retain strong photoluminescence, highlighting their superior emission stability even after prolonged incubation in water. (**d**) A magnified TEM image provides a detailed view of a single polymer capsule. The embedded sketches illustrate the random distribution of $CsPbBr_3$ nanocrystals within the capsule matrix, as well as the presence of aggregated structures. These aggregates, identified as nanocrystal clusters, are highlighted with yellow frames, emphasizing their spatial arrangement within the polymer shell. (**e**) The graph plots the PLQY of polymer-encapsulated $CsPbBr_3$ nanocrystals dispersed in water over several months. Dotted lines represent the PLQY of un-encapsulated nanocrystals in toluene prior to encapsulation. (**f**) Confocal fluorescence images show cells incubated for 24 h in a medium containing $CsPbBr_3$ nanocrystals encapsulated with Cs–oleate coating. The images were captured using an excitation wavelength of 400 nm. Cell nuclei were stained with DAPI dye, which emits a blue fluorescence signal, providing a clear contrast against the bright green fluorescence emitted by the encapsulated nanocrystals. Reprinted with permission from ref. [91]. Copyright 2022, American Chemical Society.

In a recent investigation, $CsPbBr_3$ perovskite nanocrystals were selected due to their proven ability to attenuate X-Rays, positioning them as effective scintillators in conventional X-Ray detection panels [100]. The nanocrystals are rapidly co-synthesized and double-encapsulated within SiO_2, forming $CsPbBr_3$–SiO_2@SiO_2 nanoparticles that efficiently prevent aggregation, degradation, or leakage. These nanoparticles are subsequently conjugated with antibodies (Ab), yielding $CsPbBr_3$–SiO_2@SiO_2–Ab nanoparticles, which are specifically designed to target pancreatic cancer cells. The underlying methodology for attenuation-based in vivo cancer imaging via X-Ray inspection is illustrated in Figure 7a. Upon introduction of the antibody-conjugated nanoparticles into xenograft mice, the X-Ray scintillating nanocrystals within the particles significantly attenuate the incident X-Ray photons, resulting in pronounced contrast at the cancerous site (Figure 7b). An innovative and streamlined synthetic route for encapsulating $CsPbBr_3$ nanocrystals within

SiO_2 nanoparticles has been developed. This process involves the rapid injection of a nanocrystal precursor solution, containing a base catalyst for SiO_2 synthesis, into a silicon precursor solution. This triggers the base-catalyzed formation of SiO_2 nanoparticles while simultaneously facilitating the ligand-assisted re-precipitation of $CsPbBr_3$ nanocrystals within a remarkably short time frame. The resulting $CsPbBr_3$–SiO_2 nanoparticles exhibit a spherical morphology, with diameters ranging from 20 to 30 nm, as confirmed by TEM images (Figure 7c–e).

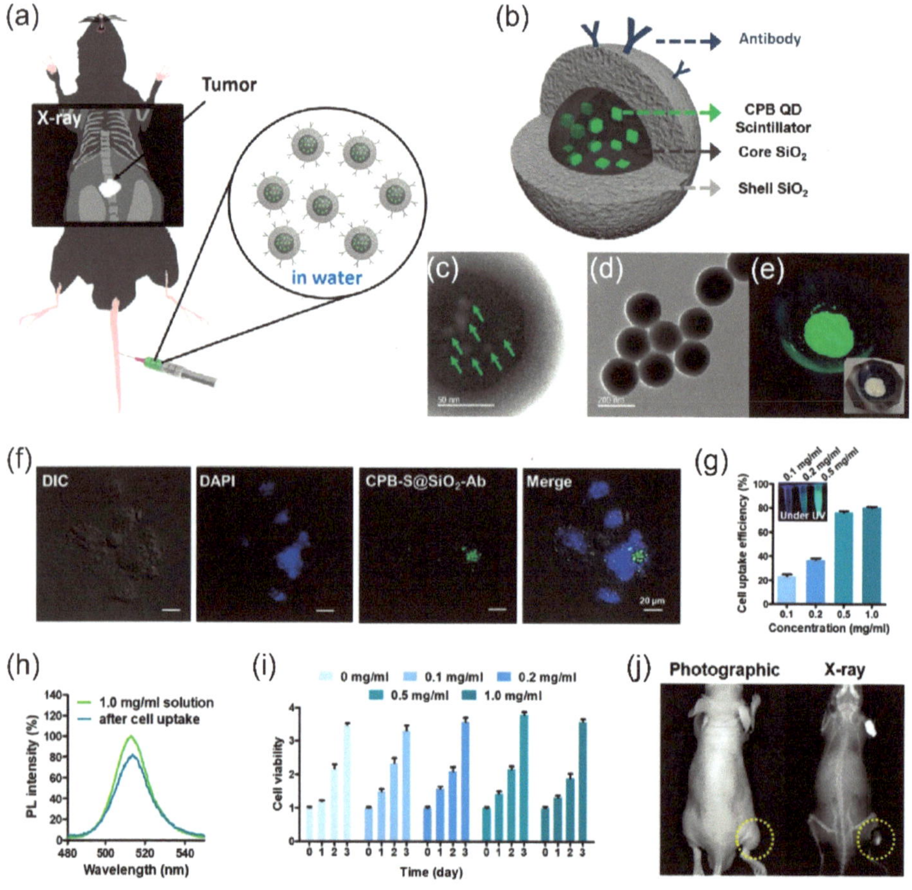

Figure 7. Schematic representation of (**a**) the cancer detection process and (**b**) the structure of $CsPbBr_3$–SiO_2@SiO_2–Ab nanoparticles. TEM images of $CsPbBr_3$–SiO_2@SiO_2 nanoparticles captured at (**c**) high and (**d**) low magnifications. (**e**) Photographs of $CsPbBr_3$–SiO_2@SiO_2 nanoparticle powder under UV light and ambient daylight. (**f**) Confocal laser scanning microscopy images of Panc-1 cells treated with $CsPbBr_3$–SiO_2@SiO_2–Ab nanoparticles. Cell nuclei are stained with blue DAPI, while the nanoparticles are visualized in green. (**g**) Cellular uptake efficiencies assessed via photoluminescence of Panc-1 cells treated with varying concentrations of nanoparticles. (**h**) Photoluminescence spectra of nanoparticle solutions and Panc-1 cells treated with the same solution. (**i**) Cell viability, as determined by the WST-1 assay, following treatment with different concentrations of nanoparticle solutions. (**j**) In vivo photographic and X-Ray images of pancreatic tumor-bearing mice taken 2 h after intravenous injection of $CsPbBr_3$–SiO_2@SiO_2–Ab nanoparticles. Reprinted with permission from ref. [100]. Copyright 2021, Wiley-VCH GmbH.

To target the CD44 surface adhesion receptor on pancreatic cells, the surfaces of the synthesized $CsPbBr_3$–SiO_2@SiO_2 nanoparticles were functionalized with anti-CD44 antibodies. Given that Cluster of Differentiation 44 (CD44) expression is often linked to poor prognosis, it serves as a critical prognostic marker and therapeutic target for pancreatic cancer. The uptake of $CsPbBr_3$–SiO_2@SiO_2–Ab nanoparticles in Panc-1 cells was assessed using confocal laser-scanning microscopy (Figure 7f). The nuclei of the cells were stained with DAPI, which emits in the blue spectral range. The PL intensities of the Panc-1 cells, reflecting the uptake of $CsPbBr_3$ nanocrystals, steadily increased as the concentration of nanoparticles rose but plateaued thereafter with no further significant increase (Figure 7g). For the 1 mg mL^{-1} solution, the maximum uptake reached 76.8% (Figure 7h). Figure 7i depicts the effect of SiO_2@SiO_2–Ab nanoparticles on the viability of Panc-1 cells, as determined by the water-soluble tetrazolium salt assay. No cytotoxicity was observed under the cell culture conditions, nor was there any noticeable effect on cell proliferation or differentiation. The non-toxicity of the $CsPbBr_3$–SiO_2@SiO_2–Ab nanoparticles clearly indicates that minimal Pb was released from the nanoparticles.

Once the transplanted Panc-1 cells had reached a sufficient size in the xenograft mouse model, a solution of $CsPbBr_3$–SiO_2@SiO_2–Ab nanoparticles was intravenously administered. As shown in Figure 7j, distinct white signals were observed at the tumor site under X-Ray irradiation, indicating that the antibody-conjugated nanoparticles successfully targeted the CD44 prognostic markers. More importantly, this result demonstrates the feasibility of detecting cancer in vivo using straightforward plain X-Ray imaging. Additionally, the real-time biodistribution of the injected $CsPbBr_3$–SiO_2@SiO_2–Ab nanoparticles was monitored via X-Ray imaging. The signal intensity increased steadily over time, peaking 2 h post-injection, and subsequently declined gradually. These findings suggest that the approach outlined in this study holds significant potential for real-time in vivo cancer detection. Note that implanting pancreatic tumors in the mouse leg (Figure 7j) is primarily for the convenience of experimental procedures and the reliability of data. Although this differs from the tumor's natural growth environment, it remains a widely used experimental method [101–104].

While halide perovskite nanocrystals show promise for deep-tissue in vivo bioimaging under X-Ray excitation, radiation exposure remains an inescapable challenge in practical applications [105–108]. As a result, researchers are increasingly exploring alternative, non-damaging excitation sources that can enable the use of halide perovskites for bioimaging without causing harm to tissues. NIR laser technology, which can be precisely focused within a localized region, offers numerous advantages, including high spatial resolution, substantial tissue penetration depth, minimal background fluorescence, and reduced photodamage to biological tissues [109–111]. Recent advances in fluorescence imaging using NIR light—particularly in the second near-infrared window (NIR-II, 900–1700 nm)—have spurred significant progress, owing to its superior tissue penetration (1–10 cm) and enhanced spatial resolution compared to visible or UV light (0.1–2 mm) [112–114]. This is primarily due to its reduced tissue absorption, scattering, and auto-fluorescence. Nevertheless, fluorescence-based imaging techniques still face considerable challenges, including the limited penetration depth of both absorption and emission light, as well as the trade-off between imaging depth and resolution.

Two-photon fluorescence imaging, which utilizes the simultaneous absorption of two NIR photons as the excitation source to generate visible light emission, is considered a highly effective technique for monitoring cellular dynamics and detecting early pathological changes [115,116]. In recent years, biological imaging under NIR illumination using lanthanide-doped nanoparticles, aggregation-induced emission nanoparticles, and II-VI quantum dots has attracted considerable attention [117–119]. In the field of halide perovskite nanocrystals, this bioimaging approach has also been demonstrated [120–122]. In a recent study, individual $CsPbBr_3$ perovskite nanocrystals were encapsulated within silica nanoshells to effectively prevent the leakage of Pb^{2+} ions (Figure 8a) [123]. TEM images confirmed the successful coating of SiO_2 protective shells on the $CsPbBr_3$ nanocrystals (Figure 8b). The two-photon and one-photon photoluminescence properties of these nanocrystals, along with their potential applications in bioimaging, were comprehensively investigated through optical characterization and two-photon cellular imaging experiments. Figure 8c illustrates the fundamental mechanisms of light absorption and photoluminescence. In the two-photon process, a single NIR photon (800 nm) with an energy of 1.55 eV is insufficient for a direct inter-band transition, given that the calculated bandgap of $CsPbBr_3$ is 2.34 eV. Consequently, two photons are required to excite an electron from the valence band (VB) to the conduction band (CB). Following relaxation within the bands, the electrons and holes reach their respective band edges, resembling the process in single-photon excitation above the bandgap, resulting in green photoluminescence with a peak at 526 nm. Simultaneously, some electrons and holes recombine non-radiatively.

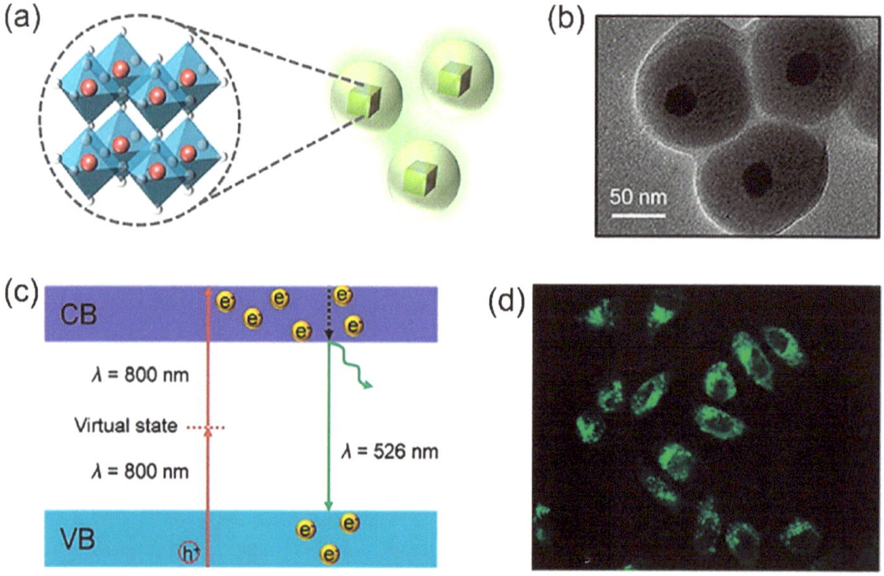

Figure 8. (**a**) Schematic illustration of $CsPbBr_3@SiO_2$ core–shell nanocrystals engineered for two-photon bioimaging applications. (**b**) TEM images of the as-prepared $CsPbBr_3@SiO_2$ nanocrystals. (**c**) Schematic representation depicting the two-photon absorption and subsequent emission process in $CsPbBr_3@SiO_2$ nanocrystals. (**d**) Fluorescence image of HepG2 cells from the experimental group under 800 nm NIR illumination, demonstrating enhanced luminescent activity. Reprinted with permission from ref. [123]. Copyright 2022, Elsevier.

To evaluate the potential application of $CsPbBr_3@SiO_2$ nanocrystals in two-photon fluorescence biological imaging, HepG2 cells were incubated with $CsPbBr_3@SiO_2$ for 24 h. Under 800-nm laser scanning, the control group of HepG2 cells exhibited minimal fluorescence, while the experimental group emitted bright green fluorescence due to two-photon absorption (Figure 8d). This suggests that the cellular luminescence signals originated from $CsPbBr_3@SiO_2$ nanocrystals that were internalized into the cells via endocytosis. These findings highlight that $CsPbBr_3@SiO_2$ nanocrystals, as two-photon imaging nanoprobes, could effectively minimize interference from cellular auto-fluorescence, ensuring high-quality imaging.

Other studies have also explored the use of halide perovskite nanocrystals for two-photon up-conversion bioimaging. For instance, it has been demonstrated that antibody-functionalized 1D $CsPbBr_3$ nanowires, 2D $CsPbBr_3$ nanoplatelets, and 3D $CsPbBr_3$ nanocubes can be employed for selective and simultaneous two-photon imaging of heterogeneous breast cancer cells within the NIR biological window [124–126]. These biocompatible nanostructures can be fabricated by coating them with amine-functionalized poly(ethylene glycol)-propionic acid. Remarkably, over 86% of the photoluminescence intensity is retained by these nanocrystals after 35 days submerged in water, showcasing exceptional photostability by maintaining 99% of the photoluminescence intensity after 3 h of UV light exposure. Furthermore, antibody-conjugated 1D and 2D perovskite nanomaterials have demonstrated the ability to perform simultaneous two-photon imaging of triple-negative breast cancer cells and human epidermal growth factor receptor 2 (HER2)-positive breast cancer cells.

3. Halide Perovskite Nanocrystals for Chemical Sensing Applications

Halide perovskite nanocrystals have not only led to pioneering advancements in the field of bioimaging but have also achieved notable outcomes in chemical sensing, particularly in the analysis and detection of metabolic products [127–129]. As illustrated in Figure 9, a recent study reports the synthesis of $CsPbCl_3$:Mn^{2+}@SiO_2 core–shell nanoparticles in water solutions, which was accomplished through the usage of capping ligand, surface modification, and crystal engineering strategies [130]. The SN_2 reaction between 4-chlorobutyric acid (CBA) and oleylamine (OAm) generates a zwitterionic ligand that promotes the dispersion of $CsPbCl_3$:Mn^{2+}@SiO_2 nanoparticles, while the robust SiO_2 shell significantly improves their overall stability. Furthermore, Mn^{2+} doping introduces a secondary emission center while enabling potential postsynthetic modifications. This transformation results in the formation of MnO_2@$CsPbCl_3$:Mn^{2+}@SiO_2 nanoparticles, which exhibit remarkable oxidase (OXD)-like activity. Theoretical calculations suggest that electron transfer from $CsPbCl_3$ to in situ MnO_2, combined with the adsorption–desorption process of 3,3′,5,5′-tetramethylbenzidine (TMB), significantly enhances the OXD-like activity. Mn^{4+} in MnO_2@$CsPbCl_3$:Mn^{2+}@SiO_2 (fluorescent nanozyme) is reduced to Mn^{2+} and dissociates in the presence of ascorbic acid (AA), thus inhibiting OXD-like activity and triggering fluorescence "turn-on/off" behavior, enabling dual-mode recognition. Ultimately, a biomarker detection platform based on the MnO_2@$CsPbCl_3$:Mn^{2+}@SiO_2 fluorescent nanozyme is established, with AA serving as the reporter molecule, enabling the precise detection of human serum alkaline phosphatase (ALP). This demonstrates the considerable potential of perovskites in biosensing applications.

The sensitive detection of *Staphylococcus aureus* enterotoxins (SEs) is essential for food inspection because of their high pathogenicity, which can be achieved through surface-enhanced Raman scattering (SERS) utilizing SERS-active nanostructures [126,131,132]. As illustrated in Figure 10, a recent study developed an Au–Ag Janus nanoparticle/$CsPbBr_3$ composite-based SERS immunoassay for SE detection [133]. The plasmonic Au–Ag Janus nanoparticles exhibited inherent SERS activity facilitated by 2-mercaptobenzoimidazole-5-carboxylic acid ligands. $CsPbBr_3$@mesoporous silica nanomaterials (MSNs) were synthesized and subsequently converted into $CsPb_2Br_5$@MSNs in the aqueous phase. Paired SE antibody–antigen-driven plasmonic Au–Ag Janus nanoparticle–$CsPb_2Br_5$@MSN composites were then prepared. These composites displayed enhanced SERS activity, which was attributed to the suppression of plasmonic decay through electromagnetic field enhancement and electron transfer mechanisms. A positive correlation was established between the

SERS signals and SE concentrations of the composites. An additive-free SERS immunoassay was developed for the straightforward, sensitive, and reproducible detection of SEs. This study lays the groundwork for the development of multiple additive-free SERS-active plasmonic nanoparticle/perovskite composites, opening new avenues for the exploration of SERS-based detection probes in food safety monitoring.

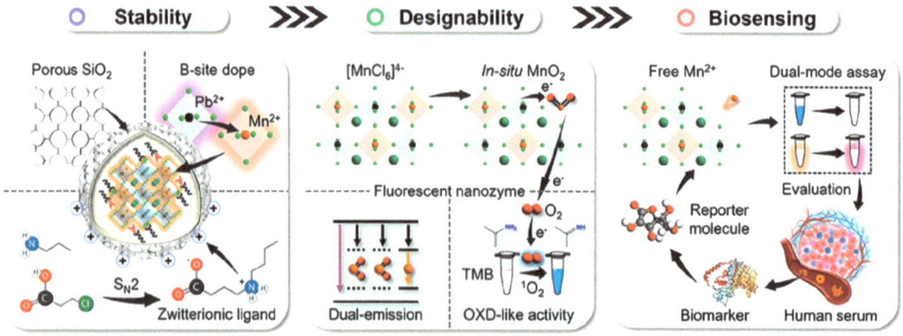

Figure 9. Schematic representation of the trinity strategy that empowers $CsPbCl_3$ nanocrystals to function as hydrophilic and highly efficient fluorescent nanozymes for the development of biomarker reporting platforms. Reprinted with permission from ref. [130]. Copyright 2024, American Chemical Society.

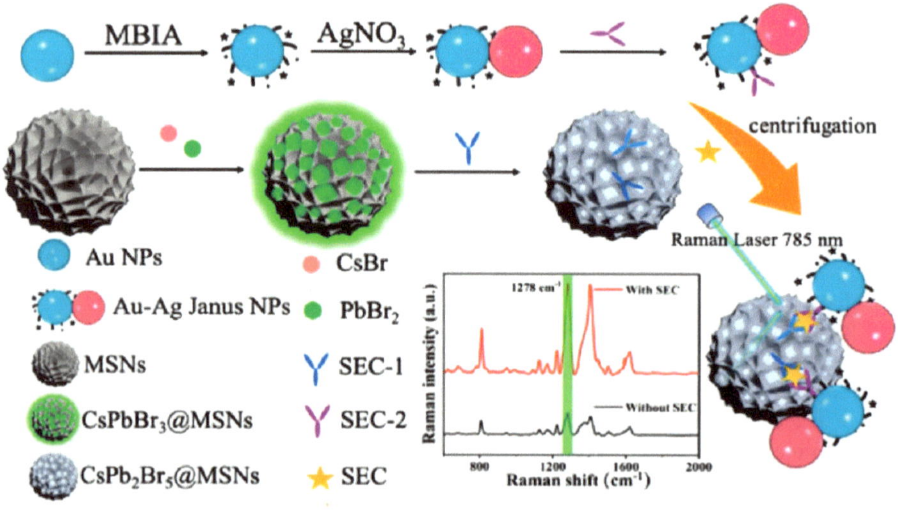

Figure 10. Schematic representation of SERS-active composites featuring Au–Ag Janus nanoparticles/$CsPbBr_3$ for immunoassays targeting Staphylococcus aureus enterotoxins. Reprinted with permission from ref. [133]. Copyright 2022, American Chemical Society.

Dual-functional perovskite nanomaterials with both hydrogen sulfide (H_2S)-sensing and antibacterial properties hold significant promise. As shown in Figure 11, chiral halide perovskite nanomaterials (R/S-α-PEA)$_2$CsPb$_2$Br$_7$ were synthesized and evenly dispersed in aqueous solution [134]. The encapsulation of methoxypolyethylene glycol amine (mPEG-NH_2) enhanced the water stability and uniformity of the particle size. The incorporation of chiral cations successfully induced circular dichroism (CD) signals. As probes, these perovskites demonstrated a rapid and sensitive luminescence quenching response to H_2S, with imaging detection being performed at the *Escherichia coli* (*E. coli*) level. Moreover, their strong positive charge properties as antibacterial agents promotes membrane lysis, resulting in E. coli cell death and showcasing significant antibacterial effects. This work represents a preliminary exploration into the use of chiral perovskites in biological applications, providing valuable insights into the development of bi-functional perovskite nanomaterials for biological sensing and antibacterial treatments.

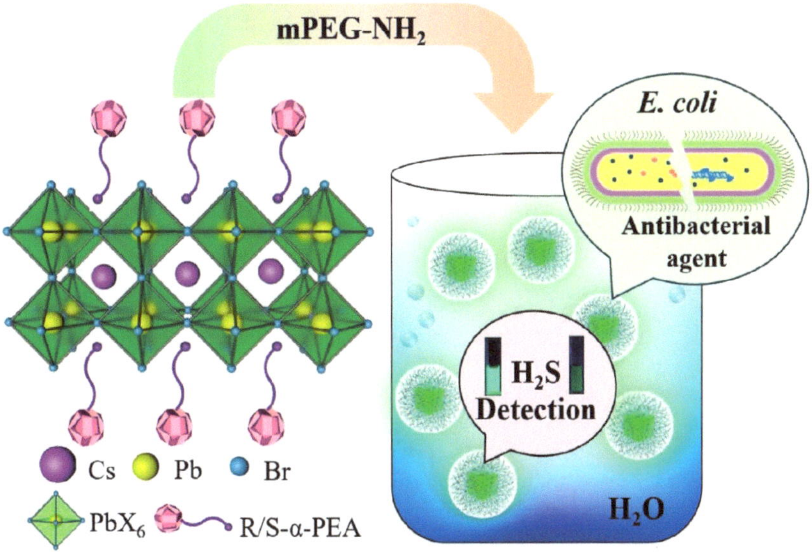

Figure 11. Liquid-phase dual-functional chiral perovskites have been developed for both hydrogen sulfide (H_2S) detection and antibacterial applications against *Escherichia coli*. Reprinted with permission from ref. [134]. Copyright 2024, Elsevier.

Similarly, $CsPbBr_3$ nanocrystals have been innovatively encapsulated within nanomicelles formed from a polyethylene glycol–polycaprolactone (PEG-PCL) block copolymer, resulting in biocompatible $CsPbBr_3$ nanocrystals with exceptional water resistance [135]. Thanks to their superior water resistance and biocompatibility, these $CsPbBr_3$ nanocrystals are ideal for real-time, long-term quantitative monitoring of H_2S levels in living cell and zebrafish models.

In another study, phospholipid membrane (PM)-coated $CsPbX_3$ nanocrystals exhibiting intrinsic bio-catalytic activity were discovered [136]. Unlike other peroxidase-like nanozymes that rely on external chromogenic reagents, the PM-$CsPbX_3$ nanocrystals function as self-reporting nanoprobes, enabling an "add-to-answer" detection model. Remarkably, the fluorescence of PM-$CsPbX_3$ nanocrystals is rapidly quenched upon the addition of H_2O_2 and can be restored by removing excess H_2O_2 (Figure 12a). This unexpected observation opens new possibilities for using PM-$CsPbX_3$ nanocrystals in the development of multicolor bio-inks and metabolite-responsive paper analytical devices, showcasing their significant potential in bioanalysis (Figure 12b). This is the first report of the nanozyme-like properties of $CsPbX_3$ perovskite nanocrystals, marking a major advancement in the nanozyme field and paving the way for novel in vitro disease diagnostic applications.

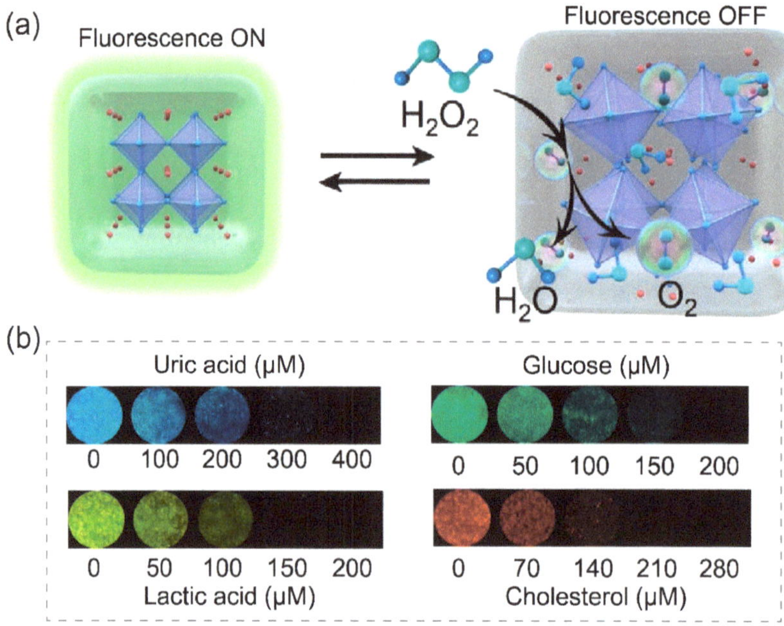

Figure 12. (**a**) Schematic of the bio-catalytic activity of PM-$CsPbX_3$ nanocrystals. (**b**) Photographs of oxidase/PM-$CsPbX_3$ nanocrystals materials with different substrates under UV light. Reprinted with permission from ref. [136]. Copyright 2021, Wiley-VCH GmbH.

4. Halide Perovskite Nanocrystals for Other Novel Applications

Quantitative fluorescence immunoassays are vital for point-of-care testing in food safety and clinical healthcare [137]. However, existing labeling materials, ranging from organic luciferins to quantum dots, often fall short in meeting the increasing demand for multi-target detection due to their broad emission spectra [138,139]. In a recent study, researchers introduced halide perovskite nanocrystals as novel labeling materials, which feature a narrower photoluminescence spectrum [140]. A sensitive and quantitative immunoassay method, the perovskite nanocrystal fluorescence-linked immunosorbent assay, was also demonstrated. To achieve stable and water-dispersible halide perovskite nanocrystals, they were functionalized with hydroxyl groups. Perovskite antibodies as fluorescent probes were created through electrostatic adsorption with a narrow FWHM of 18 nm, laying the groundwork for future multi-target detection applications. Both indirect competitive and sandwich immunoassays were successfully demonstrated. This work presents a promising labeling material for point-of-care tests and paves the way for the next generation of immunoassays.

Single-molecule localization microscopy is crucial for visualizing cellular microstructures [141,142]. However, existing microscopy imaging materials, such as organic fluorophores and quantum dots, often fail to meet the increasing demand for multi-target detection due to their broad emission spectra [143]. In a recent study, a core–shell nanocomposite, $CsPb(ClBr)_3$@PEG, was introduced [144]. This probe was synthesized through a one-step encapsulation of hydrophilic $CsPbBr_3$ nanocrystals into functionalized PEG. The resulting material exhibits wavelength tunability with a narrow FWHM of 11 nm. The PEG shell provides broad spectral tunability, superior photostability lasting several years, and the potential for further surface functionalization. $CsPb(ClBr)_3$@PEG achieves high localization precision and a tenfold improvement in spatial resolution. When applied to imaging exosomes, this probe successfully distinguished two adjacent exosomes. Furthermore, after biotin modification, $CsPb(ClBr)_3$@PEG proved versatile for microscopy imaging of cellular microstructures. Its remarkable photostability and narrow FWHM suggest that this $CsPbBr_3$-based nanoprobe holds significant potential as a commercial dye for multi-target super-resolution bioimaging.

5. Conclusions and Perspectives

Halide perovskite nanocrystals have garnered considerable attention in the biomedical field due to their excellent optical properties, including high quantum yields, tunable emission wavelengths, and narrow emission spectra. These characteristics make them ideal candidates for applications such as bioimaging and chemical sensing. However, despite their promising potential, several significant challenges remain, particularly concerning their metabolic pathways, which limit their widespread use in biomedical applications. Notably, the metabolic behavior of halide perovskite nanocrystals within the body, their eventual degradation products, and their potential systemic toxicity are not yet well understood. Addressing these issues will be essential for ensuring the safe and effective use of these materials in clinical settings. Below are some of the challenges and future research directions regarding the use of halide perovskite nanocrystals in biomedical applications.

Unclear Metabolic Pathways and Degradation Products: One of the most critical concerns when using nanomaterials in biomedical applications is understanding how these materials are metabolized by the body. In the case of halide perovskite nanocrystals, there is currently a lack of clarity regarding their metabolic pathways in vivo [145]. It remains unclear how these nanocrystals are processed by the body, where they accumulate, and how they are ultimately excreted. This uncertainty complicates the prediction of long-term effects and the optimization of perovskite-based materials for in vivo use. Therefore, future research should focus on understanding the metabolic pathways of halide perovskite nanocrystals in vivo. Studies that track the distribution, metabolism, and excretion of these nanocrystals will be essential for assessing their safety and efficacy. Advanced imaging techniques, such as in vivo fluorescence tracking and mass spectrometry, could be employed to monitor the fate of perovskite nanocrystals in living organisms. Additionally, identifying the degradation products of perovskite nanocrystals and assessing their toxicity will be critical.

Bioaccumulation and Clearance: Another major issue is the potential for the bioaccumulation of halide perovskite nanocrystals in the body. If the nanocrystals are not efficiently cleared by the body after use, they could accumulate in various organs, leading to chronic toxicity. The bioaccumulation of nanomaterials depends on factors such as particle size, surface charge, and surface functionalization. However, these factors are not well understood in the context of halide perovskite nanocrystals. Moreover, research on the clearance mechanisms of these nanocrystals is still in its infancy, and the routes by which these materials are eliminated from the body remain unclear. Therefore, extensive studies are needed to understand how halide perovskite nanocrystals are cleared from the body. Investigating the bioaccumulation potential of these nanocrystals in various organs, as well as their clearance pathways, will provide crucial information on their long-term safety. Nanomaterials that are easily cleared from the body are less likely to pose risks, so research into enhancing the excretion of perovskite nanocrystals (through renal or hepatic pathways) should be prioritized. Additionally, the optimization of particle size, surface charge, and functionalization for efficient renal or lymphatic clearance could significantly reduce the risk of bioaccumulation.

Funding: This research was funded by Shandong Provincial Natural Science Foundation, grant number ZR2024QE319.

Data Availability Statement: No new data were created or analyzed in this study.

Conflicts of Interest: The authors declare no conflicts of interest.

References

1. Protesescu, L.; Yakunin, S.; Bodnarchuk, M.I.; Krieg, F.; Caputo, R.; Hendon, C.H.; Yang, R.X.; Walsh, A.; Kovalenko, M.V. Nanocrystals of Cesium Lead Halide Perovskites ($CsPbX_3$, X = Cl, Br, and I): Novel Optoelectronic Materials Showing Bright Emission with Wide Color Gamut. *Nano Lett.* **2015**, *15*, 3692–3696. [CrossRef] [PubMed]
2. Nedelcu, G.; Protesescu, L.; Yakunin, S.; Bodnarchuk, M.I.; Grotevent, M.J.; Kovalenko, M.V. Fast Anion-Exchange in Highly Luminescent Nanocrystals of Cesium Lead Halide Perovskites ($CsPbX_3$, X = Cl, Br, I). *Nano Lett.* **2015**, *15*, 5635–5640. [CrossRef] [PubMed]
3. Dey, A.; Ye, J.; De, A.; Debroye, E.; Ha, S.K.; Bladt, E.; Kshirsagar, A.S.; Wang, Z.; Yin, J.; Wang, Y.; et al. State of the Art and Prospects for Halide Perovskite Nanocrystals. *ACS Nano* **2021**, *15*, 10775–10981. [CrossRef] [PubMed]
4. Zhang, F.; Zhong, H.; Chen, C.; Wu, X.; Hu, X.; Huang, H.; Han, J.; Zou, B.; Dong, Y. Brightly Luminescent and Color-Tunable Colloidal CH_3 NH_3 PbX_3 (X = Br, I, Cl) Quantum Dots: Potential Alternatives for Display Technology. *ACS Nano* **2015**, *9*, 4533–4542. [CrossRef]
5. Song, P.; Qiao, B.; Song, D.; Cao, J.; Shen, Z.; Xu, Z.; Zhao, S.; Wageh, S.; Al-Ghamdi, A. Modifying the Crystal Field of $CsPbCl_3$:Mn^{2+} Nanocrystals by Co-Doping to Enhance Its Red Emission by a Hundredfold. *ACS Appl. Mater. Interfaces* **2020**, *12*, 30711–30719. [CrossRef]
6. Hu, M.; Jia, S.; Liu, Y.; Cui, J.; Zhang, Y.; Su, H.; Cao, S.; Mo, L.; Chu, D.; Zhao, G.; et al. Large and Dense Organic–Inorganic Hybrid Perovskite CH_3 NH_3 PbI_3 Wafer Fabricated by One-Step Reactive Direct Wafer Production with High X-Ray Sensitivity. *ACS Appl. Mater. Interfaces* **2020**, *12*, 16592–16600. [CrossRef]
7. Yuan, R.; Shen, L.; Shen, C.; Liu, J.; Zhou, L.; Xiang, W.; Liang, X. $CsPbBr_3$: *X* Eu^{3+} Perovskite QD Borosilicate Glass: A New Member of the Luminescent Material Family. *Chem. Commun.* **2018**, *54*, 3395–3398. [CrossRef]
8. Raja, S.N.; Bekenstein, Y.; Koc, M.A.; Fischer, S.; Zhang, D.; Lin, L.; Ritchie, R.O.; Yang, P.; Alivisatos, A.P. Encapsulation of Perovskite Nanocrystals into Macroscale Polymer Matrices: Enhanced Stability and Polarization. *ACS Appl. Mater. Interfaces* **2016**, *8*, 35523–35533. [CrossRef]
9. Mendoza-Carreño, J.; Molet, P.; Otero-Martínez, C.; Alonso, M.I.; Polavarapu, L.; Mihi, A. Nanoimprinted 2D-Chiral Perovskite Nanocrystal Metasurfaces for Circularly Polarized Photoluminescence. *Adv. Mater.* **2023**, *35*, 2210477. [CrossRef]
10. Song, J.; Li, J.; Li, X.; Xu, L.; Dong, Y.; Zeng, H. Quantum Dot Light-Emitting Diodes Based on Inorganic Perovskite Cesium Lead Halides ($CsPbX_3$). *Adv. Mater.* **2015**, *27*, 7162–7167. [CrossRef]
11. Meinardi, F.; Akkerman, Q.A.; Bruni, F.; Park, S.; Mauri, M.; Dang, Z.; Manna, L.; Brovelli, S. Doped Halide Perovskite Nanocrystals for Reabsorption-Free Luminescent Solar Concentrators. *ACS Energy Lett.* **2017**, *2*, 2368–2377. [CrossRef] [PubMed]
12. Sun, R.; Lu, P.; Zhou, D.; Xu, W.; Ding, N.; Shao, H.; Zhang, Y.; Li, D.; Wang, N.; Zhuang, X.; et al. Samarium-Doped Metal Halide Perovskite Nanocrystals for Single-Component Electroluminescent White Light-Emitting Diodes. *ACS Energy Lett.* **2020**, *5*, 2131–2139. [CrossRef]
13. Gualdrón-Reyes, A.F.; Masi, S.; Mora-Seró, I. Progress in Halide-Perovskite Nanocrystals with near-Unity Photoluminescence Quantum Yield. *Trends Chem.* **2021**, *3*, 499–511. [CrossRef]
14. De Trizio, L.; Infante, I.; Manna, L. Surface Chemistry of Lead Halide Perovskite Colloidal Nanocrystals. *Acc. Chem. Res.* **2023**, *56*, 1815–1825. [CrossRef]
15. Pan, G.; Bai, X.; Yang, D.; Chen, X.; Jing, P.; Qu, S.; Zhang, L.; Zhou, D.; Zhu, J.; Xu, W.; et al. Doping Lanthanide into Perovskite Nanocrystals: Highly Improved and Expanded Optical Properties. *Nano Lett.* **2017**, *17*, 8005–8011. [CrossRef]
16. Hong, W.; Huang, Y.; Chang, C.; Zhang, Z.; Tsai, H.; Chang, N.; Chao, Y. Efficient Low-Temperature Solution-Processed Lead-Free Perovskite Infrared Light-Emitting Diodes. *Adv. Mater.* **2016**, *28*, 8029–8036. [CrossRef]
17. Zhu, F.-P.; Yong, Z.-J.; Liu, B.-M.; Chen, Y.-M.; Zhou, Y.; Ma, J.-P.; Sun, H.-T.; Fang, Y.-Z. Superbroad Near-Infrared Photoluminescence from Bismuth-Doped CsPbI_3 Perovskite Nanocrystals. *Opt. Express* **2017**, *25*, 33283. [CrossRef]
18. Liang, J.; Wang, C.; Zhao, P.; Lu, Z.; Ma, Y.; Xu, Z.; Wang, Y.; Zhu, H.; Hu, Y.; Zhu, G.; et al. Solution Synthesis and Phase Control of Inorganic Perovskites for High-Performance Optoelectronic Devices. *Nanoscale* **2017**, *9*, 11841–11845. [CrossRef]
19. Zhang, B.; Guo, F.; Xue, J.; Yang, L.; Zhao, Y.; Ge, M.; Cai, Q.; Liu, B.; Xie, Z.; Chen, D.; et al. Photoluminescence Study of the Photoinduced Phase Separation in Mixed-Halide Hybrid Perovskite $CH_3NH_3Pb(Br_xI_{1-x})3$ Crystals Synthesized via a Solvothermal Method. *Sci. Rep.* **2017**, *7*, 17695. [CrossRef]
20. Zhao, H.; Hu, Z.; Wei, L.; Zeng, P.; Kuang, C.; Liu, X.; Bai, S.; Gao, F.; Liu, M. Efficient and High-Luminance Perovskite Light-Emitting Diodes Based on $CsPbBr_3$ Nanocrystals Synthesized from a Dual-Purpose Organic Lead Source. *Small* **2020**, *16*, 2003939. [CrossRef]

21. Liu, Z.; Qin, X.; Chen, Q.; Jiang, T.; Chen, Q.; Liu, X. Metal–Halide Perovskite Nanocrystal Superlattice: Self-Assembly and Optical Fingerprints. *Adv. Mater.* **2023**, *35*, 2209279. [CrossRef] [PubMed]
22. Zhou, Z.-R.; Liao, Z.-H.; Wang, F. Shape-Controlled Synthesis of One-Dimensional Cesium Lead Halide Perovskite Nanocrystals: Methods and Advances. *J. Mater. Chem. C* **2023**, *11*, 3409–3427. [CrossRef]
23. Mohammed, O.F. Outstanding Challenges of Zero-Dimensional Perovskite Materials. *J. Phys. Chem. Lett.* **2019**, *10*, 5886–5888. [CrossRef]
24. Smock, S.R.; Chen, Y.; Rossini, A.J.; Brutchey, R.L. The Surface Chemistry and Structure of Colloidal Lead Halide Perovskite Nanocrystals. *Acc. Chem. Res.* **2021**, *54*, 707–718. [CrossRef]
25. Otero-Martínez, C.; García-Lojo, D.; Pastoriza-Santos, I.; Pérez-Juste, J.; Polavarapu, L. Dimensionality Control of Inorganic and Hybrid Perovskite Nanocrystals by Reaction Temperature: From No-Confinement to 3D and 1D Quantum Confinement. *Angew. Chem.* **2021**, *133*, 26881–26888. [CrossRef]
26. Liu, J.; Zheng, X.; Mohammed, O.F.; Bakr, O.M. Self-Assembly and Regrowth of Metal Halide Perovskite Nanocrystals for Optoelectronic Applications. *Acc. Chem. Res.* **2022**, *55*, 262–274. [CrossRef]
27. Mondal, N.; De, A.; Das, S.; Paul, S.; Samanta, A. Ultrafast Carrier Dynamics of Metal Halide Perovskite Nanocrystals and Perovskite-Composites. *Nanoscale* **2019**, *11*, 9796–9818. [CrossRef]
28. Yang, B.; Han, K. Charge-Carrier Dynamics of Lead-Free Halide Perovskite Nanocrystals. *Acc. Chem. Res.* **2019**, *52*, 3188–3198. [CrossRef]
29. Kamat, P.V.; Kuno, M. Halide Ion Migration in Perovskite Nanocrystals and Nanostructures. *Acc. Chem. Res.* **2021**, *54*, 520–531. [CrossRef]
30. Ramasamy, P.; Lim, D.-H.; Kim, B.; Lee, S.-H.; Lee, M.-S.; Lee, J.-S. All-Inorganic Cesium Lead Halide Perovskite Nanocrystals for Photodetector Applications. *Chem. Commun.* **2016**, *52*, 2067–2070. [CrossRef]
31. Yan, F.; Tan, S.T.; Li, X.; Demir, H.V. Light Generation in Lead Halide Perovskite Nanocrystals: LEDs, Color Converters, Lasers, and Other Applications. *Small* **2019**, *15*, 1902079. [CrossRef] [PubMed]
32. Liu, C.; Zeng, Q.; Wei, H.; Yu, Y.; Zhao, Y.; Feng, T.; Yang, B. Metal Halide Perovskite Nanocrystal Solar Cells: Progress and Challenges. *Small Methods* **2020**, *4*, 2000419. [CrossRef]
33. Kim, Y.-H.; Wolf, C.; Kim, Y.-T.; Cho, H.; Kwon, W.; Do, S.; Sadhanala, A.; Park, C.G.; Rhee, S.-W.; Im, S.H.; et al. Highly Efficient Light-Emitting Diodes of Colloidal Metal–Halide Perovskite Nanocrystals beyond Quantum Size. *ACS Nano* **2017**, *11*, 6586–6593. [CrossRef] [PubMed]
34. Akkerman, Q.A.; Rainò, G.; Kovalenko, M.V.; Manna, L. Genesis, Challenges and Opportunities for Colloidal Lead Halide Perovskite Nanocrystals. *Nat. Mater.* **2018**, *17*, 394–405. [CrossRef]
35. Ochsenbein, S.T.; Krieg, F.; Shynkarenko, Y.; Rainò, G.; Kovalenko, M.V. Engineering Color-Stable Blue Light-Emitting Diodes with Lead Halide Perovskite Nanocrystals. *ACS Appl. Mater. Interfaces* **2019**, *11*, 21655–21660. [CrossRef]
36. Liu, D.; Guo, Y.; Que, M.; Yin, X.; Liu, J.; Xie, H.; Zhang, C.; Que, W. Metal Halide Perovskite Nanocrystals: Application in High-Performance Photodetectors. *Mater. Adv.* **2021**, *2*, 856–879. [CrossRef]
37. Surendran, A.; Yu, X.; Begum, R.; Tao, Y.; Wang, Q.J.; Leong, W.L. All Inorganic Mixed Halide Perovskite Nanocrystal–Graphene Hybrid Photodetector: From Ultrahigh Gain to Photostability. *ACS Appl. Mater. Interfaces* **2019**, *11*, 27064–27072. [CrossRef]
38. Zhang, Q.; Yin, Y. All-Inorganic Metal Halide Perovskite Nanocrystals: Opportunities and Challenges. *ACS Cent. Sci.* **2018**, *4*, 668–679. [CrossRef]
39. Zhong, F.; He, Y.; Sun, Y.; Dong, F.; Sheng, J. Metal-Doping of Halide Perovskite Nanocrystals for Energy and Environmental Photocatalysis: Challenges and Prospects. *J. Mater. Chem. A* **2022**, *10*, 22915–22928. [CrossRef]
40. Shyamal, S.; Pradhan, N. Halide Perovskite Nanocrystal Photocatalysts for CO_2 Reduction: Successes and Challenges. *J. Phys. Chem. Lett.* **2020**, *11*, 6921–6934. [CrossRef]
41. Zhou, L.; Xu, Y.; Chen, B.; Kuang, D.; Su, C. Synthesis and Photocatalytic Application of Stable Lead-Free Cs_2 $AgBiBr_6$ Perovskite Nanocrystals. *Small* **2018**, *14*, 1703762. [CrossRef] [PubMed]
42. Wang, J.; Shi, Y.; Wang, Y.; Li, Z. Rational Design of Metal Halide Perovskite Nanocrystals for Photocatalytic CO_2 Reduction: Recent Advances, Challenges, and Prospects. *ACS Energy Lett.* **2022**, *7*, 2043–2059. [CrossRef]
43. Cortés-Villena, A.; Bellezza, D.; Cunha, C.; Rosa-Pardo, I.; Seijas-Da Silva, Á.; Pina, J.; Abellán, G.; Seixas De Melo, J.S.; Galian, R.E.; Pérez-Prieto, J. Engineering Metal Halide Perovskite Nanocrystals with BODIPY Dyes for Photosensitization and Photocatalytic Applications. *J. Am. Chem. Soc.* **2024**, *146*, 14479–14492. [CrossRef] [PubMed]
44. Zhang, L.; Miao, J.; Li, J.; Li, Q. Halide Perovskite Materials for Energy Storage Applications. *Adv. Funct. Mater.* **2020**, *30*, 2003653. [CrossRef]
45. Kostopoulou, A.; Brintakis, K.; Nasikas, N.K.; Stratakis, E. Perovskite Nanocrystals for Energy Conversion and Storage. *Nanophotonics* **2019**, *8*, 1607–1640. [CrossRef]
46. Narayanan, S.; Parikh, N.; Tavakoli, M.M.; Pandey, M.; Kumar, M.; Kalam, A.; Trivedi, S.; Prochowicz, D.; Yadav, P. Metal Halide Perovskites for Energy Storage Applications. *Eur. J. Inorg. Chem.* **2021**, *2021*, 1201–1212. [CrossRef]

47. Ngo, L.T.; Huang, W.; Chan, M.; Su, T.; Li, C.; Hsiao, M.; Liu, R. Comprehensive Neurotoxicity of Lead Halide Perovskite Nanocrystals in Nematode *Caenorhabditis elegans*. *Small* **2024**, *20*, 2306020. [CrossRef]
48. Infante, I.; Manna, L. Are There Good Alternatives to Lead Halide Perovskite Nanocrystals? *Nano Lett.* **2021**, *21*, 6–9. [CrossRef]
49. Babayigit, A.; Ethirajan, A.; Muller, M.; Conings, B. Toxicity of Organometal Halide Perovskite Solar Cells. *Nat. Mater.* **2016**, *15*, 247–251. [CrossRef]
50. Ren, M.; Qian, X.; Chen, Y.; Wang, T.; Zhao, Y. Potential Lead Toxicity and Leakage Issues on Lead Halide Perovskite Photovoltaics. *J. Hazard. Mater.* **2022**, *426*, 127848. [CrossRef]
51. Patsiou, D.; Del Rio-Cubilledo, C.; Catarino, A.I.; Summers, S.; Mohd Fahmi, A.; Boyle, D.; Fernandes, T.F.; Henry, T.B. Exposure to Pb-Halide Perovskite Nanoparticles Can Deliver Bioavailable Pb but Does Not Alter Endogenous Gut Microbiota in Zebrafish. *Sci. Total Environ.* **2020**, *715*, 136941. [CrossRef] [PubMed]
52. Dong, Y.; Zhao, Y.; Zhang, S.; Dai, Y.; Liu, L.; Li, Y.; Chen, Q. Recent Advances toward Practical Use of Halide Perovskite Nanocrystals. *J. Mater. Chem. A* **2018**, *6*, 21729–21746. [CrossRef]
53. Zhang, J.; Zhu, Y. Exploiting the Photo-Physical Properties of Metal Halide Perovskite Nanocrystals for Bioimaging. *ChemBioChem* **2024**, *25*, e202300683. [CrossRef]
54. Yang, D.; Li, X.; Zeng, H. Surface Chemistry of All Inorganic Halide Perovskite Nanocrystals: Passivation Mechanism and Stability. *Adv. Mater. Interfaces* **2018**, *5*, 1701662. [CrossRef]
55. Li, J.; Cao, H.-L.; Jiao, W.-B.; Wang, Q.; Wei, M.; Cantone, I.; Lü, J.; Abate, A. Biological Impact of Lead from Halide Perovskites Reveals the Risk of Introducing a Safe Threshold. *Nat. Commun.* **2020**, *11*, 310. [CrossRef]
56. Liu, M.; Matuhina, A.; Zhang, H.; Vivo, P. Advances in the Stability of Halide Perovskite Nanocrystals. *Materials* **2019**, *12*, 3733. [CrossRef]
57. Luo, B.; Naghadeh, S.B.; Zhang, J.Z. Lead Halide Perovskite Nanocrystals: Stability, Surface Passivation, and Structural Control. *ChemNanoMat* **2017**, *3*, 456–465. [CrossRef]
58. Boote, B.W.; Andaraarachchi, H.P.; Rosales, B.A.; Blome-Fernández, R.; Zhu, F.; Reichert, M.D.; Santra, K.; Li, J.; Petrich, J.W.; Vela, J.; et al. Unveiling the Photo- and Thermal-Stability of Cesium Lead Halide Perovskite Nanocrystals. *ChemPhysChem* **2019**, *20*, 2647–2656. [CrossRef]
59. Chen, B.; Wang, S.; Song, Y.; Li, C.; Hao, F. A Critical Review on the Moisture Stability of Halide Perovskite Films and Solar Cells. *Chem. Eng. J.* **2022**, *430*, 132701. [CrossRef]
60. Li, J.; Wang, L.; Yuan, X.; Bo, B.; Li, H.; Zhao, J.; Gao, X. Ultraviolet Light Induced Degradation of Luminescence in CsPbBr3 Perovskite Nanocrystals. *Mater. Res. Bull.* **2018**, *102*, 86–91. [CrossRef]
61. Chatterjee, S.; Biswas, S.; Sourav, S.; Rath, J.; Akhil, S.; Mishra, N. Strategies To Achieve Long-Term Stability in Lead Halide Perovskite Nanocrystals and Its Optoelectronic Applications. *J. Phys. Chem. Lett.* **2024**, *15*, 10118–10137. [CrossRef] [PubMed]
62. Getachew, G.; Huang, W.-W.; Chou, T.-H.; Rasal, A.S.; Chang, J.-Y. Brightly Luminescent (NH4)xCs1-xPbBr3 Quantum Dots for in Vitro Imaging and Efficient Photothermal Ablation Therapy. *J. Colloid Interface Sci.* **2022**, *605*, 500–512. [CrossRef]
63. Cao, Y.; Shao, Y.; Zhang, J.; Chen, C.; Wang, Q. The Photothermal Stability Study of Silica-Coated CsPbBr3 Perovskite Nanocrystals. *J. Solid State Chem.* **2022**, *311*, 123086. [CrossRef]
64. Getachew, G.; Tien, Y.-C.; Kan, T.-C.; Dirersa, W.B.; Wibrianto, A.; Orchirbat, S.; Chang, J.; Rasal, A.S.; Gurav, V.; Kizhepat, S.; et al. Defect-Passivated Metal Halide Perovskite Quantum Dots Stabilized into Biodegradable Porous Polydopamine Nanoparticles for Photothermal/Chemodynamic/Gas Therapy of Cancer. *Chem. Eng. J.* **2023**, *467*, 143560. [CrossRef]
65. Zheng, X.; Hou, Y.; Sun, H.-T.; Mohammed, O.F.; Sargent, E.H.; Bakr, O.M. Reducing Defects in Halide Perovskite Nanocrystals for Light-Emitting Applications. *J. Phys. Chem. Lett.* **2019**, *10*, 2629–2640. [CrossRef]
66. Ye, J.; Byranvand, M.M.; Martínez, C.O.; Hoye, R.L.Z.; Saliba, M.; Polavarapu, L. Defect Passivation in Lead-Halide Perovskite Nanocrystals and Thin Films: Toward Efficient LEDs and Solar Cells. *Angew. Chem.* **2021**, *133*, 21804–21828. [CrossRef]
67. Jellicoe, T.C.; Richter, J.M.; Glass, H.F.J.; Tabachnyk, M.; Brady, R.; Dutton, S.E.; Rao, A.; Friend, R.H.; Credgington, D.; Greenham, N.C.; et al. Synthesis and Optical Properties of Lead-Free Cesium Tin Halide Perovskite Nanocrystals. *J. Am. Chem. Soc.* **2016**, *138*, 2941–2944. [CrossRef]
68. Chen, L.-J.; Lee, C.-R.; Chuang, Y.-J.; Wu, Z.-H.; Chen, C. Synthesis and Optical Properties of Lead-Free Cesium Tin Halide Perovskite Quantum Rods with High-Performance Solar Cell Application. *J. Phys. Chem. Lett.* **2016**, *7*, 5028–5035. [CrossRef]
69. Liu, T.; Zhao, X.; Li, J.; Liu, Z.; Liscio, F.; Milita, S.; Schroeder, B.C.; Fenwick, O. Enhanced Control of Self-Doping in Halide Perovskites for Improved Thermoelectric Performance. *Nat. Commun.* **2019**, *10*, 5750. [CrossRef]
70. Zhong, Q.; Cao, M.; Zhang, Q. Encapsulation of Lead Halide Perovskite Nanocrystals (NCs) at the Single-Particle Level: Strategies and Properties. *Nanoscale* **2021**, *13*, 19341–19351. [CrossRef]
71. Lv, W.; Li, L.; Xu, M.; Hong, J.; Tang, X.; Xu, L.; Wu, Y.; Zhu, R.; Chen, R.; Huang, W. Improving the Stability of Metal Halide Perovskite Quantum Dots by Encapsulation. *Adv. Mater.* **2019**, *31*, 1900682. [CrossRef] [PubMed]
72. Kitazawa, N. Preparation and Optical Properties of Nanocrystalline $(C_6H_5C_2H_4NH_3)_2PbI_4$-Doped PMMA Films. *J. Mater. Sci.* **1998**, *33*, 1441–1444. [CrossRef]

73. Papavassiliou, G.C.; Patsis, A.P.; Lagouvardos, D.J.; Koutselas, I.B. Spectroscopic studies of $(C_{10}H_{21}NH_3)_2PbI_4$,$(CH_3NH_3)(C_{10}H_{21}NH_3)$ $2Pb_2I_7$,$(CH_3NH_3)PbI_3$, and similar compounds. *Synth. Met.* **1993**, *57*, 3889–3894. [CrossRef]
74. Li, S.; Lei, D.; Ren, W.; Guo, X.; Wu, S.; Zhu, Y.; Rogach, A.L.; Chhowalla, M.; Jen, A.K.-Y. Water-Resistant Perovskite Nanodots Enable Robust Two-Photon Lasing in Aqueous Environment. *Nat. Commun.* **2020**, *11*, 1192. [CrossRef] [PubMed]
75. Huang, S.; Li, Z.; Kong, L.; Zhu, N.; Shan, A.; Li, L. Enhancing the Stability of CH_3 NH_3 $PbBr_3$ Quantum Dots by Embedding in Silica Spheres Derived from Tetramethyl Orthosilicate in "Waterless" Toluene. *J. Am. Chem. Soc.* **2016**, *138*, 5749–5752. [CrossRef]
76. Zheng, J.; Zhang, W.; Huang, Y.; Shao, J.; Khan, M.S.; Chi, Y. Encapsulation of Pure Water-Stable Perovskite Nanocrystals (PNCs) into Biological Environment-Stable PNCs for Cell Imaging. *Inorg. Chem.* **2024**, *63*, 5623–5633. [CrossRef]
77. Dirin, D.N.; Benin, B.M.; Yakunin, S.; Krumeich, F.; Raino, G.; Frison, R.; Kovalenko, M.V. Microcarrier-Assisted Inorganic Shelling of Lead Halide Perovskite Nanocrystals. *ACS Nano* **2019**, *13*, 11642–11652. [CrossRef]
78. Jia, X.; Shi, T.; Zhang, Q.; Huang, H.; Chu, P.K.; Yu, X.-F.; He, R. Perovskite Nanocrystals Photo-Initiated in Situ Encapsulation for Optical Tracking. *Chem. Eng. J.* **2024**, *500*, 156756. [CrossRef]
79. Zhang, H.; Wang, X.; Liao, Q.; Xu, Z.; Li, H.; Zheng, L.; Fu, H. Embedding Perovskite Nanocrystals into a Polymer Matrix for Tunable Luminescence Probes in Cell Imaging. *Adv. Funct. Mater.* **2017**, *27*, 1604382. [CrossRef]
80. Talianov, P.M.; Yakubova, A.A.; Bukreeva, A.; Masharin, M.; Eliseev, I.E.; Zelenkov, L.; Muslimov, A.R.; Bukatin, A.; Gordeeva, A.; Kudryavtseva, V.; et al. Incorporation of Perovskite Nanocrystals into Polymer Matrix for Enhanced Stability in Biological Media: In Vitro and In Vivo Studies. *ACS Appl. Bio Mater.* **2022**, *5*, 2411–2420. [CrossRef]
81. Chen, B.; Wang, F. Emerging Frontiers of Upconversion Nanoparticles. *Trends Chem.* **2020**, *2*, 427–439. [CrossRef]
82. Wang, F.; Banerjee, D.; Liu, Y.; Chen, X.; Liu, X. Upconversion Nanoparticles in Biological Labeling, Imaging, and Therapy. *The Analyst* **2010**, *135*, 1839. [CrossRef] [PubMed]
83. Wen, S.; Zhou, J.; Zheng, K.; Bednarkiewicz, A.; Liu, X.; Jin, D. Advances in Highly Doped Upconversion Nanoparticles. *Nat. Commun.* **2018**, *9*, 2415. [CrossRef] [PubMed]
84. He, Y.; Zhang, L.; Chen, G.; Liu, Y.; Shi, S.; Jiang, P.; Ding, J.; Xu, S.; Geng, C. ZnO/SiO_2 Encapsulation of Perovskite Nanocrystals for Efficient and Stable Light-Emitting Diodes. *Appl. Surf. Sci.* **2023**, *611*, 155724. [CrossRef]
85. Miao, Y.; Xie, R.; Kan, Q.; Yu, Y.; Dong, S.; Wang, S.; Mao, L. Silica-Stabilized Lead Halide Perovskite Nanocrystals: Advantages, Progress, and Future Directions. *Adv. Opt. Mater.* **2024**, *13*, 2402320. [CrossRef]
86. Lou, S.; Zhou, Z.; Xuan, T.; Li, H.; Jiao, J.; Zhang, H.; Gautier, R.; Wang, J. Chemical Transformation of Lead Halide Perovskite into Insoluble, Less Cytotoxic, and Brightly Luminescent $CsPbBr_3$ /$CsPb_2$ Br_5 Composite Nanocrystals for Cell Imaging. *ACS Appl. Mater. Interfaces* **2019**, *11*, 24241–24246. [CrossRef]
87. Wei, Y.; Deng, X.; Xie, Z.; Cai, X.; Liang, S.; Ma, P.; Hou, Z.; Cheng, Z.; Lin, J. Enhancing the Stability of Perovskite Quantum Dots by Encapsulation in Crosslinked Polystyrene Beads via a Swelling–Shrinking Strategy toward Superior Water Resistance. *Adv. Funct. Mater.* **2017**, *27*, 1703535. [CrossRef]
88. Kim, Y.-H.; Kim, S.; Kakekhani, A.; Park, J.; Park, J.; Lee, Y.-H.; Xu, H.; Nagane, S.; Wexler, R.B.; Kim, D.-H.; et al. Comprehensive Defect Suppression in Perovskite Nanocrystals for High-Efficiency Light-Emitting Diodes. *Nat. Photonics* **2021**, *15*, 148–155. [CrossRef]
89. Seth, S.; Ahmed, T.; De, A.; Samanta, A. Tackling the Defects, Stability, and Photoluminescence of $CsPbX_3$ Perovskite Nanocrystals. *ACS Energy Lett.* **2019**, *4*, 1610–1618. [CrossRef]
90. Ten Brinck, S.; Zaccaria, F.; Infante, I. Defects in Lead Halide Perovskite Nanocrystals: Analogies and (Many) Differences with the Bulk. *ACS Energy Lett.* **2019**, *4*, 2739–2747. [CrossRef]
91. Avugadda, S.K.; Castelli, A.; Dhanabalan, B.; Fernandez, T.; Silvestri, N.; Collantes, C.; Baranov, D.; Imran, M.; Manna, L.; Pellegrino, T.; et al. Highly Emitting Perovskite Nanocrystals with 2-Year Stability in Water through an Automated Polymer Encapsulation for Bioimaging. *ACS Nano* **2022**, *16*, 13657–13666. [CrossRef] [PubMed]
92. Tedetti, M.; Sempéré, R. Penetration of Ultraviolet Radiation in the Marine Environment. A Review. *Photochem. Photobiol.* **2006**, *82*, 389–397. [CrossRef] [PubMed]
93. Bucci, P.; Prieto, M.J.; Milla, L.; Calienni, M.N.; Martinez, L.; Rivarola, V.; Alonso, S.; Montanari, J. Skin Penetration and UV-damage Prevention by Nanoberries. *J. Cosmet. Dermatol.* **2018**, *17*, 889–899. [CrossRef] [PubMed]
94. Kim, F.H.; Penumadu, D.; Kardjilov, N.; Manke, I. High-Resolution X-Ray and Neutron Computed Tomography of Partially Saturated Granular Materials Subjected to Projectile Penetration. *Int. J. Impact Eng.* **2016**, *89*, 72–82. [CrossRef]
95. Villarraga-Gómez, H.; Morse, E.P.; Smith, S.T. Assessing the Effect of Penetration Length Variations on Dimensional Measurements with X-Ray Computed Tomography. *Precis. Eng.* **2023**, *79*, 146–163. [CrossRef]
96. Slavíková, M.; Krejčí, F.; Žemlička, J.; Pech, M.; Kotlík, P.; Jakůbek, J. X-Ray Radiography and Tomography for Monitoring the Penetration Depth of Consolidants in Opuka—The Building Stone of Prague Monuments. *J. Cult. Herit.* **2012**, *13*, 357–364. [CrossRef]
97. Hao, S.; Daemi, S.R.; Heenan, T.M.M.; Du, W.; Tan, C.; Storm, M.; Rau, C.; Brett, D.J.L.; Shearing, P.R. Tracking Lithium Penetration in Solid Electrolytes in 3D by In-Situ Synchrotron X-Ray Computed Tomography. *Nano Energy* **2021**, *82*, 105744. [CrossRef]

98. Bjärngard, B.E.; Shackford, H. Attenuation in High-Energy X-Ray Beams. *Med. Phys.* **1994**, *21*, 1069–1073. [CrossRef]
99. Archer, B.R.; Fewell, T.R.; Conway, B.J.; Quinn, P.W. Attenuation Properties of Diagnostic X-ray Shielding Materials. *Med. Phys.* **1994**, *21*, 1499–1507. [CrossRef]
100. Ryu, I.; Ryu, J.; Choe, G.; Kwon, H.; Park, H.; Cho, Y.; Du, R.; Yim, S. In Vivo Plain X-Ray Imaging of Cancer Using Perovskite Quantum Dot Scintillators. *Adv. Funct. Mater.* **2021**, *31*, 2102334. [CrossRef]
101. Tseng, W.W.; Winer, D.; Kenkel, J.A.; Choi, O.; Shain, A.H.; Pollack, J.R.; French, R.; Lowy, A.M.; Engleman, E.G. Development of an Orthotopic Model of Invasive Pancreatic Cancer in an Immunocompetent Murine Host. *Clin. Cancer Res.* **2010**, *16*, 3684–3695. [CrossRef]
102. Kim, M.P.; Evans, D.B.; Wang, H.; Abbruzzese, J.L.; Fleming, J.B.; Gallick, G.E. Generation of Orthotopic and Heterotopic Human Pancreatic Cancer Xenografts in Immunodeficient Mice. *Nat. Protoc.* **2009**, *4*, 1670–1680. [CrossRef] [PubMed]
103. Hoffman, R.M. Orthotopic Metastatic Mouse Models for Anticancer Drug Discovery and Evaluation: A Bridge to the Clinic. *Investig. New Drugs* **1999**, *17*, 343–360. [CrossRef] [PubMed]
104. Provenzano, P.P.; Cuevas, C.; Chang, A.E.; Goel, V.K.; Von Hoff, D.D.; Hingorani, S.R. Enzymatic Targeting of the Stroma Ablates Physical Barriers to Treatment of Pancreatic Ductal Adenocarcinoma. *Cancer Cell* **2012**, *21*, 418–429. [CrossRef] [PubMed]
105. Brix, G.; Nissen-Meyer, S.; Lechel, U.; Nissen-Meyer, J.; Griebel, J.; Nekolla, E.A.; Becker, C.; Reiser, M. Radiation Exposures of Cancer Patients from Medical X-Rays: How Relevant Are They for Individual Patients and Population Exposure? *Eur. J. Radiol.* **2009**, *72*, 342–347. [CrossRef] [PubMed]
106. Damilakis, J.; Adams, J.E.; Guglielmi, G.; Link, T.M. Radiation Exposure in X-Ray-Based Imaging Techniques Used in Osteoporosis. *Eur. Radiol.* **2010**, *20*, 2707–2714. [CrossRef] [PubMed]
107. Knight, S.P. A Paediatric X-Ray Exposure Chart. *J. Med. Radiat. Sci.* **2014**, *61*, 191–201. [CrossRef]
108. Hwang, S.-Y.; Choi, E.-S.; Kim, Y.-S.; Gim, B.-E.; Ha, M.; Kim, H.-Y. Health Effects from Exposure to Dental Diagnostic X-Ray. *Environ. Health Toxicol.* **2018**, *33*, e2018017. [CrossRef]
109. Lammertyn, J.; Peirs, A.; De Baerdemaeker, J.; Nicolaı, B. Light Penetration Properties of NIR Radiation in Fruit with Respect to Non-Destructive Quality Assessment. *Postharvest Biol. Technol.* **2000**, *18*, 121–132. [CrossRef]
110. Xiong, X.; Xu, Z.; Huang, H.; Wang, Y.; Zhao, J.; Guo, X.; Zhou, S. A NIR Light Triggered Disintegratable Nanoplatform for Enhanced Penetration and Chemotherapy in Deep Tumor Tissues. *Biomaterials* **2020**, *245*, 119840. [CrossRef]
111. Sun, A.; Guo, H.; Gan, Q.; Yang, L.; Liu, Q.; Xi, L. Evaluation of Visible NIR-I and NIR-II Light Penetration for Photoacoustic Imaging in Rat Organs. *Opt. Express* **2020**, *28*, 9002. [CrossRef] [PubMed]
112. Xu, W.; Wang, D.; Tang, B.Z. NIR-II AIEgens: A Win–Win Integration towards Bioapplications. *Angew. Chem. Int. Ed.* **2021**, *60*, 7476–7487. [CrossRef] [PubMed]
113. Shaw, P.A.; Forsyth, E.; Haseeb, F.; Yang, S.; Bradley, M.; Klausen, M. Two-Photon Absorption: An Open Door to the NIR-II Biological Window? *Front. Chem.* **2022**, *10*, 921354. [CrossRef]
114. Cao, J.; Zhu, B.; Zheng, K.; He, S.; Meng, L.; Song, J.; Yang, H. Recent Progress in NIR-II Contrast Agent for Biological Imaging. *Front. Bioeng. Biotechnol.* **2020**, *7*, 487. [CrossRef]
115. Yao, S.; Belfield, K.D. Two-Photon Fluorescent Probes for Bioimaging. *Eur. J. Org. Chem.* **2012**, *2012*, 3199–3217. [CrossRef]
116. Bestvater, F.; Spiess, E.; Stobrawa, G.; Hacker, M.; Feurer, T.; Porwol, T.; Berchner-Pfannschmidt, U.; Wotzlaw, C.; Acker, H. Two-photon Fluorescence Absorption and Emission Spectra of Dyes Relevant for Cell Imaging. *J. Microsc.* **2002**, *208*, 108–115. [CrossRef]
117. Jethva, P.; Momin, M.; Khan, T.; Omri, A. Lanthanide-Doped Upconversion Luminescent Nanoparticles—Evolving Role in Bioimaging, Biosensing, and Drug Delivery. *Materials* **2022**, *15*, 2374. [CrossRef]
118. Yi, Z.; Luo, Z.; Qin, X.; Chen, Q.; Liu, X. Lanthanide-Activated Nanoparticles: A Toolbox for Bioimaging, Therapeutics, and Neuromodulation. *Acc. Chem. Res.* **2020**, *53*, 2692–2704. [CrossRef]
119. Ma, Q.; Wang, J.; Li, Z.; Lv, X.; Liang, L.; Yuan, Q. Recent Progress in Time-Resolved Biosensing and Bioimaging Based on Lanthanide-Doped Nanoparticles. *Small* **2019**, *15*, 1804969. [CrossRef]
120. Zheng, W.; Huang, P.; Gong, Z.; Tu, D.; Xu, J.; Zou, Q.; Li, R.; You, W.; Bünzli, J.-C.G.; Chen, X. Near-Infrared-Triggered Photon Upconversion Tuning in All-Inorganic Cesium Lead Halide Perovskite Quantum Dots. *Nat. Commun.* **2018**, *9*, 3462. [CrossRef]
121. Liang, T.; Fu, J.; Li, M.; Li, H.; Hao, Y.; Ma, W. Application of Upconversion Photoluminescent Materials in Perovskite Solar Cells: Opportunities and Challenges. *Mater. Today Energy* **2021**, *21*, 100740. [CrossRef]
122. Ruan, L.; Zhang, Y. NIR-Excitable Heterostructured Upconversion Perovskite Nanodots with Improved Stability. *Nat. Commun.* **2021**, *12*, 219. [CrossRef] [PubMed]
123. Zhong, C.-Y.; Xiao, L.; Zhou, J.; Chen, Z.; Chen, Y.; Liu, Z.-Q.; Zhang, J.Z. Two-Photon Photoluminescence and Bio-Imaging Application of Monodispersed Perovskite-in-Silica Nanocrystals with High Biocompatibility. *Chem. Eng. J.* **2022**, *431*, 134110. [CrossRef]
124. Chan, K.K.; Giovanni, D.; He, H.; Sum, T.C.; Yong, K.-T. Water-Stable All-Inorganic Perovskite Nanocrystals with Nonlinear Optical Properties for Targeted Multiphoton Bioimaging. *ACS Appl. Nano Mater.* **2021**, *4*, 9022–9033. [CrossRef]

125. Talianov, P.M.; Peltek, O.O.; Masharin, M.; Khubezhov, S.; Baranov, M.A.; Drabavičius, A.; Timin, A.S.; Zelenkov, L.E.; Pushkarev, A.P.; Makarov, S.V.; et al. Halide Perovskite Nanocrystals with Enhanced Water Stability for Upconversion Imaging in a Living Cell. *J. Phys. Chem. Lett.* **2021**, *12*, 8991–8998. [CrossRef]
126. Pramanik, A.; Patibandla, S.; Gao, Y.; Gates, K.; Ray, P.C. Water Triggered Synthesis of Highly Stable and Biocompatible 1D Nanowire, 2D Nanoplatelet, and 3D Nanocube $CsPbBr_3$ Perovskites for Multicolor Two-Photon Cell Imaging. *JACS Au* **2021**, *1*, 53–65. [CrossRef]
127. Yin, W.; Li, H.; Chesman, A.S.R.; Tadgell, B.; Scully, A.D.; Wang, M.; Huang, W.; McNeill, C.R.; Wong, W.W.H.; Medhekar, N.V.; et al. Detection of Halomethanes Using Cesium Lead Halide Perovskite Nanocrystals. *ACS Nano* **2021**, *15*, 1454–1464. [CrossRef]
128. Jia, D.; Xu, M.; Mu, S.; Ren, W.; Liu, C. Recent Progress of Perovskite Nanocrystals in Chem/Bio Sensing. *Biosensors* **2022**, *12*, 754. [CrossRef]
129. Kim, J.; John, A.T.; Li, H.; Huang, C.; Chi, Y.; Anandan, P.R.; Murugappan, K.; Tang, J.; Lin, C.; Hu, L.; et al. High-Performance Optoelectronic Gas Sensing Based on All-Inorganic Mixed-Halide Perovskite Nanocrystals with Halide Engineering. *Small Methods* **2024**, *8*, 2300417. [CrossRef]
130. Wang, S.; Wei, Z.; Xu, Q.; Yu, L.; Xiao, Y. Trinity Strategy: Enabling Perovskite as Hydrophilic and Efficient Fluorescent Nanozyme for Constructing Biomarker Reporting Platform. *ACS Nano* **2024**, *18*, 1084–1097. [CrossRef]
131. Temur, E.; Zengin, A.; Boyacı, İ.H.; Dudak, F.C.; Torul, H.; Tamer, U. Attomole Sensitivity of Staphylococcal Enterotoxin B Detection Using an Aptamer-Modified Surface-Enhanced Raman Scattering Probe. *Anal. Chem.* **2012**, *84*, 10600–10606. [CrossRef] [PubMed]
132. Wang, J.; Wu, X.; Wang, C.; Shao, N.; Dong, P.; Xiao, R.; Wang, S. Magnetically Assisted Surface-Enhanced Raman Spectroscopy for the Detection of *Staphylococcus Aureus* Based on Aptamer Recognition. *ACS Appl. Mater. Interfaces* **2015**, *7*, 20919–20929. [CrossRef] [PubMed]
133. Xu, Y.; Shi, L.; Jing, X.; Miao, H.; Zhao, Y. SERS-Active Composites with Au–Ag Janus Nanoparticles/Perovskite in Immunoassays for Staphylococcus Aureus Enterotoxins. *ACS Appl. Mater. Interfaces* **2022**, *14*, 3293–3301. [CrossRef]
134. Zhang, J.; Yuan, Z.; Wang, C.; Liu, L.; Wang, Y.; Guo, Y.; Zhao, G. Aqueous-Phase Dual-Functional Chiral Perovskites for Hydrogen Sulfide (H2S) Detection and Antibacterial Applications in Escherichia Coli. *J. Colloid Interface Sci.* **2024**, *661*, 740–749. [CrossRef]
135. Luo, F.; Li, S.; Cui, L.; Zu, Y.; Chen, Y.; Huang, D.; Weng, Z.; Lin, Z. Biocompatible Perovskite Quantum Dots with Superior Water Resistance Enable Long-Term Monitoring of the H_2 S Level in Vivo. *Nanoscale* **2021**, *13*, 14297–14303. [CrossRef]
136. Li, M.; Zeng, Y.; Qu, X.; Jalalah, M.; Alsareii, S.A.; Li, C.; Harraz, F.A.; Li, G. Biocatalytic $CsPbX_3$ Perovskite Nanocrystals: A Self-Reporting Nanoprobe for Metabolism Analysis. *Small* **2021**, *17*, 2103255. [CrossRef]
137. Fu, L.; Qian, Y.; Zhou, J.; Zheng, L.; Wang, Y. Fluorescence-based Quantitative Platform for Ultrasensitive Food Allergen Detection: From Immunoassays to DNA Sensors. *Compr. Rev. Food Sci. Food Saf.* **2020**, *19*, 3343–3364. [CrossRef]
138. Andrásfalvy, B.K.; Galiñanes, G.L.; Huber, D.; Barbic, M.; Macklin, J.J.; Susumu, K.; Delehanty, J.B.; Huston, A.L.; Makara, J.K.; Medintz, I.L. Quantum Dot–Based Multiphoton Fluorescent Pipettes for Targeted Neuronal Electrophysiology. *Nat. Methods* **2014**, *11*, 1237–1241. [CrossRef]
139. Huang, L.; Liao, T.; Wang, J.; Ao, L.; Su, W.; Hu, J. Brilliant Pitaya-Type Silica Colloids with Central–Radial and High-Density Quantum Dots Incorporation for Ultrasensitive Fluorescence Immunoassays. *Adv. Funct. Mater.* **2018**, *28*, 1705380. [CrossRef]
140. Dong, Y.; Tang, X.; Zhang, Z.; Song, J.; Niu, T.; Shan, D.; Zeng, H. Perovskite Nanocrystal Fluorescence-Linked Immunosorbent Assay Methodology for Sensitive Point-of-Care Biological Test. *Matter* **2020**, *3*, 273–286. [CrossRef]
141. Sauer, M.; Heilemann, M. Single-Molecule Localization Microscopy in Eukaryotes. *Chem. Rev.* **2017**, *117*, 7478–7509. [CrossRef] [PubMed]
142. Lelek, M.; Gyparaki, M.T.; Beliu, G.; Schueder, F.; Griffié, J.; Manley, S.; Jungmann, R.; Sauer, M.; Lakadamyali, M.; Zimmer, C. Single-Molecule Localization Microscopy. *Nat. Rev. Methods Primer* **2021**, *1*, 39. [CrossRef] [PubMed]
143. Jin, D.; Xi, P.; Wang, B.; Zhang, L.; Enderlein, J.; Van Oijen, A.M. Nanoparticles for Super-Resolution Microscopy and Single-Molecule Tracking. *Nat. Methods* **2018**, *15*, 415–423. [CrossRef] [PubMed]
144. Yang, Z.; Zong, S.; Yang, K.; Zhu, K.; Li, N.; Wang, Z.; Cui, Y. Wavelength Tunable Aqueous $CsPbBr_3$ -Based Nanoprobes with Ultrahigh Photostability for Targeted Super-Resolution Bioimaging. *ACS Appl. Mater. Interfaces* **2022**, *14*, 17109–17118. [CrossRef]
145. Mei, L.; Xie, R.; Zhu, S.; Deng, S.; Xu, H.; Fan, X.; Yin, W.; Gu, Z. Neurotoxicity Study of Lead-Based Perovskite Nanoparticles. *Nano Today* **2023**, *50*, 101830. [CrossRef]

inorganics

MDPI

Article

A Novel TiO_2-Cuttlebone Photocatalyst for Highly Efficient Catalytic Degradation of Tetracycline Hydrochloride

Qing Li [1,2], Penghui Liu [1,2], Huizhen Lin [1,2], Hun Xue [1,2,*] and Jingyun Mao [1,2,*]

[1] College of Environmental and Resource Sciences, College of Carbon Neutral Modern Industry, Fujian Normal University, Fuzhou 350007, China; liqing0909fujian@163.com (Q.L.); ppenghui1996@163.com (P.L.); 15260937515@163.com (H.L.)

[2] Fujian Key Laboratory of Pollution Control & Resource Reuse, Fujian Normal University, Fuzhou 350007, China

* Correspondence: xuehun@fjnu.edu.cn (H.X.); mjy19960427@163.com (J.M.)

Abstract: The harmful effects of antibiotics on aquatic environments have become a growing concern of modern society. Developing high-performance photocatalysts capable of degrading antibiotics under solar light is, therefore, crucial. In this study, TiO_2-cuttlebone composites are prepared via the sol–gel method, to produce carbonate radicals ($\bullet CO_3^-$) under solar light irradiation. The $\bullet CO_3^-$ radicals exhibit high selectivity for the degradation of tetracycline hydrochloride (TC). Compared to TiO_2 alone, the TiO_2-cuttlebone composite demonstrates excellent solar-driven photocatalytic activity for TC degradation in both freshwater and seawater. The reaction pathways of TC degradation in seawater are elucidated using HPLC-MS/MS analysis. Moreover, a TiO_2-cuttlebone self-suspending photocatalyst device is fabricated using 3D printing technology and low-temperature deposition methods, with aluminum–plastic (AP) as a substrate. This innovative device is easily recyclable from photocatalytic solutions while maintaining high stability, making it highly desirable for practical applications.

Keywords: TiO_2-cuttlebone; carbonate radical; photocatalysis; tetracycline hydrochloride; seawater; self-suspending; photocatalyst device

1. Introduction

Over the past few decades, the rapid development of the aquaculture industry has led to the accumulation of significant amounts of antibiotics in seawater [1–4]. Tetracycline hydrochloride (TC), a common antimicrobial agent used in aquaculture, presents significant environmental hazards due to its low biodegradability and high stability [5–7]. The concentration of TC in aquaculture water, including freshwater and seawater, is generally in the range of $\mu g/L^{-1}$ to mg/L^{-1}. Although these levels are relatively low, the entry of TC into the food chain through ecological cycles results in bioaccumulation, posing great risks to the environment and to human health [8]. Therefore, the development of efficient and sustainable technologies to remove TC from both freshwater and seawater is urgently needed.

The application of photocatalysis for environmental remediation has attracted increasing research attention over the past decades [9–12]. Among these efforts, the photocatalytic degradation of TC has been demonstrated over a series of TiO_2 [13,14], WO_3 [15,16], and C_3N_4 [17,18] nanocomposites. Nevertheless, their photocatalytic efficiency is commonly limited by the complex molecular structure of TC. In a previous work, we reported a $CaCO_3/TiO_2$ composite photocatalyst that demonstrates excellent activity regarding the degradation of TC [19]. This high performance is attributable to the reaction between photocatalytically generated $\bullet OH$ and holes with CO_3^{2-}, which produces carbonate radicals ($\bullet CO_3^-$) that exhibit unique activity and selectivity toward TC degradation. Cuttlefish bone is an abundant, low-cost, environmentally friendly, and non-toxic material in the ocean that is particularly rich in $CaCO_3$. Therefore, combining TiO_2 with cuttlebone is a promising approach for generating a good deal of $\bullet CO_3^-$ free radicals under light irradiation, thereby accelerating TC degradation. However, no reports currently exist in the open literature on this topic.

Moreover, most catalysts developed for photocatalytic degradation are powder-based, which makes them difficult to recover in practical aquaculture applications and may potentially cause secondary pollution. Anchoring nanometer photocatalysts onto bulk substrates is considered to be an effective strategy to address these challenges [20–25]. In this regard, we have designed and fabricated self-suspending aluminum–plastic (AP)/semiconductor photocatalyst devices for the degradation of TC [26]. Compared to powder systems, the device maximizes light utilization, quickly replenishes oxygen during the photocatalytic process, and facilitates the generation of active $\bullet OH$ and $\bullet O_2^-$ species. Furthermore, the conductive aluminum accelerates electron separation and reduces the recombination of charge carriers [26,27]. In this context, the fabrication of TiO_2-cuttlebone composite and further immobilization of the designated catalysts on the AP substrate are anticipated to enhance both photocatalytic activity and cyclic stability.

Bearing these considerations in mind, for the present study, we first synthesized TiO_2-cuttlebone (TCb) composites via the sol–gel method for the degradation of TC (5 and 30 mg/L) in freshwater and seawater under solar light and natural light irradiation, respectively. It is clear that the combination of cuttlebone with TiO_2 remarkably enhances TC degradation efficiency compared to pure TiO_2, primarily due to the generation of a large number of $\bullet CO_3^-$ radicals under light irradiation. A self-suspending photocatalyst device was then prepared using an AP substrate, through 3D printing technology and low-temperature deposition. TCb nanoparticles were immobilized on the surface of the AP plate. This floating device improves light utilization and the oxygenation rate of the photocatalyst. The catalytic activity of the self-suspending AP-TCb device is notably higher than that of TCb powder, under the condition of equal TCb mass. This work provides new insights into the preparation of selective oxidation photocatalytic materials for TC degradation and the fabrication of photocatalyst devices, with promising applications in real-world seawater treatment.

2. Results and Discussion

Figure 1a shows the XRD partners of the prepared catalysts. TiO_2 exhibits diffraction peaks at 25.3°, 37.8°, 48.1°, 53.9°, and 55.1°, corresponding to the (101), (004), (200), (105), and (211) crystal planes of anatase TiO_2 (JCPDS No. 21-1272), respectively [28]. The XRD pattern for cuttlebone powder shows diffraction peaks at 26.3°, 27.3°, 33.2°, 36.3°, 38.0°, 38.5°, 41.3°, 43.0°, 45.9°, 48.5°, 50.3°, and 52.5°, which are assignable to orthorhombic $CaCO_3$ (JCPDS No. 05-0453) [29]. Additionally, the XRD pattern of the resulting TCb nanocomposite reveals diffraction peaks corresponding to both anatase TiO_2 and orthorhombic $CaCO_3$, indicating that the sol–gel method, followed by calcination, leads to the successful hybridization of TiO_2 with cuttlebone powder.

Figure 1b displays the FT-IR spectra of the cuttlebone, TiO_2, and TCb samples. In the TiO_2 spectrum, a peak at 3422 cm^{-1} corresponding to the stretching and bending vibrations of the O-H bond is observed. In addition, a wide band in the range of 400–800 cm^{-1} is attributable to the typical vibration of the O–Ti–O bond within TiO_2. A peak at 1383 cm^{-1} can be assigned to the $-OCH_2CH_3$ groups, which might result from the incomplete hydrolysis of titanium tetraisopropoxide during the synthetic processing of TiO_2 sol [30]. Pure cuttlebone exhibits four characteristic peaks at 1475 cm^{-1}, 1083 cm^{-1}, 875 cm^{-1}, and 712 cm^{-1}, all of which can be attributed to the stretching vibrations of C–O bonds [31]. In the TCb composite, the main peaks from both TiO_2 and cuttlebone are apparent, demonstrating the coexistence of the two components.

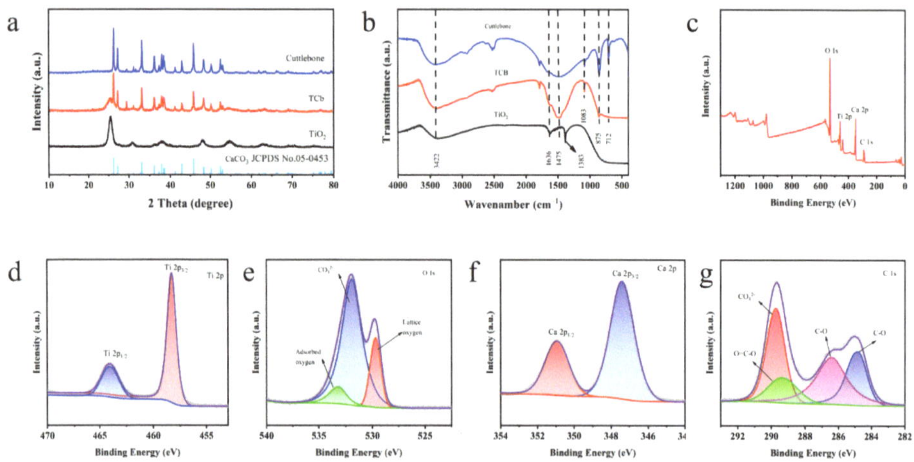

Figure 1. (**a**) XRD patterns and (**b**) FTIR spectra of cuttlebone, TiO_2, and TCb; (**c**) XPS figures of TCb: (**d**) Ti 2p, (**e**) O 1s, (**f**) Ca 2p, and (**g**) C 1s.

The presence of Ti, O, Ca, and C elements in the TCb nanocomposite is confirmed by XPS analysis (Figure 1c). In Figure 1d, the high-resolution Ti 2p XPS spectrum shows Ti $2p_{1/2}$ and Ti $2p_{3/2}$ peaks at 465.0 and 459.3 eV, respectively. The O 1s XPS peak can be deconvoluted into three components at 532.8, 531.9, and 530.7 eV (Figure 1e), which are ascribable to surface-adsorbed oxygen species, carbonate-like species, and lattice oxygen, respectively [32]. In Figure 1f, the peaks at 350.9 and 347.3 eV are attributable to Ca $2p_{1/2}$ and Ca $2p_{3/2}$, confirming the +2 oxidation state of Ca in the composite [19]. The XPS spectrum in the C 1s region is deconvoluted into peaks at 289.7, 289.3, 286.3, and 284.8 eV, attributable to CO_3^{2-}, O=C-O, C-O, and C-C bonds, respectively (Figure 1g). The existence of the C–C group is consistent with the incomplete hydrolysis of titanium tetraisoproxide during the synthesis process of the TiO_2 sol. This finding aligns with the results from the FT-IR analysis. Overall, these results provide solid evidence for the construction of the TiO_2-cuttlebone hybrid composite.

The micromorphology of TCb was examined using SEM. As shown in Figure 2a, TCb is mainly composed of clusters. To further investigate the detailed structural and morphological characteristics, transmission electron microscopy (TEM) analysis was performed (Figure 2b,c). The result is consistent with the SEM image. Moreover, the nanoparticles are distributed in clusters with a width and length of approximately 100 nm and 1 μm. HR-TEM analysis was further conducted on the TCb sample (Figure 2c–e). The measured lattice fringe distances of 0.35 and 0.19 nm correspond to the (101) and (200) crystal planes of anatase TiO_2, respectively. The size of the TiO_2 nanoparticles was measured to be about 7 nm. Furthermore, the elemental mapping analysis of the TCb sample revealed the presence of Ca, Ti, C, and O, as shown in Figure 2f,g. Ca and Ti elements were evenly distributed throughout the composite. The EDS result (Figure S1) shows that the content of Ti and Ca elements is 10.55% and 12.08%, which is close to the feed ratio of the synthetic process. These findings further confirm the immobilization of TiO_2 nanoparticles on the cuttlebone component.

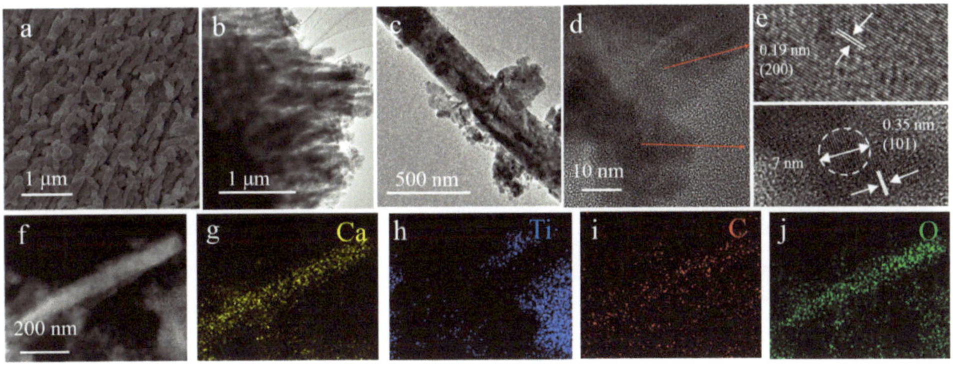

Figure 2. (**a**) SEM, (**b**,**c**) TEM, and (**d**–**f**) HRTEM images of TCb; (**g**–**j**) elemental mapping images of TCb.

The specific surface area and pore structure of the synthesized samples were evaluated via the N_2 adsorption–desorption technique, as shown in Figure 3a. Both the TiO_2 and TCb samples exhibit type IV curves, indicative of the typical adsorption behavior of mesoporous structures. Based on the Brunauer–Emmett–Teller (BET) isotherms, the specific surface areas of the cuttlebone, TiO_2, and TCb samples were calculated to be 8.0, 167.4, and 107.5 m^2/g, respectively. Furthermore, the BJH patterns (Figure 3b) revealed that TiO_2 has an average pore size of 7.2 nm, with a narrow pore size distribution similar to that observed for the TCb sample.

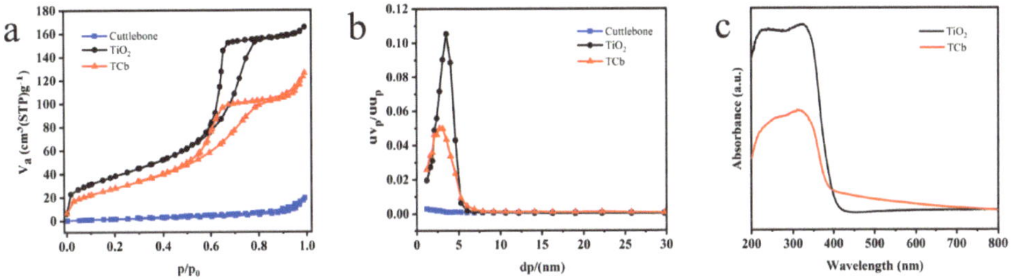

Figure 3. (**a**) N_2 adsorption-desorption isotherms and (**b**) pore size distributions of the cuttlebone, TiO_2, and TCb samples; (**c**) UV-vis DRS spectra of TiO_2 and TCb nanocomposite.

Figure 3c displays the UV-vis DRS spectra of the prepared samples. Bare TiO_2 shows strong absorption in the region of 200–398 nm, which is consistent with its band gap of 3.1 eV [33]. In comparison, the UV-vis DRS spectrum of the TCb nanocomposite exhibits enhanced light absorption in the 400~800 nm region, which could be favorable for photocatalytic reactions.

The photocatalytic activities of the synthesized catalyst were initially investigated by the degradation of TC in freshwater, as shown in Figure 4. Before photocatalytic degradation, the catalyst was stored in the dark to adsorb TC for 30 min to establish absorption–desorption equilibrium. After 5 min of light irradiation, 96.4% of TC (5 mg/L) was degraded over the TCb catalyst (Figure 4a). In contrast, no degradation of TC was observed in the dark, and only 15.4% of TC was eliminated under light irradiation without a catalyst. Bare TiO_2 showed inferior photocatalytic performance, with 71.9% of TC degradation under the same conditions. The obviously higher photocatalytic activity of the TCb nanocomposite compared to bare TiO_2 cannot be ascribed to differences in the specific surface areas since the TCb nanocomposite has a smaller surface area (107.5 m^2/g)

than bare TiO_2 (167.4 m^2/g). Instead, the enhanced photocatalytic activity of the TCb nanocomposite suggests that cuttlebone plays a key role in promoting TC degradation.

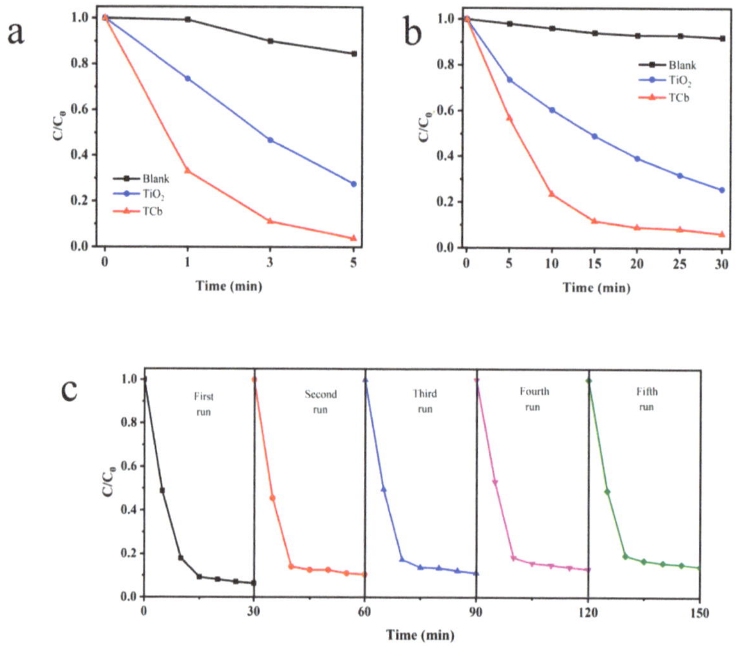

Figure 4. Degradation curves of (**a**) 5 mg L^{-1} and (**b**) 30 mg L^{-1} TC under solar light illumination over the synthesized catalysts; (**c**) cycling tests for the photocatalytic degradation of TC (30 mg L^{-1}) over TCb.

The ratio of TC to photocatalyst in the reaction system also influences performance. When the concentration of TC in the reaction system increased to 30 mg L^{-1} while the amount of photocatalyst remained constant, 79.7% and 92.0% of the TC was degraded within a light irradiation of 30 min over TiO_2 and TCb, respectively (Figure 4b). This is less efficient compared to the degradation of 5 mg L^{-1} TC. The TCb nanocomposite also exhibited high stability during the photocatalytic degradation of TC (30 mg L^{-1}). Recent progress in the photocatalytic degradation of TC by TiO_2-based materials in recent years is shown in Table S1 [19,29,34–38], indicating that TCB has a high degradation efficiency for TC. As presented in Figure 4c, the photocatalytic degradation of TC in freshwater was successfully repeated five times without obvious changes in activity.

LC-QTOF-MS was employed to investigate the intermediates generated in the system, as shown in Figure 5a–c. A possible degradation pathway of TC is proposed in Scheme 1. Firstly, CO_3^- initiates an oxidative attack on the target molecule, generating the intermediate product of M_1 (m/z 431) [39]. Then, M_1 undergoes deamination, demethylation, and addition reactions to form M_2 (m/z 417) and M_3 (m/z 305) [40]. Subsequently, M_3 undergoes a series of ring-opening reactions, gradually decomposing into smaller molecules, including M_4 and M_5. These smaller molecules eventually decompose further, yielding H_2O, CO_2, and other inorganic substances. Concurrently, CO_3^- selectively oxidizes the target molecule, producing another intermediate product of M_6 (m/z 459), as described in previous research [41]. M_6 undergoes demethylation, deamination, deprotonation, and addition reactions, and a six-membered ring is opened to generate M_7 (m/z 411) and M_8 (m/z 344). M_7 and M_8 continue to undergo ring opening and demethylation reactions, further decomposing into smaller molecular compounds. Finally, these smaller molecules undergo a series of oxidative degradations, ultimately decomposing the target molecule and completing the degradation process.

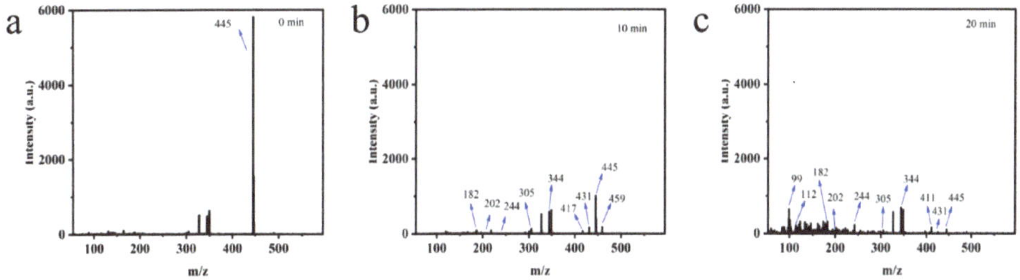

Figure 5. HRESI-TOF-MS spectra of the transformation products after light irradiation of (**a**) 0, (**b**) 10, and (**c**) 30 min in the TCb system.

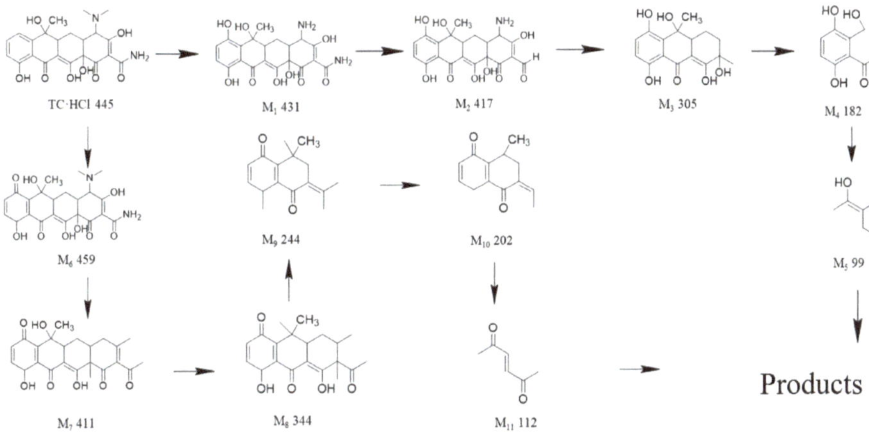

Scheme 1. Proposed degradation pathways of TC in the TCb system.

To explore the effect of the photocatalytic mechanism of TC over TCb, BQ, AO, IPA, and 4-cp were employed as scavengers of $\bullet O_2^-$, h^+, $\bullet OH$, and $\bullet CO_3^-$ and added into a TC aqueous solution. As shown in Figure 6a, $\bullet O_2^-$ and h^+ play an important role in the degradation of TC over TiO_2 instead of $\bullet OH$ and $\bullet CO_3^-$ radicals. This result is consistent with our previous research [29].

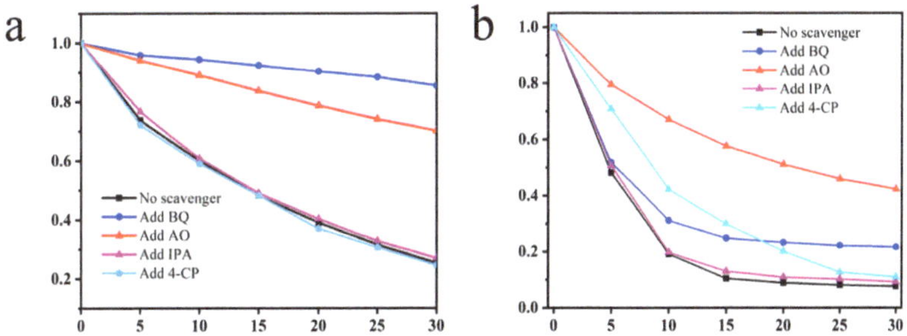

Figure 6. Effects of different scavengers on TC degradation over (**a**) TiO_2, and (**b**) TCB (BQ: 0.5 mM, AO: 100 mM, IPA: 100 mM, 4-CP: 0.1 mM).

Compared with TiO_2, the TCb composite shows a different mechanism of TC degradation (Figure 6b). An obvious decrease in TC degradation efficiency occurs after the AO scavenger is added to the system. When the BQ scavenger is added, there is a certain effect on degradation. However, the negative effect of BQ and AO on TC degradation is weaker than that of TiO_2. This may be due to the consumption of a portion of the species of $\bullet O_2^-$ or h^+ to facilitate the conversion of CO_3^{2-} into $\bullet CO_3^-$. In the case of IPA, there are no obvious effects on the catalytic activity of TCb. Furthermore, TC degradation efficiency is remarkably reduced when 4-CP is added, indicating that the $\bullet CO_3^-$ radical is significant in this reaction [19]. Furthermore, ESR measurements (Figure S2) demonstrate the existence of the $\bullet CO_3^-$ radical in the photochemical system of TCb.

The above observations clearly indicate that the degradation of TC is a photocatalytic process over TiO_2, with cuttlebone acting as a promoter to enhance photocatalytic efficiency. Under solar light irradiation, TiO_2 is excited and produces electrons (e^-) and holes (h^+). The e^- can activate O_2 to generate $\bullet O_2^-$ species, while the h^+ contribute to the formation of $\bullet OH$ radicals. The generated $\bullet OH$, $\bullet O_2^-$, or h^+ species can then react with CO_3^{2-}/HCO_3^- to produce selective $\bullet CO_3^-$ radicals [29]. This leads to the synergistic degradation of TC over the TCb hybrid composite, resulting in improved photocatalytic performance.

Based on the excellent photocatalytic performance regarding TC degradation in freshwater, the photocatalytic activity for the degradation of TC (5 mg/L) in seawater over TiO_2 and TCb has also been investigated, due to its great significance for environmental protection. As displayed in Figure 7a, the TC degradation efficiency reaches 88.5% and 97.8% within 20 min with TiO_2 and TCb, respectively. Moreover, the photocatalytic degradation of TC in seawater can be cycled five times without a noticeable loss in activity (Figure 7b). These results highlight the great potential of TCb for the elimination of TC in seawater.

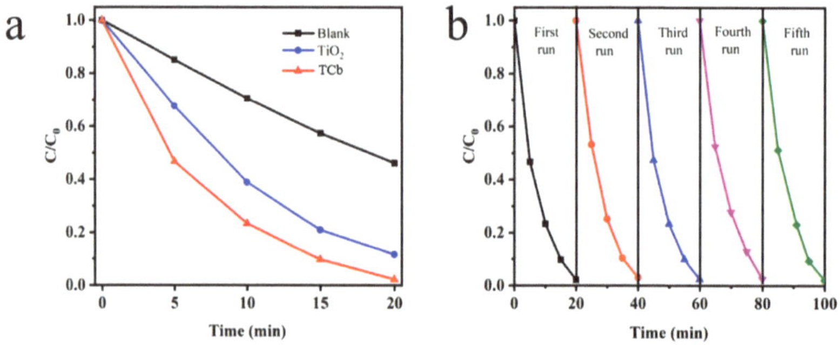

Figure 7. (**a**) Photocatalytic degradation of TC (5 mg L^{-1}) in seawater over the obtained samples, and (**b**) cycling tests for the photocatalytic degradation of TC (5 mg L^{-1}) over TCb in seawater.

To promote the practical application of the prepared TCb, we further immobilized the catalyst on an AP substrate to create a self-suspending AP-TCb photocatalyst device (for more details, please refer to the experimental section). Figure S3 shows the SEM image of the cross-section of AP-TCb. As depicted, TCb is distributed over the surface of AP, with a thickness of about 3 μm.

The photocatalytic performance of AP-TCb has been evaluated for TC degradation in freshwater and seawater under natural light, and an indicative scheme of experiment arrangement is proposed (Scheme 2). As illustrated in Figure 8a,b, after light irradiation for 20 min, the degradation efficiency of TC was 98.34% in freshwater and 96.4% in seawater. Compared to TCb powder, the AP-TCb device exhibited significant enhancements for TC degradation in both freshwater and seawater, which can be attributed to the following three major factors: (a) increased light-capture efficiency, leading to greater light utilization; (b) the timely replenishment of oxygen consumed in the photocatalytic process; and (c) the presence of Al in the AP, which accelerates electron migration, promoting the separation of electrons and holes [26]. These results demonstrate the great potential of this AP-TCb self-suspending device for practical water treatment in both freshwater and seawater environments.

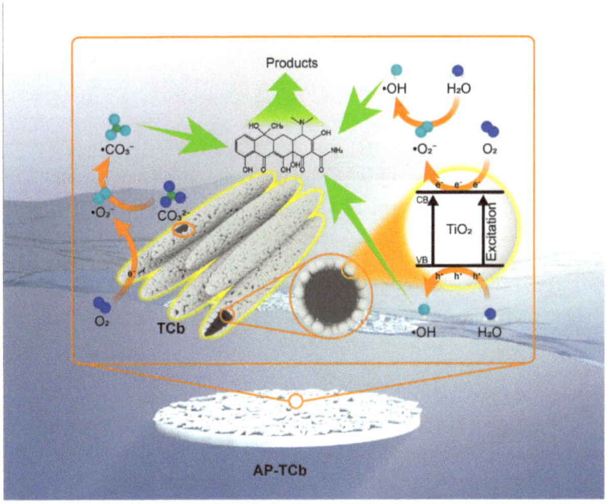

Scheme 2. Possible photocatalytic TC degradation mechanism over AP-TCb under natural light irradiation.

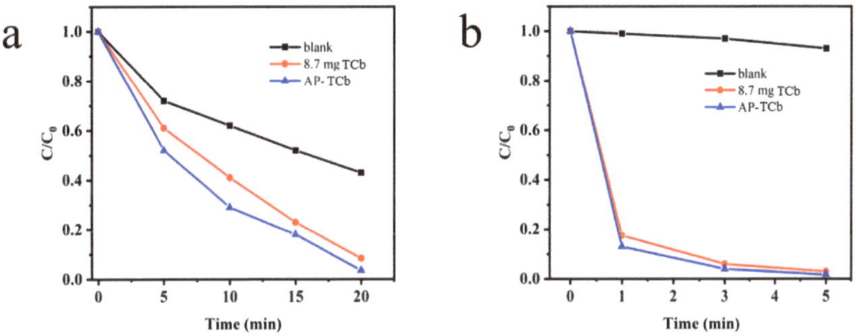

Figure 8. Photocatalytic degradation of TC (5 mg/L) in (**a**) seawater and (**b**) freshwater under natural light irradiation through the use of TCb and AP-TCb.

3. Methods and Materials

3.1. Chemical and Reagents

Tetracycline hydrochloride (96%), benzoquinone (BQ, 99%), ammonium oxalate (AO, 99.8%), isopropyl alcohol (IPA, ≥99.5%), 4-chlorophenol (4-CP, 99%), and 5,5-dimethyl-1-pyrroline-N-oxide (DMPO) were purchased from Aladdin (Shanghai, China). Titanium tetraisopropoxide ($C_{12}H_{28}O_4Ti$, 97%) was obtained from Sigma-Aldrich (St. Louis, MO, USA). Waste cuttlebone was obtained from a seafood market in Fuzhou, China. Seawater was obtained from Pingtan, China.

3.2. Synthesis Process

The raw cuttlebone powder was prepared via a grinding method, followed by sieving through a 100-mesh screen. TiO_2-cuttlebone nanocomposites (TCb) were synthesized via the sol–gel method, followed by a calcining process. In brief, titanium tetraisopropoxide was first hydrolyzed under acidic conditions. The resulting suspension was dialyzed to a pH of ~4. According to our previous work, the calcination of 100 mL of titanium glue can obtain 3 g TiO_2. Then, the raw cuttlebone powder (1.8 g) was added to the obtained transparent TiO_2 sol (40 mL), dehydrated in a microwave oven, and calcinated at 300 °C for 5 h. The resulting product was washed with ethanol and dried overnight at 60 °C. For comparison, bare TiO_2 and cuttlebone were synthesized using the same procedure without adding raw cuttlebone or TiO_2 sol.

The self-suspending AP substrate was fabricated by 3D printing, according to our previous work [26]. For TCb immobilization, the AP substrate was washed with ethanol several times and dried. Its initial mass was recorded as m_1. TCb (0.10 g) was dispersed in 50 mL of anhydrous ethanol and ultrasonicated for 30 min to obtain a TCb suspension with a concentration of 2 g/L. The suspension was then poured into a Petri dish containing the AP substrate and dried in an air-circulating oven at 130 °C for 5 h. Finally, the impregnated AP substrate was ultrasonically cleaned in deionized water for 60 min to remove any loose TCb from the AP surface, followed by drying at 60 °C to obtain the AP-TCb device. The final mass was recorded as m_2. The amount of TCb loaded onto the AP substrate was determined by calculating the difference between m_2 and m_1.

3.3. Characterization and Analytical Methods

X-ray diffraction (XRD) patterns were studied with a Bruker D8 Advance X-ray diffractometer using Cu K_α radiation (40 kV, 40 mA, λ = 1.5418 Å) in a 2θ range of 5–90° with a 0.02° step size. X-ray photoelectron spectroscopy (XPS) was conducted with a PHI Quantum 2000 system, using a monochromatic Al K_α source. The specific surface areas were obtained at −196 °C via a BELSORP-mini II nitrogen adsorption–desorption apparatus. The Fourier transform infrared (FT-IR) spectra were determined in transmittance mode with a resolution of 4 cm^{-1}, using a Nicolet Is 10 FT-IR spectrometer. The surface morphology of the samples was recorded with scanning electron microscopy (SEM, Regulus 8100, Tokyo, Japan) and transmission electron microscopy (TEM) (FEI, Talos F200X, Boston, MA, USA). UV-vis diffuse reflectance spectroscopy (DRS) measurements were determined using a Varian Cary 500 spectrometer equipped with an integrating sphere, employing $BaSO_4$ as a reference. A Bruker ESP 300E spectrometer was employed to measure the electron spin resonance (ESR) signals of the radicals that were spin-trapped by DMPO.

3.4. Photochemical Measurements

The photocatalytic reactions were performed under simulated solar light irradiation. The irradiance was determined to be 269 mW·cm^{-2}, using a PL-MW 2000 photoradiometer (Beijing Perfect Light Co.; Beijing, China) at a distance of 10 cm from the 300 W Xenon lamp to liquid level and an optical power of 0.845 W, equipped with an AM 1.5 G simulated solar filter (300 W xenon lamp, PLS-SXE300, Beijing Perfect Light Co.; Beijing, China) and optical equalizer (PLS-LA320A, Beijing Perfect Light Co.; Beijing, China). Typically, a glass vessel with a water-cooling system (25 °C) was used to mix 50 mg of photocatalyst and 100 mL of TC solution at a specific concentration in the dark for 30 min, to achieve absorption-desorption equilibrium. Subsequently, the suspension was exposed to Xe lamp irradiation within the wavelength range of 200–1200 nm. A 4 mL sample from the reactor was filtered through a 0.45 μm filter membrane for analysis, using either an Agilent high-performance liquid chromatography spectrometer (1260 infinity LC) (Hong Kong, China) or a Shimadzu UV-1750 UV-vis spectrophotometer (Kyoto, Japan). The wavelength for TC detection was set at 357 nm.

The AP model was designed to be a disc-shaped structure with a radius of r_1 = 35 mm, a height of h_1 = 1.2 mm, a wall thickness of t_1 = 0.4 mm, and an infill ratio of η = 58.71% for the central infill layer. The model featured 102 circular holes with a diameter of 1 mm each to facilitate water flow. Using a densitometer, the density of the prepared AP filament was measured to be ρ_0 = 1.155 g·cm^{-3}. If the model was filled to 100% infill, the device would fail to achieve self-suspension, due to its density being greater than that of water

(1 g·cm^{-3}). Therefore, the central part of the model needed to be filled at a specific infill ratio to bring the overall density of the device close to that of water. We set the density of the model to be $\rho_1 \approx 1$ g·cm^{-3}, calculated using the following equation:

$$\rho_1 = \frac{m_1 + m_2}{V_0} \quad (1)$$

where ρ_1 is the density of the model, m_1 is the quality of the model wall, m_2 is the quality of the middle fill layer, and V_0 is the volume of the model.

According to the mass formula, $m = \rho V$, the volume formulae $V = h\pi r^2$, $m_1 = 3.522$ g, $V_0 = 4.520$ cm^3, and the volume of the central infill layer, $V_2 = 1.472$ cm^3. Therefore, using Equation (1), the mass of the central infill layer m_2 can be calculated to be 0.998 g. In the following equation:

$$\eta = \frac{m_2}{\rho_0 V_2} \times 100\% \quad (2)$$

η is the filling rate of the middle filling layer.

According to our calculations, $\eta = 58.71\%$.

4. Conclusions

In summary, we have successfully prepared a novel cuttlebone-TiO_2 (TCb) hybrid composite, as presented in the current work. Compared to pure TiO_2, the obtained TCb composite exhibits superior TC degradation efficiency under solar light and natural light irradiation in freshwater and seawater. The enhanced degradation rate is attributable to the production of highly selective $\bullet CO_3{}^-$ radicals in the reaction system. The degradation pathway of TC was explored by LC-MS analysis. To facilitate the recovery of the TCb photocatalyst in practical applications, an AP material was utilized as a substrate to prepare an AP-TCb self-suspending device. The device not only realizes the easy separation and recovery of the catalyst but also demonstrates obviously improved photocatalytic activity compared to the powder sample. This study offers a promising and convenient approach for treating low concentrations of antibiotics in real-world aquaculture wastewater.

Supplementary Materials: The following supporting information can be downloaded at: https://www.mdpi.com/article/10.3390/inorganics12120319/s1, Figure S1: The element mapping spectrum; Table S1: The comparison of efficiency of TC degradation in pure water by different photocatalysts; Figure S2: ESR signals of the DMPO-$\bullet CO_3{}^-$ in TCb photochemical system; Figure S3: SEM image of the cross section of AP-TCb.

Author Contributions: Q.L. and J.M. performed the research, analyzed the data, and wrote the paper; Q.L., P.L., H.L., H.X. and J.M. revised the paper. J.M. and H.X. were supervisors and revised the paper. All authors have read and agreed to the published version of the manuscript.

Funding: This work was supported financially by the National Key Research and Development Program of China (2023YFC3906300, 2016YFB0302303, and 2019YFC1908203), the National Natural Science Foundation of China (NSFC 21875037 and 51502036), and the National Natural Science Foundation of Fujian (2022J01650).

Institutional Review Board Statement: Not applicable.

Informed Consent Statement: Not applicable.

Data Availability Statement: The data presented in this study are available upon request from the corresponding author.

Acknowledgments: The authors of this study gratefully acknowledge the Fujian Key Laboratory of Pollution Control (Fuzhou, China) for providing the facility for sample characterizations, and gratefully acknowledge Zixin Chen for her contribution to the experiments in this paper.

Conflicts of Interest: The authors declare no conflicts of interest. The funders had no role in the design of the study; in the collection, analyses, or interpretation of data; in the writing of the manuscript; or in the decision to publish the results.

References

1. Kafaei, R.; Papari, F.; Seyedabadi, M.; Sahebi, S.; Tahmasebi, R.; Ahmadi, M.; Sorial, G.A.; Asgari, G.; Ramavandi, B. Occurrence, distribution, and potential sources of antibiotics pollution in the water-sediment of the northern coastline of the persian gulf, Iran. *Sci. Total Environ.* **2018**, *627*, 703–712. [CrossRef] [PubMed]
2. Gambino, D.; Savoca, D.; Sucato, A.; Gargano, V.; Gentile, A.; Pantano, L.; Vicari, D.; Alduina, R. Occurrence of antibiotic resistance in the mediterranean sea. *Antibiotics* **2022**, *11*, 332. [CrossRef] [PubMed]
3. Tian, L.; Xu, X.F.; Zhang, Z.L.; Ding, Y.Z.; Zhang, K.Q.; Zhi, S.L. A Comprehensive Contamination Investigation of Bohai Bay Seawater: Antibiotics Occurrence, Distribution, Ecological Risks and Their Interactive Factors. *Int. J. Environ. Res. Public Health* **2023**, *20*, 1599. [CrossRef] [PubMed]
4. Qian, G.S.; Shao, J.Y.; Hu, P.; Tang, W.T.; Xiao, Y.H.; Hao, T.W. From micro to macro: The role of seawater in maintaining structural integrity and bioactivity of granules in treating antibiotic-laden mariculture wastewater. *Water Res.* **2023**, *246*, 120702. [CrossRef]
5. Chen, Y.C.; Liu, J.T.; Zeng, Q.B.; Liang, Z.X.; Ye, X.X.; Lv, Y.C.; Liu, M.H. Preparation of Eucommia ulmoides lignin-based high-performance biochar containing sulfonic group: Synergistic pyrolysis mechanism and tetracycline hydrochloride adsorption. *Bioresour. Technol.* **2021**, *329*, 124432. [CrossRef]
6. Ni, J.W.; Wang, W.; Liu, D.M.; Zhu, Q.; Jia, J.L.; Tian, J.Y.; Li, Z.Y.; Wang, X.; Xing, Z.P. Oxygen vacancy-mediated sandwich-structural TiO_{2-x}/ultrathin g-C_3N_4/TiO_{2-x} direct Z-scheme heterojunction visible-light-driven photocatalyst for efficient removal of high toxic tetracycline antibiotics. *J. Hazard. Mater.* **2021**, *408*, 124432. [CrossRef]
7. Yang, H.C.; Yu, H.; Wang, J.H.; Ning, T.; Chen, P.; Yu, J.; Di, S.Y.; Zhu, S.K. Magnetic porous biochar as a renewable and highly effective adsorbent for the removal of tetracycline hydrochloride in water. *Environ. Sci. Pollut. Res.* **2021**, *28*, 61513–61525. [CrossRef]
8. Song, X.L.; He, J.L.; Wang, Y.; Wang, J.L.; Zhang, S.W. A novel MIL-125(Ti)-based nanocomposite for enhanced adsorption and catalytic degradation of tetracycline hydrochloride: Synergetic mechanism of calcination and the nitrogen-containing reticulated surface layer. *J. Colloid Interface Sci.* **2023**, *645*, 918–932. [CrossRef]
9. Bhoyar, T.; Vidyasagar, D.; Umare, S.S. Mitigating phytotoxicity of tetracycline by metal-free 8-hydroxyquinoline functionalized carbon nitride photocatalyst. *J. Environ. Sci.* **2023**, *125*, 37–46. [CrossRef]
10. Li, H.B.; Li, H.J.; Zhou, Z.N.; Tong, H.; Long, B.; Liu, W.; Li, W.H. Tailoring hydrophily and composition of BiOI for an ultrafast photodegradation of tetracycline hydrochloride. *J. Environ. Chem. Eng.* **2021**, *9*, 106292. [CrossRef]
11. Yu, G.L.; Yang, K.; Yang, Y.; Li, Y.F.; Sun, Q.F.; Li, P.Y.; Wang, W.M.; Song, F.M.; Ling, T.; Peng, X.J.; et al. Efficient removal of tetracycline hydrochloride through novel Fe/BiOBr/Bi_2WO_6 photocatalyst prepared by dual-strategy under visible-light irradiation. *J. Environ. Sci.* **2024**, *138*, 46–61. [CrossRef] [PubMed]
12. Hou, C.T.; Yuan, X.P.; Niu, M.M.; Li, Y.J.; Wang, L.P.; Zhang, M.Y. In situ composite of Co-MOF on a Ti-based material for visible light multiphase catalysis: Synthesis and the photocatalytic degradation mechanism. *New J. Chem.* **2022**, *46*, 11341–11349. [CrossRef]
13. Chen, M.G.; Chang, W.Y.; Zhang, J.; Zhao, W.; Chen, Z. Preparation of a Hybrid TiO_2 and 1T/2H-MoS2 Photocatalyst for the Degradation of Tetracycline Hydrochloride. *Acs Omega* **2023**, *8*, 15458–15466. [CrossRef] [PubMed]
14. Li, D.; Li, H.M.; Long, M.Y.; Bai, X.J.; Zhao, Q.Q.; Wen, Q.; Song, F. Synergetic effect of photocatalysis and peroxymonosulfate activation by MIL-53Fe@TiO_2 on efficient degradation of tetracycline hydrochloride under visible light irradiation. *Crystengcomm* **2022**, *24*, 4283–4293. [CrossRef]
15. Han, W.M.; Wu, T.; Wu, Q.S. Fabrication of WO_3/Bi_2MoO_6 heterostructures with efficient and highly selective photocatalytic degradation of tetracycline hydrochloride. *J. Colloid Interface Sci.* **2021**, *602*, 544–552. [CrossRef]
16. Wang, J.; Jiang, L.X.; Liu, F.Y.; Jia, M.; Liu, M.; Li, J.; Lai, Y.Q. Enhanced photoelectrochemical degradation of tetracycline hydrochloride with FeOOH and Au nanoparticles decorated WO_3. *Chem. Eng. J.* **2021**, *407*, 127195. [CrossRef]
17. Liang, J.C.; Li, X.Q.; Zuo, J.L.; Lin, J.; Liu, Z.L. Hybrid 0D/2D heterostructures: In-situ growth of 0D g-C_3N_4 on 2D BiOI for efficient photocatalyst. *Adv. Compos. Hybrid Mater.* **2021**, *4*, 1122–1136. [CrossRef]
18. Liu, S.Y.; Zada, A.; Yu, X.Y.; Liu, F.Z.; Jin, G. $NiFe_2O_4$/g-C_3N_4 heterostructure with an enhanced ability for photocatalytic degradation of tetracycline hydrochloride and antibacterial performance. *Chemosphere* **2022**, *307*, 135717. [CrossRef]
19. Wang, J.Q.; Qian, Q.R.; Chen, Q.H.; Liu, X.P.; Luo, Y.J.; Xue, H.; Li, Z.H. Significant role of carbonate radicals in tetracycline hydrochloride degradation based on solar light-driven TiO_2-seashell composites: Removal and transformation pathways. *Chin. J. Catal.* **2020**, *41*, 1511–1521. [CrossRef]
20. Yu, Z.Y.; Mielczarski, E.; Mielczarski, J.; Laub, D.; Buffat, P.; Klehm, U.; Albers, P.; Lee, K.; Kulik, A.; Kiwi-Minsker, L.; et al. Preparation, stabilization and characterization of TiO_2 on thin polyethylene films (LDPE). Photocatalytic applications. *Water Res.* **2007**, *41*, 862–874.
21. Balakrishnan, A.; Appunni, S.; Gopalram, K. Immobilized TiO_2/chitosan beads for photocatalytic degradation of 2,4-dichlorophenoxyacetic acid. *Int. J. Biol. Macromol.* **2020**, *161*, 282–291. [CrossRef] [PubMed]
22. Yadav, M.; Garg, S.; Chandra, A.; Hernadi, K. Immobilization of green BiOX (X = cl, Br and I) photocatalysts on ceramic fibers for enhanced photocatalytic degradation of recalcitrant organic pollutants and efficient regeneration process. *Ceram. Int.* **2019**, *45*, 17715–17722. [CrossRef]
23. Chang, C.J.; Chen, J.K.; Lin, K.S.; Wei, Y.H.; Chao, P.Y.; Huang, C.Y. Enhanced visible-light-driven photocatalytic degradation by metal wire-mesh supported Ag/flower-like Bi_2WO_6 photocatalysts. *J. Alloys Compd.* **2020**, *813*, 152186. [CrossRef]

24. Martin de Vidales, M.J.; Nieto-Marquez, A.; Morcuende, D.; Atanes, E.; Blaya, F.; Soriano, E.; Fernandez-Martinez, F. 3D printed floating photocatalysts for wastewater treatment. *Catal. Today* **2019**, *328*, 157–163. [CrossRef]
25. Liu, Z.; Xia, X.S.; Li, W.; Xiao, L.R.; Sun, X.L.; Luo, F.B.; Chen, Q.H.; Qian, Q.R. In Situ Growth of Ca^{2+}-Based Metal–Organic Framework on $CaSiO_3$/ABS/TPU 3D Skeleton for Methylene Blue Removal. *Materials* **2020**, *13*, 4403. [CrossRef]
26. Zhao, Z.A.; Mao, J.Y.; Lu, C.Y.; Yang, S.Q.; Qian, Q.R.; Chen, Q.H.; Xue, H.; Sun, X.L.; Yang, M.Q. Design and fabrication of self-suspending aluminum-plastic/semiconductor photocatalyst devices for solar energy conversion. *J. Environ. Sci.* **2024**, *136*, 615–625. [CrossRef]
27. Wu, W.; Song, H.W.; Gan, Q.Q.; Liu, D.X. Decomposition of dimethyl methylphosphonate vapor on ultrathin-film titania photocatalytic light absorber. *Chemosphere* **2021**, *274*, 129719. [CrossRef]
28. Torane, A.P.; Ubale, A.B.; Kanade, K.G.; Pagare, P.K. Photocatalytic dye degradation study of TiO_2 material. *Mater. Today Proc.* **2020**, *43*, 2738–2741. [CrossRef]
29. Huang, Z.Q.; Wang, J.Q.; Yang, M.Q.; Qian, Q.R.; Liu, X.P.; Xiao, L.R.; Xue, H. Construction of TiO_2-Eggshell for Efficient Degradation of Tetracycline Hydrochloride: Sunlight Induced In-situ Formation of Carbonate Radical. *Materials* **2021**, *14*, 1598. [CrossRef]
30. Xue, H.; Chen, Y.; Liu, X.; Qian, Q.; Luo, Y.; Cui, M.; Chen, Y.; Yang, D.P.; Chen, Q. Visible light-assisted efficient degradation of dye pollutants with biomass-supported TiO_2 hybrids. *Mater. Sci. Eng. C Mater. Biol. Appl.* **2018**, *82*, 197–203. [CrossRef]
31. Dai, Y.Z.; Zou, H.F.; Zhu, H.F.; Zhou, X.Q.; Song, Y.H.; Zheng, K.Y.; Shi, Z.; Sheng, Y. Facile surfactant- and template-free synthesis and luminescence properties of needle-like calcite $CaCO_3$:Eu^{3+} phosphors. *Crystengcomm* **2018**, *20*, 496–504. [CrossRef]
32. Li, Q.H.; Dong, M.; Li, R.; Cui, Y.Q.; Xie, G.X.; Wang, X.X.; Long, Y.Z. Enhancement of Cr(VI) removal efficiency via adsorption/photocatalysis synergy using electrospun chitosan/g-C_3N_4/TiO_2 nanofibers. *Carbohydr. Polym.* **2021**, *253*, 117200. [CrossRef] [PubMed]
33. Tao, R.H.; Wu, J.M.; Xiao, J.Z.; Zhao, Y.P.; Dong, W.W.; Fang, X.D. Conformal growth of ZnO on TiO_2 nanowire array for enhanced photocatalytic activity. *Appl. Surf. Sci.* **2013**, *279*, 324–328. [CrossRef]
34. Wang, C.; Liu, Y.X.; Li, Y.Y.; Sun, X.; Xu, L.M.; Huang, W.X. Facile defect construction of TiO_2 nanotube for excellent photocatalytic degradation of tetracycline under visible light. *J. Photochem. Photobiol. A* **2023**, *437*, 114475. [CrossRef]
35. Tian, J.L.; Wei, L.X.; Ren, Z.Q.; Lu, J.F.; Ma, J. The facile fabrication of Z-scheme Bi_2WO_6-P25 heterojunction with enhanced photodegradation of antibiotics under visible light. *J. Environ. Chem. Eng.* **2021**, *9*, 106167. [CrossRef]
36. Liu, H.J.; Hou, M.C.; Fu, H.; Hu, A.J.; Zhai, Y.L.; Wang, L.W.; Zhai, D.; Zhang, S.L.; Wang, S.P. 2D/1D BiOBr/TiO_2 flexible nanofibrous film heterojunction photocatalyst for tetracycline degradation. *Surf. Interfaces* **2024**, *44*, 103795. [CrossRef]
37. Li, H.X.; Liang, L.; Niu, X.H.; Zhang, D.Y.; Fan, H.Y.; Wang, K.J. Construction of a Bi_2WO_6/TiO_2 heterojunction and its photocatalytic degradation performance. *New J. Chem.* **2022**, *46*, 8185–8194. [CrossRef]
38. Zhang, X.L.; Han, D.F.; Dai, M.J.; Chen, K.; Han, Z.Y.; Fan, Y.Y.; He, Y.; Han, D.X.; Niu, L. Enhanced photocatalytic degradation of tetracycline by constructing a controllable Cu_2O-TiO_2 heterojunction with specific crystal facets. *Catal. Sci. Technol.* **2021**, *11*, 6248–6256. [CrossRef]
39. Zhang, H.; Zeng, Y.X.; Wang, X.Q.; Zhan, X.Y.; Xu, J.C.; Jin, A.; Hong, B. Sea-Urchin carbon nitride with carbon vacancies (C-v) and oxygen substitution (O-s) for photodegradation of Tetracycline: Performance, mechanism insight and pathways. *Chem. Eng. J.* **2022**, *446*, 137053. [CrossRef]
40. Yan, J.T.; Chai, B.; Liu, Y.Y.; Fan, G.Z.; Song, G.S. Construction of 3D/2D $ZnFe_2O_4$/g-C_3N_4 S-scheme heterojunction for efficient photo-Fenton degradation of tetracycline hydrochloride. *Appl. Surf. Sci.* **2023**, *607*, 155088. [CrossRef]
41. Burns, J.M.; Cooper, W.J.; Ferry, J.L.; King, D.W.; DiMento, B.P.; McNeill, K.; Miller, C.J.; Miller, W.L.; Peake, B.M.; Rusak, S.A.; et al. Methods for reactive oxygen species (ROS) detection in aqueous environments. *Aquat. Sci.* **2012**, *74*, 683–734. [CrossRef]

inorganics

MDPI

Article

Synthesis and Characterization of Broccoli-like Ag/Cu_2O Nanostructures on ZnO Nanowires Using the Plasma–Liquid Interaction Method

Phung Thi Thu [1], Ta Ngoc Bach [2], Le Thi Hong Phong [2], Do Hoang Tung [3], Vu Hong Ky [2], Do Khanh Tung [2], Vu Dinh Lam [4], Do Hung Manh [2], Nguyen Huy Dan [2], Trinh Xuan Anh [5] and Ngo Thi Hong Le [2,4,*]

1 University of Science and Technology of Hanoi, Vietnam Academy of Science and Technology, 18 Hoang Quoc Viet, Cau Giay, Hanoi 100000, Vietnam; thuphungi2@gmail.com

2 Institute of Materials Science, Vietnam Academy of Science and Technology, 18 Hoang Quoc Viet, Cau Giay, Hanoi 100000, Vietnam; bachtn.ims@gmail.com (T.N.B.); lephong.ims@gmail.com (L.T.H.P.); kyvuhong@gmail.com (V.H.K.); dokhanhtung@gmail.com (D.K.T.); manhdh@ims.vast.ac.vn (D.H.M.); dannh@ims.vast.ac.vn (N.H.D.)

3 Institute of Physics, Vietnam Academy of Science and Technology, 18 Hoang Quoc Viet, Cau Giay, Hanoi 100000, Vietnam; dhtung@gmail.com

4 Graduate University of Science and Technology, Vietnam Academy of Science and Technology, 18 Hoang Quoc Viet, Cau Giay, Hanoi 100000, Vietnam; lamvd@gust-edu.vast.vn

5 Department of Chemistry, School of Chemistry and Life Sciences, Hanoi University of Science and Technology, 1 Dai Co Viet, Hanoi 100000, Vietnam; trinhxuananh@gmail.com

* Correspondence: hongle@ims.vast.ac.vn or hongle2009@gmail.com

Abstract: We have designed an excellent visible-light-driven and high-performance photocatalyst with a Ag-Cu_2O-ZnO nanowire heterostructure in our work by combining the hydrothermal approach with plasma–liquid technology. The structural and morphological characteristics and optical properties of the samples were evaluated using X-ray diffraction, field-emission scanning electron microscopy, and spectrophotometry, respectively. The results show that the Ag nanoparticles are mainly positioned on the Cu_2O nanoclusters compared with the ZnO nanowire surface, forming broccoli-like Ag-Cu_2O nanoclusters during the Ar gas plasma treatment process in an aqueous solution. The diameter of the Ag/Cu_2O nanoclusters ranges from 150 to 180 nm. The Ag-Cu_2O-ZnO nanowires exhibited improved photocatalytic performance, decomposing approximately 98% methyl orange dye in 30 min. This is a consequence of the synergistic interactions between the p-n heterojunction formed at the Cu_2O-ZnO interfaces and the localized surface plasmon resonance (LSPR) effect of the Ag nanoparticles, which broaden the visible light absorption range and effectively separate the photogenerated charge carriers.

Keywords: ZnO nanowires; plasma–liquid; photocatalysis; methyl orange; localized surface plasmon resonance; hydrothermal method

1. Introduction

In recent decades, the photocatalytic approach has demonstrated exceptional benefits for both environmental protection and the rehabilitation of polluted water [1]. Breakthrough achievements in nanomaterial synthesis technology have contributed to accelerating the catalytic process to satisfy practical needs. Semiconductors have drawn a lot of interest as a key contender among the representative catalysts [1,2]. However, a challenging problem nowadays is the quick recombination rate of the electron–hole pairs produced in photocatalytic reactions [1,2]. It is imperative to address this bottleneck to produce a high-performance photocatalyst.

In parallel with the well-known typical semiconductors, thanks to their high catalytic activity, affordable cost, nontoxicity, and facile preparation, both ZnO and copper(I) oxide

(Cu_2O) have emerged as prominent candidates [3–6]. Typically, ZnO nanowires have a highly promising shape among ZnO 1D structures because of effective electron transport in narrow dimensions and their anisotropy [3,7]. Nevertheless, ZnO nanowires only absorb 3–5% of visible light because they have a large band gap (>3 eV) [7]. Therefore, the exploitation of a ZnO/Cu_2O heterojunction is expected to slow down the charge carrier recombination rate—since there exists an irreversible electric field at the p-n junction of the ZnO/Cu_2O system—and expand the absorption spectrum toward the visible light range, as shown in previous articles [8–11]. Furthermore, the concordance between the energy level and the electronic structure in the ZnO and Cu_2O forms a superior catalytic p-n system compared with other p-n systems [8–13]. On the other hand, the formation of a Schottky barrier at the precious metal–semiconductor interface can efficiently prevent the recombination of electron–hole pairs [14–16]. Combined with the surface plasmonic resonance effect of noble metal nanoparticles, which provides a considerable adsorption ability in the visible light region, precious metal/semiconductor-based catalysts greatly increase the photo-decomposing process [14–19]. In addition to mono- and binary structures, ternary structures have been recently developed and offer highly promising potential. The synergistic effects between the interfaces in ternary structures bring out the possibility of boosting the charge carrier separation and triggering the absorption range in the visible light region [20–28]. The $ZnO/Au/Cu_2O$ heterojunction system, as described by Yuan et al., demonstrates a better Z-scheme charge separation and deep oxidative decomposition for organic dye, resulting in enhanced catalytic performance of over 78% compared with pure ZnO [27]. Tsai et al. showed that $Ag/Cu_2O/ZnO$ NRs with remarkable kinetic constant enhancements degrade RhB more effectively than Ag/ZnO and Cu_2O/ZnO [23]. Numerous well-known fabrication methods for ZnO nanowires have been applied, such as carbo-thermal evaporation [29], metal–organic vapor phase epitaxy [30], and molecular beam epitaxy [31]. Unfortunately, these methods have disadvantages, such as being time-consuming, costly, and complicated experimental systems. Studies have demonstrated that the fabrication of ZnO nanowires using the hydrothermal method is simple and economical. Additionally, various fabrication techniques for mono- and binary structures, as well as ternary structures—including co-precipitation [32], hydrothermal [27], sol–gel [14], solution combustion [33], and plasma technology [28,34,35]—have been applied. Moreover, plasma technology has increasingly satisfied the urgent requirement for the development of a quick, easy, and eco-friendly approach. The interaction between plasma and aqueous solutions at the interface has been found to offer a strong solution for the synthesis and functionalization of nanomaterials [36,37]. In this case, the plasma is generated either inside the solution or close to its surface. Importantly, more active species, such as electrons, radicals, ions, and produced UV light, are introduced to promote reactions in the solution [36,37]. To the best of our knowledge, the use of the plasma–liquid approach to construct ternary heterostructures and simultaneously study their underlying formation mechanisms is still limited.

Therefore, we recently synthesized a CuO-Ag-ZnO nanowire photocatalyst in which Ag nanoparticles were intercalated between the CuO nanoparticles and the ZnO nanowire using the plasma–liquid method in alkaline media [28]. However, that work did not mention the formation mechanism of the nanoparticles on the ZnO nanowires during the plasma treatment. Additionally, to understand the comprehensive scenario of the influence of the media on the formation of nanoparticles on ZnO nanowires under plasma conditions and to study the effect of the composite order, from semiconductor–metal–semiconductor to metal–semiconductor–semiconductor, on photocatalytic activity, we designed and prepared a ternary heterostructure made of Ag-Cu_2O-ZnO nanowire in this work. In acidic media, Cu_2O particles were formed instead of CuO particles under the same plasma conditions. Furthermore, the Ag nanoparticles were preferentially located in Cu_2O nanoclusters instead of the ZnO nanowires, forming a broccoli-shaped structure attached to the ZnO nanowires. The results showed that the presence of Ag and Cu_2O partially quenched the photoluminescence in both the UV and visible light regions. Consequently, the photocatalytic activity decomposing methyl orange in the ternary Ag-Cu_2O-ZnO structure took place strongly within 30 min with an efficiency of approximately 98%. The stability and photo-reusability were dramatically improved, merely decreasing by about 3% in terms of performance after 20 cycles. The formation of the Ag-Cu_2O-ZnO heterostructure within the plasma–liquid treatment is also discussed in detail.

2. Results and Discussion

2.1. Decorating Ag NPs and Cu_2O NPs on ZnO NWs via the Plasma–Liquid Interaction Method

ZnO nanowires (NWs) were prepared using the hydrothermal method [28]. The ZnO-nanowire-growing procedure involves a mixed aqueous solution as a precursor, which contains the alkaline reagent Na_2CO_3 and Zn^{2+} salt. The formation process of the ZnO was provided by Hu et.al. [38].

The interaction between the Ar gas plasma and the aqueous solution can generate strongly reactive species such as atomic hydrogen (H), H*, H^+, and OH radicals, caused by the dissociation of water molecules (H_2O), hydrated electrons (e^-_{aq}), and UV light—which has a strong reducing ability—at the same time [39–42]. These species take part in and push up the reactions in the solution. The decoration of the Ag and Cu_2O NPs on the ZnO NWs using the plasma–liquid treatment includes the following stages: Firstly, the H^+ species participate in the bombarding process on the surface of the ZnO NW, resulting in the removal of oxygen atoms and, thus, the creation of oxygen vacancies on the surface. Consequently, the nanoparticle adherence to the ZnO surface is greatly enhanced. This process was considered for a Pt-attached MoO_x system [41] by the Li group and activated TiO_2 [39] by the Yu group. Secondly, the reduction reaction of metal ions, in order to synthesize nanoparticles in the plasma–liquid process with the reducing agents e^-_{aq} and H*, is performed as per the following equations [37,43]:

$$Ag^+ + ne^-{}_{aq} \rightarrow Ag^0 \quad (1)$$

$$Ag^+ + H^* \rightarrow Ag^0 + H^+ \quad (2)$$

The hydroxyl OH radicals produced at the plasma–liquid interface location react swiftly with Cu^{2+} together with ascorbic acid in order to form Cu_2O nanoparticles, as per the reaction below [44]:

$$Cu^{2+} + OH^- + C_6H_8O_6 \rightarrow Cu_2O + C_6H_6O_6 + H_2O \quad (3)$$

Lastly, the ZnO nanowires can make electron–hole pairs on their surface by absorbing the UV light generated within the plasma process. On the ZnO surface, these electrons and holes are utilized for direct reduction or to produce radicals that further reduce Ag^+ and Cu^{2+} ions [34,45]. Notably, Ag and Cu_2O nanoparticles preferentially agglomerate with each other to form a broccoli-like hierarchical structure because of their energy priority [46]. With the benefits of being rapid and easy to use, the plasma–liquid method can effectively produce a Ag-Cu_2O-ZnO ternary structure without adding chemical surfactants as other methods require.

2.2. Structural and Morphological Characteristics

Figure 1 introduces the XRD patterns of as-prepared samples: ZnO and Ag-Cu_2O-ZnO NWs determined by a Bruker D8 advanced X-ray power diffractometer device. The diffraction peaks characterize the hexagonal wurtzite structure of the ZnO NWs located at 31.9°, 34.5°, 36.4°, 47.66°, 56.69°, and 63.1° corresponding to crystal planes of (100), (002), (101), (102), (110), and (103), in excellent consistency with the standard card of ZnO (JCPDS: 36-1451). Remarkably, these peaks are all shown in Ag-Cu_2O-ZnO X-ray diffraction patterns without any shifting, indicating the advantage of the plasma–liquid method, which does not destroy the ZnO crystalline structure. The presence of diffraction peaks with Miller indices (011), (002), and (022), matching the standard card of Cu_2O (JCPDS: 05-0667), confirms the presence of Cu_2O, as pointed out in other published articles [9,11,44]. A peak at around 43° can be due to the overlap contribution of the Ag (200) and Cu_2O (002) planes. A peak at around 38° can be attributed to the (111) plane of Ag nanoparticles, corresponding to the standard card of Ag (JCPDS: 04–0783). From the X-ray diffraction pattern, a ternary Ag-Cu_2O-ZnO composite was successfully synthesized using the plasma–liquid method. Furthermore, no other X-ray peaks could be observed, confirming the purity of the as-prepared samples.

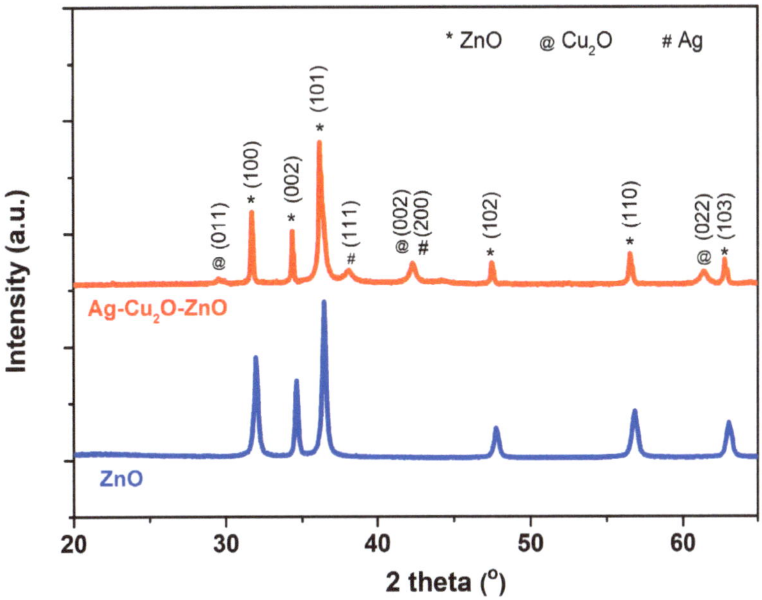

Figure 1. XRD patterns of ZnO and Ag-Cu_2O-ZnO synthesized using the hydrothermal approach combined with the plasma–liquid method.

The surface morphologies of ZnO and Ag-Cu_2O-ZnO NWs measured by FESEM are displayed in Figure 2. The ZnO NWs prepared using the hydrothermal technique have diameters of 20–70 nm and lengths of approximately 10 μm (Figure 2a,b). Under the impact of plasma current, the Cu_2O and Ag nanoparticles were decorated on the surfaces of the ZnO NWs (Figure 2c) in a broccoli-like shape. As a result, broccolis with 150–180 nm diameters are observed. The particle size distribution histogram of the Ag NPs and Cu_2O NPs is shown in Figure 2d. The average particle size of the nanoparticles was calculated to be 10.3 nm from the original data. The broccoli shape is an overlay result of the consecutive formation of Cu_2O crystals together with the deposition of Ag nanoparticles. To elucidate this, we implemented EDX-STEM mapping to measure the Ag-Cu_2O-ZnO composite (Figure 2e). Obviously, four elements—Zn (blue), Cu (red), Ag (green), and O (yellow)—were detected, and one cannot observe any other elements. It is worth mentioning that the majority of Ag elements are located in the broccoli of Cu_2O NPs. HRTEM images robustly demonstrated the existence of distinguishable lattice fringes of ZnO, Cu_2O, and Ag with d-spacings of 0.25 nm, 0.24 nm, and 0.23 nm, respectively, which are assigned to crystal planes of (101) (ZnO), (111) (Cu_2O), and (111) (Ag), as shown in Figure 3. Thus, both Ag and Cu_2O NPs were successfully decorated on the ZnO NW surfaces.

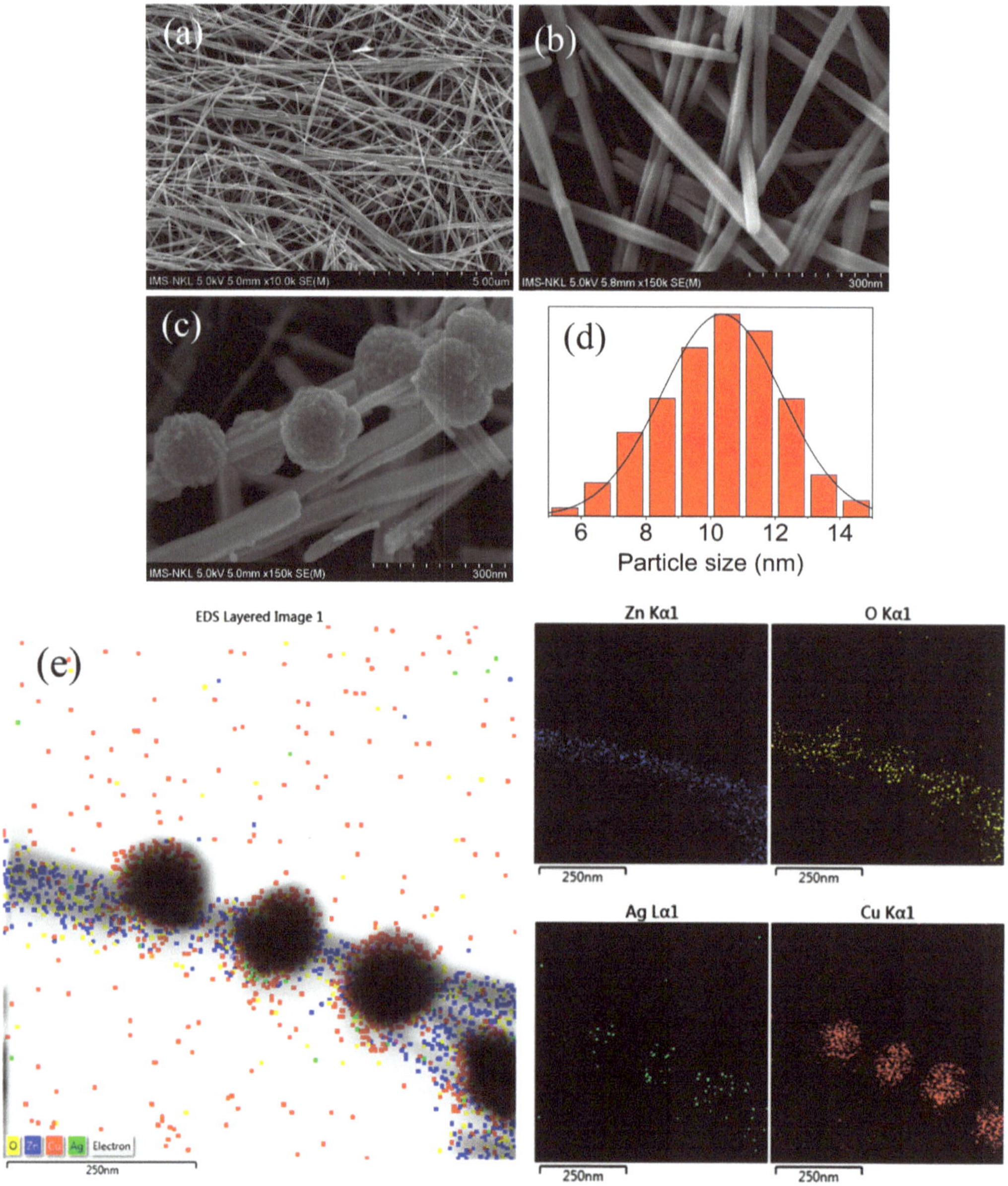

Figure 2. Field-emission scanning electron microscopy images of ZnO (**a**,**b**) and Ag-Cu_2O-ZnO NWs (**c**); particle size distribution histogram of Ag NPs and Cu_2O NPs (**d**); and EDX-STEM mapping images of Zn, O, Ag, and Cu elements (**e**).

Analysis of the room-temperature PL spectra of the ZnO and Ag-Cu_2O-ZnO NWs showed that the spectrum intensity gradually declined at both the UV and VIS regions when the Cu_2O and Ag NPs were attached (Figure 4a). As reported in previous articles [28,46], the PL spectrum of ZnO NWs consists of an emission peak in the 380–390 nm region, which stems from free electron–hole pair recombination, and a broad emission peak in the visible region, which arises from defect levels [46]. Therefore, the suppression of the PL spectrum of ZnO NWs is a crucial task in gaining an excellent photocatalyst. As expected, a part of the PL spectrum is quenched upon the decoration of Cu_2O and Ag NPs on the ZnO NWs surface, indicating an inhibition in electron–hole pair recombination and an increase in charge separation efficiency [47]. The reason for the decrease in the PL spectrum intensity can be ascribed to (1) the heterostructure formation between the ZnO and Cu_2O, improving the charge transfer route, and (2) the formation of an 'electron reservoir' by the Ag NPs to catch the photogenerated electrons.

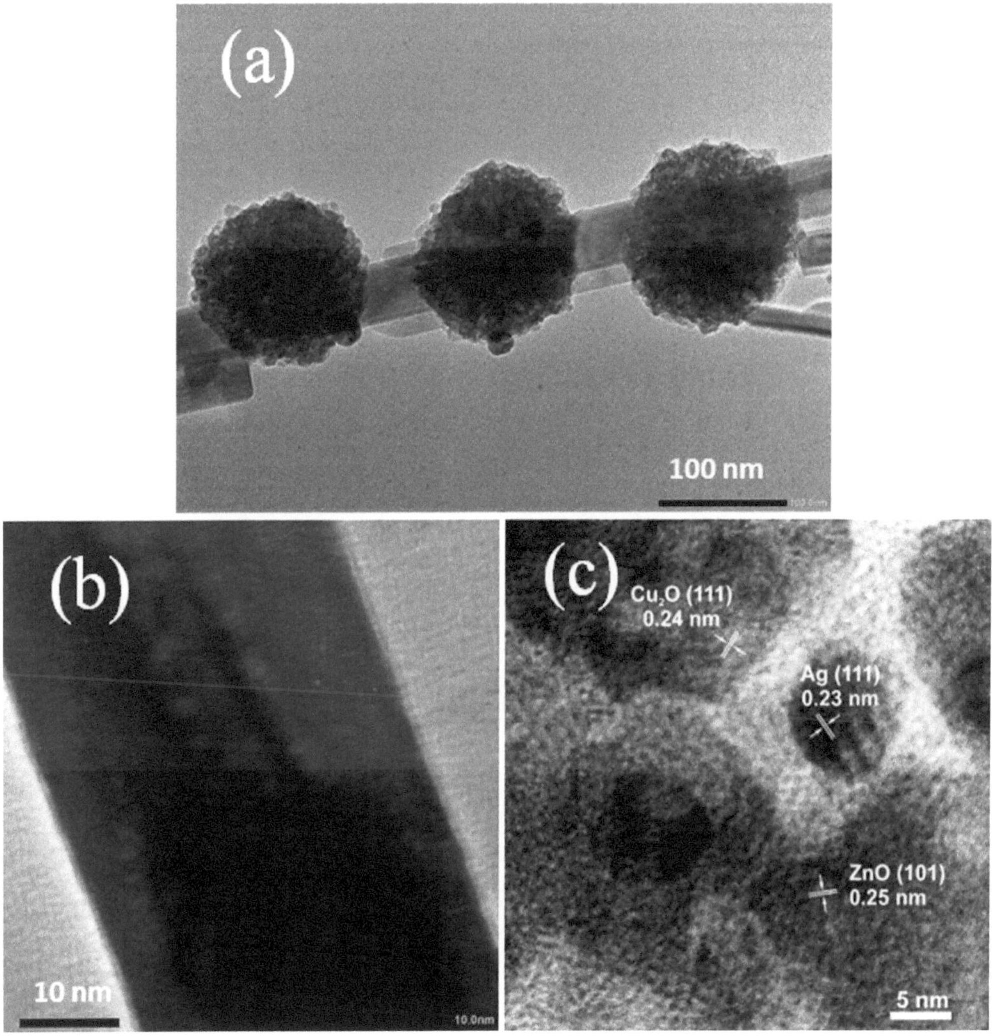

Figure 3. TEM (**a**) and HR-TEM images of ZnO NW (**b**) and Ag-Cu_2O-ZnO (**c**).

The optical absorption spectra of ZnO and Ag-Cu_2O-ZnO samples are studied and depicted in Figure 4b. Obviously, the ZnO NWs only absorb in the UV region with an absorption edge at ~385 nm. In the ternary Ag-Cu_2O-ZnO NWs, the absorption edge moves forward along the wavelength side, and a broad absorption range is found in the visible light region. This is a consequence of the narrower bandgap of Cu_2O [48] and the localized surface plasmon resonance (LSPR) effect of Ag NPs, which forms an LSPR peak at ~445 nm. Therefore, the enhanced absorbance in the visible light region produces more photogenerated charge carriers. These findings also confirm the formation of a Ag-Cu_2O-ZnO heterojunction.

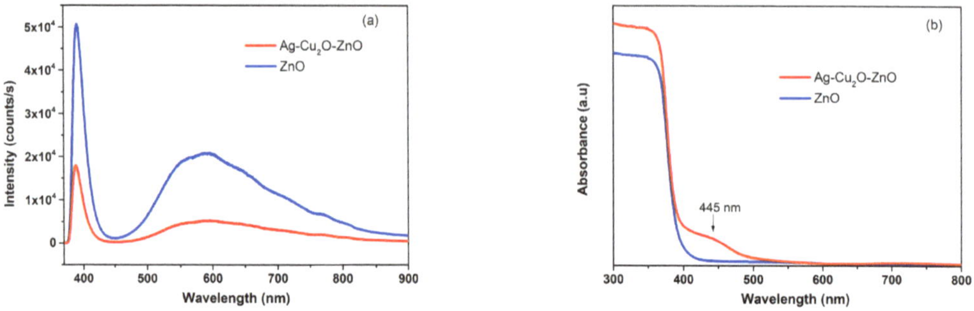

Figure 4. PL spectra (**a**) and absorbance spectra (**b**) of ZnO and Ag-Cu_2O-ZnO.

2.3. Photocatalytic Activity

Electrochemical impedance spectroscopy (EIS) is exploited as a useful tool for learning more information about charge separation/transfer and charge recombination in photocatalysts. Interface layer resistance on the electrode surface is exhibited by an arc radius on a Nyquist plot. As shown in Figure 5a, the arc radius for the Ag-Cu_2O-ZnO NWs is significantly smaller than that of their pure ZnO counterpart, indicating that this ternary heterostructure possesses faster interfacial charge transfer and more efficient charge separation. On the other hand, as noted in numerous articles, the higher photocurrent demonstrates the better separation efficiency of the charge carriers. Therefore, the photocurrent versus the time of the as-prepared samples was measured in three on–off irradiation cycles using solar-simulated light and then plotted in the inset of Figure 5a. It is evident that the photocurrent intensity is greatly enhanced when the Ag and Cu_2O NPs are decorated on the ZnO NW surface. As a result, the current intensity of the Ag-Cu_2O-ZnO NWs is about 600×10^{-9} A/cm^2, three times higher than that of pure ZnO NWs. Consequently, the Ag-Cu_2O-ZnO NWs unquestionably exhibit a more effective photogenerated electron–hole separation ability. Furthermore, the interfacial electric field forming in the Ag-Cu_2O-ZnO may also be responsible for these observations since they hasten the migration of charge carriers under UV–vis light.

The degradation of methyl orange (MO) dye with respect to time under solar simulator illumination is used to estimate the photocatalytic activity of the ZnO and Ag-Cu_2O-ZnO samples. The decrease in MO absorption peak intensity between 350 and 550 nm is considered a crucial indicator for this study. Figure 5b,c show the UV–vis absorption spectra of the MO dye in terms of irradiation time using the ZnO and Ag-Cu_2O-ZnO photocatalysts, respectively. As indicated, the MO dye is decolorized based on illumination time. A peak shift to a shorter wavelength side can also be found, which is explained by the formation of intermediate species during photocatalytic reaction [33,49].

As seen in Figure 5b,c, the MO dye was strongly decomposed, ~27% and 82% in the first 10 min and ~89% and 98% within 30 min, corresponding to pure ZnO and Ag-Cu_2O-ZnO NWs. This is assumed to be due to the mutual working of the active species—•OH, •O_2^- photogenerated holes, and electrons—in decomposing organic molecules in the aqueous solution. Since •OH and •O_2^- are generated by surface redox during photocatalysis, photogenerated holes and electrons can have a key impact on the degradation of organic dyes.

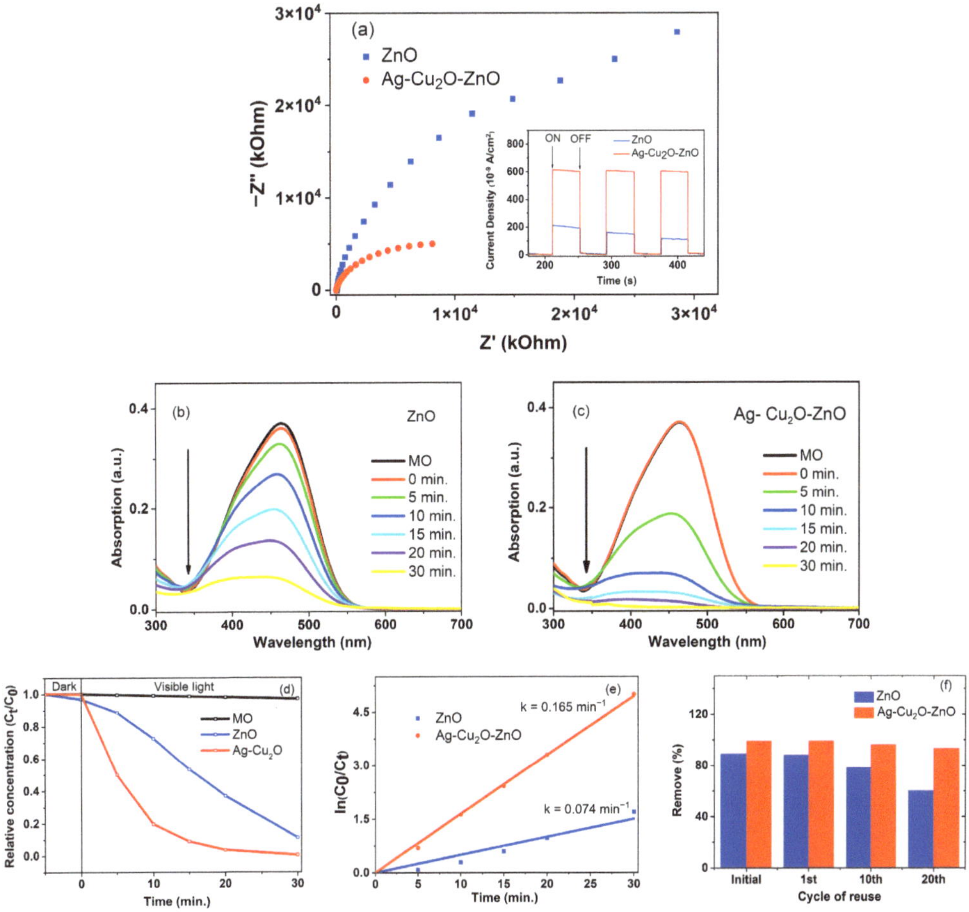

Figure 5. EIS Nyquist plot curves of ZnO and Ag-Cu_2O-ZnO samples. The inset is the photocurrent-time curves of as-prepared samples under on–off solar-simulated light (**a**). The UV–vis absorption spectra of the MO dye in terms of time under solar-simulated light using the photocatalysts of ZnO (**b**) and Ag-Cu_2O-ZnO (**c**). C_t/C_0 curves as a function of the time of the MO degradation process (**d**). Kinetics of the MO degradation process (**e**). Photocatalytic activity reuse cycle of Ag-Cu_2O-ZnO NWs compared with ZnO NWs (**f**).

For further information, the kinetics (k) of the MO decomposing reaction was calculated with the Langmuir–Hinshelwood equation [28]. The k-values of the ZnO and the Ag-Cu_2O-ZnO NWs were 0.074 min^{-1} and 0.165 min^{-1}, respectively. Evidently, these findings indicate that the Ag/Cu_2O broccolis contribute to accelerating the MO photo-decomposing process of ZnO host material more quickly. Therefore, the necessary decomposing time is much shorter in comparison with pure ZnO NWs.

It is worth mentioning that the k-value of the Ag-Cu_2O-ZnO NWs is greater than the compounds used in many previous papers, as seen in Table 1. The k-values of the binary composites, TiO_2-ZnO (0.011 min^{-1}) and ZnO-WO_3 (0.0521 min^{-1}), are ~15 and 3 times lower than that of Ag-Cu_2O-ZnO, respectively. In the Ag-Cu_2O-ZnO nanorods prepared by Tsai et al. [23], their degradation rate was only 0.041 min^{-1}, equal to ¼k of our composite. Another composite, ZnO-Au-Cu_2O nanorods [24], showed excellent photocatalytic performance driven by visible light; nevertheless, the k-value (0.3495 h^{-1})

was a limitation of this structure. Similarly, WO_3-ZnO@rGO, ZnO-CdO-CuO, and ZnO-Fe_3O_4-g-C_3N_4 composites have ~6–7 times lower k-values compared with our Ag-Cu_2O-ZnO NWs. Consequently, these composites took much more time to photodegrade organic dyes. It can be confirmed that decorating Ag-Cu_2O NPs on ZnO wires is a superior strategy. This finding has also been found in a CuO-Ag-ZnO nanowire structure in our previous report [28]. In that work, the MO dye is completely decomposed in 30 min at a high k-value of 0.2007 min^{-1}. This k-value difference may originate from the main interfaces between the CuO-Ag-ZnO and Ag-Cu_2O-ZnO structures. In the CuO-Ag-ZnO NWs, the formation of the interfacial electric field in the semiconductor–metal–semiconductor was demonstrated to further improve photocatalytic efficiency [28]. The electrons in the CuO NPs excited from the valance band to the conduction band can transfer to the Fermi level of the Ag NPs and the ZnO conduction band. Then, the electrons from the Fermi level can also move into the conduction band of the ZnO; thus, more electrons participate in the photocatalytic process.

Table 1. A comparison of the photocatalytic performance of different systems.

Structure	Method	Organic Dye	Kinetics k	Yield (%)	Time (min)/Ref.
Cu_2O-ZnO/kaolinite	Co-precipitation	MB	0.098 min^{-1}	93	105 [32]
Cu_2O-ZnO	Precipitation and calcination	MO	1.30762 h^{-1}	98	240 [13]
Ag-Cu_2O-ZnO	Hydrothermal and photoreduction	RhB	0.041 min^{-1}	80	60 [23]
WO_3-ZnO@rGO	Ultrasound	MB	0.0278 min^{-1}	94	90 [50]
TiO_2-ZnO	Hydrothermal	RhB	0.011 min^{-1}	89	180 [51]
ZnO-CdO-CuO	Co-precipitation	MB	0.027 min^{-1}	94	100 [52]
ZnO-Au-Cu_2O	Electrodeposition and sputtering	MO	0.3495 h^{-1}	~75	240 [24]
ZnO-WO_3	Sol–gel	MO	0.0521 min^{-1}	100	90 [53]
ZnO-Fe_3O_4-g-C_3N_4	Sol–gel and annellation	MO	0.0243 min^{-1}	~97	140 [54]
Ag-Cu_2O-ZnO NW	Hydrothermal and plasma–liquid	MO	0.165 min^{-1}	98	30 This work

Moreover, one of the critical factors for a good photocatalyst regarding large-scale production is stability. We, hence, conducted a repeated 20-cycle photocatalytic experiment to evaluate this factor under solar simulator illumination. The results are depicted in Figure 5f. After 10 cycles, the Ag-Cu_2O-ZnO photocatalyst still maintained great activity with a performance of ~97%. Although the catalytic activity continues to decrease in subsequent experimental cycles, the MO dye still decomposed to 95% in the 20th cycle, while that of the ZnO NWs significantly reduced by ~15% after 20 cycles. This result demonstrates the high stability of Ag-Cu_2O-ZnO in the visible light region. As expected, the synergistic contribution of the Ag/Cu_2O NPs improved the reusability of the Ag-Cu_2O-ZnO heterostructure. This demonstrates the superiority of a ternary structure in addressing chemical dyes that contaminate water sources.

In this work, the bottom of the conduction band (CB) and the top of the valence band (VB) of ZnO and Cu_2O are situated at −1.06 (V vs. NHE), 1.7 (V vs. NHE) and −0.77 (V vs. NHE), 2.57 (V vs. NHE), respectively [48]. Combined with the aforementioned findings and discussion, we propose a photocatalytic mechanism in the ternary Ag-Cu_2O-ZnO heterostructure, as illustrated in Figure 6.

The photocatalytic enhancement of this structure can be explained as follows: (i) The expansion of the absorption band to the visible light region due to the LSPR effect of Ag NPs and the narrow bandgap of Cu_2O (2.76 eV [53]) improves the light-harvesting efficiency. (ii) When exposed to visible light, the electrons from the Fermi level with the appearance of an LSPR effect are excited to a higher energy state; these electrons are thus moved to the CB of the Cu_2O. In the photocatalytic process, this can help raise the concentration of charge carriers, which is crucial for separating photogenerated electron–hole pairs [53]. (iii) Because of the p-n heterojunction and tight surface contact between ZnO and Cu_2O, photogenerated electrons continue to flow readily from the CB of Cu_2O to the CB of ZnO. As a result, the electrons on the CB of ZnO and the holes on the VB of Cu_2O can be split and accumulated. The strong oxidants $\bullet O^{2-}$ and $\bullet OH$, which take part in the MO degrading process, can be produced as a result of the redox process by the excess electrons on the CB of ZnO and holes on the VB of Cu_2O.

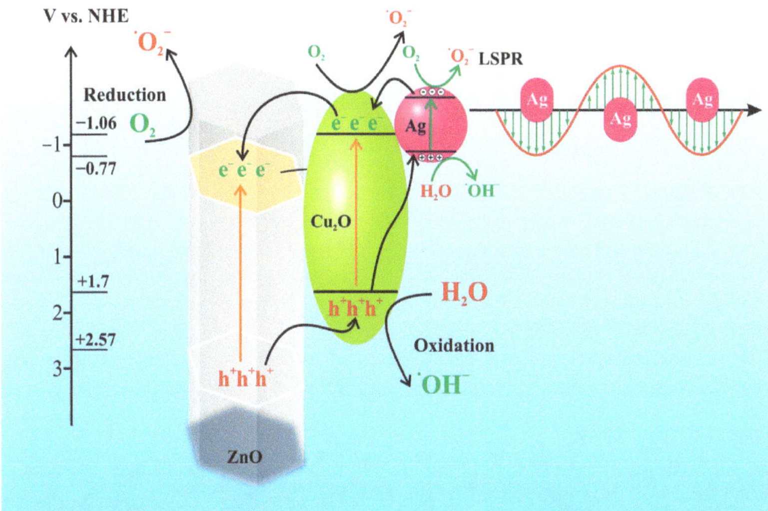

Figure 6. Proposed photocatalytic mechanism scheme of Ag−Cu_2O−ZnO ternary heterostructure.

3. Materials and Methods

Reagents: All chemicals used are analytic grade and high purity. $Zn(NO_3)_2$, Na_2CO_3, $AgNO_3$, $Cu(NO_3)_2$, NH_3, and L'-ascorbic acid were purchased from Sigma Aldrich (St. Louis, MO, USA), and deionized water was the solvent used to dissolve the applied chemicals. Distilled water and ethanol were utilized for cleaning.

3.1. Synthesis of ZnO Nanowires

ZnO nanowires (NWs) were prepared using the hydrothermal method with conditions outlined in our previous publication [28]. $Zn(NO_3)_2$ (0.05 M) and Na_2CO_3 (0.12 M) were separately dissolved in distilled water in two 100mL beakers for 1 h at room temperature. Prior to transferring the solutions to a Teflon put-in steel autoclave, these two solutions were mixed by a magnetic stirrer until the solution mixture reached a milky white color. The hydrothermal process was run at 200 °C for a full day. Eventually, after being taken out by a centrifuge, the white powder was repeatedly cleaned with distilled water and ethanol to get rid of the unreacted substances and contaminants. The powder was dried at 70 °C before being stored securely.

3.2. Decoration of Silver and Copper(I) Oxide on ZnO NWs Using the Plasma–Liquid Interaction Technique

Preparation of two precursor solutions:

(1) $AgNO_3$ salt was dissolved with deionized water in a 500 mL beaker in order to obtain a 50 mM Ag-precursor solution.

(2) We prepared a 15 mM L'ascorbic acid solution and a 50 mM $Cu(NO_3)_2$ solution in deionized water. In total, 50 mM of $Cu(NO_3)_2$ and 15 mM of L'ascorbic acid were combined in a 6:2 volumetric ratio to generate a Cu-precursor solution.

Setting up the plasma–liquid system:

A simple plasma–liquid system operating at ambient conditions comprises three main components, a DC high-voltage source, a plasma nozzle, and a gold electrode, described in detail in Refs. [34,35]. The plasma nozzle, consisting of a plasma electrode comprising a 1.6 mm diameter Tungsten rod, was connected to an Ar gas flow system that was supplied during the experiment. The gold electrode was immersed in the examined solution contained in a 50 mL beaker. The DC power source supplied a high voltage of 2.5 kV to ignite the Ar gas plasma between the plasma nozzle and the aqueous solution surface. The generated current was kept at a constant value of 5 mA. Reactions in the solution were accelerated extraordinarily quickly owing to radicals and activated chemical species generated by plasma–liquid interactions.

The following steps were used to decorate the surface of the ZnO NWs with Ag and Cu_2O nanoparticles:

- The as-prepared ZnO NW powder was uniformly dispersed in distilled water within a 50 mL beaker using ultrasonic equipment.
- Plasma was ignited for three minutes.
- To form the ternary structure of Ag-Cu_2O-ZnO NWs, 5 mL of Cu-precursor solution was added, followed by five minutes of plasma treatment. Next, 80 μL of 50 mM Ag-precursor solution was added, and the plasma ignition process was continued for an additional five minutes.
- The Ag-Cu_2O-ZnO powder was isolated using centrifugation. These samples were dried in a vacuum oven for 24 h at 50 °C and stored in sealed vessels.

3.3. Sample Characterization

The surface morphology of the as-prepared samples was observed using field-emission scanning electron microscopy (FESEM, Hitachi S-4800, Tokyo, Japan). The crystal structure, composition, and morphology of the samples were certified by a high-resolution transmission electron microscope (HRTEM; JEM 2100, JEOL, Tokyo, Japan). Their crystal structure was determined by a Bruker D8 advanced X-ray power diffractometer with Cu-ka radiation. The photoluminescence spectrum was measured using high-resolution photoluminescence spectroscopy (Horiba iHR550, Horiba, Osaka, Japan) with the excitation of a 355 nm Teem Photonic laser and then detected using a thermoelectrically cooled Si-CCD camera (Horiba Synapse). The UV–vis absorption spectrum was examined by a Carry 5000 UV–vis spectrophotometer (Agilent, Santa Clara, CA, USA).

3.4. Photocatalytic Activity Evaluation

The photocatalytic activity of the samples involving ZnO and Ag-Cu_2O-ZnO was evaluated using the photodegradation of methyl orange (MO) under solar-simulated light. In total, 25 mg of photocatalyst was mixed with 30 mL of MO (5 ppm) contained in a beaker (Schott Duran, Mainz, Germany). After that, this mixture is stirred for an hour in a dark environment to ensure that the photocatalyst surface achieves an adsorption–desorption equilibrium state. Subsequently, the incident light source was irradiated perpendicularly to the solution surface for 30 min. Three milliliters of the solution were aspirated and placed in a centrifuge tube at determined intervals. The concentration of MO, with respect to the illuminated time interval, was measured using the UV–vis spectrophotometer.

4. Conclusions

Capitalizing on the advantages and effectiveness of the plasma–liquid method, we successfully synthesized a ternary Ag-Cu_2O-ZnO catalyst and proposed an underlying formation mechanism for this material system. The results showed that the coexistence of Cu_2O and Ag NPs on ZnO NWs acts as a key intermediate species for promoting fast charge transfer and separation and impeding electron–hole recombination. Indeed, the ternary Ag-Cu_2O-ZnO composite showed superior photocatalytic activity. About 82% and 98% of the MO contents decomposed under visible light irradiation within 10 and 30 min. In addition, the catalytic performance still achieved above 95% after 20 reuse cycles, indicating the excellent stability and reusability of the Ag-Cu_2O-ZnO. The findings demonstrate the outstanding photocatalytic efficiency of semiconductor- and noble metal-based structures, paving the way for significant potential applications in pollutant treatments.

Author Contributions: Conceptualization, N.T.H.L., P.T.T. and V.D.L.; methodology, N.T.H.L. and T.X.A.; validation, N.T.H.L., P.T.T., V.H.K. and D.K.T.; formal analysis, P.T.T., N.T.H.L., T.N.B., V.H.K. and N.H.D.; investigation, P.T.T., T.N.B., L.T.H.P., V.D.L. and D.H.T.; data curation, P.T.T., L.T.H.P., V.H.K., D.K.T. and T.N.B.; writing—original draft preparation, P.T.T., D.H.M., N.H.D. and N.T.H.L.; writing—review and editing, N.T.H.L.; visualization, supervision, N.T.H.L., V.D.L. and D.H.M.; project administration, N.T.H.L.; funding acquisition, N.T.H.L. All authors have read and agreed to the published version of the manuscript.

Funding: This work was financially supported by the Ministry of Science and Technology, Vietnam, with the grant code NDT. 70.e-ASIA/19.

Data Availability Statement: The data that support the plots within this paper are available from the corresponding authors upon reasonable request.

Conflicts of Interest: The authors declare no conflicts of interest.

References

1. Al-Nuaim, M.A.; Alwasiti, A.A.; Shnain, Z.Y. The photocatalytic process in the treatment of polluted water. *Chem. Pap.* **2023**, *77*, 677–701. [CrossRef] [PubMed]
2. Ren, G.; Han, H.; Wang, Y.; Liu, S.; Zhao, J.; Meng, X.; Li, Z. Recent advances of photocatalytic application in water treatment: A review. *Nanomaterials* **2021**, *11*, 1804. [CrossRef]
3. Ong, C.B.; Ng, L.Y.; Mohammad, A.W. A review of ZnO nanoparticles as solar photocatalysts: Synthesis, mechanisms and applications. *Renew. Sustain. Energy Rev.* **2018**, *81*, 536–551. [CrossRef]
4. Qi, K.; Cheng, B.; Yu, J.; Ho, W. Review on the improvement of the photocatalytic and antibacterial activities of ZnO. *J. Alloys Compd.* **2017**, *727*, 792–820. [CrossRef]
5. Su, Q.; Zuo, C.; Liu, M.; Tai, X. A review on Cu_2O-based composites in photocatalysis: Synthesis, modification and applications. *Molecules* **2023**, *28*, 5576. [CrossRef]
6. Kuo, C.H.; Huang, M.H. Morphologically controlled synthesis of Cu_2O nanocrystals and their properties. *Nano Today* **2010**, *5*, 106–116. [CrossRef]
7. Consonni, V.; Briscoe, J.; Kärber, E.; Li, X.; Cossuet, T. ZnO nanowires for solar cells: A comprehensive review. *Nanotechnology* **2019**, *30*, 362001. [CrossRef]
8. Kandjani, A.E.; Sabri, Y.M.; Periasamy, S.R.; Zohora, N.; Amin, M.H.; Nafady, A.; Bhargava, S.K. Controlling core/shell formation of nanocubic p-Cu_2O/n-ZnO Toward Enhanced Photocatalytic Performance. *Langmuir* **2015**, *31*, 10922–10930. [CrossRef]
9. Chen, X.; Lin, P.; Yan, X.; Bai, Z.; Yuan, H.; Shen, Y.; Liu, Y.; Zhang, G.; Zhang, Z.; Zhang, Y. Three-dimensional ordered ZnO/Cu_2O nanoheterojunctions for efficient metal–oxide solar cells. *ACS Appl. Mater. Interfaces* **2015**, *7*, 3216–3223. [CrossRef]
10. Yoon, J.S.; Lee, J.W.; Sung, Y.M. Enhanced photoelectrochemical properties of Zscheme ZnO/p-n Cu_2O PV-PEC cells. *J. Alloys Compd.* **2019**, *771*, 869–876. [CrossRef]
11. Jiang, T.; Xie, T.; Chen, L.; Fu, Z.; Wang, D. Carrier concentration-dependent electron transfer in Cu_2O/ZnO nanorod arrays and their photocatalytic performance. *Nanoscale* **2013**, *5*, 2938–2944. [CrossRef]
12. Zou, X.; Tian, H.; Yan, S. Syntheis of Cu_2O/ZnO hetero-nanorod arrays with enhanced visible light-driven photocatalytic activity. *Cryst. Eng. Comm* **2014**, *16*, 1149–1156. [CrossRef]
13. Wang, X.-S.; Zhang, Y.-D.; Wang, Q.-C.; Dong, B.; Wang, Y.-J.; Feng, W. Photocatalytic activity of Cu_2O/ZnO nanocomposite for the decomposition of methyl orange under visible light irreadiation. *Sci. Eng. Compos. Mater.* **2019**, *26*, 104–113. [CrossRef]
14. Zhu, X.; Wang, J.; Yang, D.; Liu, J.; He, L.; Tang, M.; Feng, W.; Wu, X. Fabrication, characterization and high photocatalytic activity of Ag–ZnO heterojunctions under UV-visible light. *RSC Adv.* **2021**, *11*, 27257–27266. [CrossRef] [PubMed]
15. Deng, Q.; Duan, X.; Ng, D.H.L.; Tang, H.; Yang, Y.; Kong, M.; Wu, Z.; Cai, W.; Wang, G. Ag nanoparticle decorated nanoporous ZnO microrods and their enhanced photocatalytic activities. *ACS Appl. Mater. Interfaces* **2012**, *4*, 6030–6037. [CrossRef] [PubMed]
16. Trang, T.N.Q.; Phan, T.B.; Nam, N.D.; Thu, V.T.H. In situ charge transfer at the Ag@ZnO photoelectrochemical interface toward the high photocatalytic performance of H_2 evolution and RhB degradation. *ACS Appl. Mater. Interfaces* **2020**, *12*, 12195–12206. [CrossRef] [PubMed]
17. Thinh, V.D.; Lam, V.D.; Bach, T.N.; Van, N.D.; Manh, D.H.; Tung, D.H.; Lien, N.T.H.; Thuy, U.T.D.; Anh, T.X.; Tung, N.T.; et al. Enhanced optical and photocatalytic properties of Au/Ag nanoparticle-decorated ZnO films. *J. Electron. Mater.* **2020**, *49*, 2626–2632. [CrossRef]
18. He, W.; Kim, H.-K.; Wamer, W.G.; Melka, D.; Callahan, J.H.; Yin, J.-J. Photogenerated charge carrier and reactive oxygen species in ZnO/Au hybrid nanostructures with enhanced photocatalytic and antibacterial activity. *J. Am. Chem. Soc.* **2014**, *136*, 750–757. [CrossRef] [PubMed]

19. Mauro, A.D.; Zimbone, M.; Scuderi, M.; Nicotra, G.; Fragalà, M.E.; Impellizzeri, G. Effect of Pt nanoparticles on the photocatalytic activity of ZnO nanofibers. *Nanoscale Res. Lett.* **2015**, *10*, 484. [CrossRef]
20. Zhao, S.; Cheng, Z.; Kang, L.; Li, M.; Gao, Z. The facile preparation of Ag decorated TiO_2/ZnO nanotube and their potent photocatalytic degradation efficiency. *RSC Adv.* **2017**, *7*, 50064–50071. [CrossRef]
21. Mulkhopadhyay, S.; Maiti, D.; Chatterjee, S.; Devi, P.S.; Kumar, G.S. Design and application of Au decorated ZnO/TiO_2 as a stable photocatalyst for wide spectral coverage. *Phys. Chem. Chem. Phys.* **2016**, *18*, 31622–31633. [CrossRef]
22. Yin, H.; Yu, K.; Song, C.; Huang, R.; Zhu, Z. Synthesis of Au-decorated V_2O_5@ZnO heteronanostructures and enhanced plasmonic photocatalytic activity. *ACS Appl. Mater. Interfaces* **2014**, *6*, 14851–14860. [CrossRef]
23. Tsai, C.-E.; Yeh, S.-M.; Chen, C.-H.; Lin, H.-N. Flexible photocatalytic paper with Cu_2O and Ag nanoparticle-decorated ZnO nanorods for visible light photodegradation of organic dye. *Nanoscale Res. Lett.* **2019**, *14*, 204. [CrossRef] [PubMed]
24. Ren, A.; Wang, B.; Zhang, H.; Ding, P.; Wang, Q. Sandwiched ZnO@Au@Cu_2O Nanorod Films as Efficient Visible-Light-Driven Plasmonic Photocatalysts. *ACS Appl. Mater. Interfaces* **2015**, *7*, 4066–4074. [CrossRef] [PubMed]
25. Subash, B.; Krishnakumar, B.; Swaminathan, M.; Shanthi, M. ZnS–Ag–ZnO as an excellent UV-light-active photocatalyst for the degradation of AV 7, AB 1, RR 120, and RY 84 dyes: Synthesis, characterization, and catalytic applications. *Ind. Eng. Chem. Res.* **2014**, *53*, 12953–12963. [CrossRef]
26. Gavade, N.L.; Babar, S.B.; Kadam, A.N.; Gophane, A.; Garadka, K.M. Fabrication of M@CuxO/ZnO (M = Ag, Au) heterostructured nanocomposite with enhanced photocatalytic performance under sunlight. *Ind. Eng. Chem. Res.* **2017**, *56*, 14489–14501. [CrossRef]
27. Yuan, X.; Pei, F.; Luo, X.; Hu, H.; Qian, H.; Wen, P.; Miao, K.; Wang, W.; Feng, G. Fabrication of ZnO/Au@Cu2O heterojunction towards deeply oxidative photodegradation of organic dyes. *Sep. Purif. Technol.* **2021**, *262*, 118301. [CrossRef]
28. Thu, P.T.; Thinh, V.D.; Lam, V.D.; Bach, T.N.; Phong, L.T.H.; Tung, D.H.; Manh, D.H.; Van Khien, N.; Anh, T.X.; Le, N.T.H. Decorating of Ag and CuO on ZnO nanowires by plasma electrolyte oxidation method for enhanced photocatalytic efficiency. *Catalysts* **2022**, *12*, 801. [CrossRef]
29. Prete, P.; Lovergine, N.; Tapfer, L. Nanostructure size evolution during Au-catalysed growth by carbo-thermal evaporation of well-aligned ZnO nanowires on (100)Si. *Appl. Phys. A* **2007**, *88*, 21–26. [CrossRef]
30. Yatsui, T.; Lim, J.; Nakamata, T.; Kitamura, K.; Ohtsu, M.; Yi, G.-C. Low-temperature (~270 °C) growth of vertically aligned ZnO nanorods using photoinduced metal organic vapour phase epitaxy. *Nanotechnology* **2007**, *18*, 065606. [CrossRef]
31. Liang, H.W.; Lu, Y.M.; Shen, D.Z.; Li, B.H.; Zhang, Z.Z.; Shan, C.X.; Zhang, J.Y.; Fan, X.W.; Du, G.T. Growth of vertically aligned single crystal ZnO nanotubes by plasma-molecular beam epitaxy. *Solid State Commun.* **2006**, *137*, 182–186. [CrossRef]
32. Fufa, P.A.; Feysia, G.B.; Gultom, N.S.; Kuo, D.-H.; Chen, X.; Kabtamu, D.M.; Zelekew, O.A. Visible light-driven photocatalytic activity of Cu2O/ZnO/Kaolinite-based composite catalyst for the degradation of organic pollutant. *Nanotechnology* **2022**, *33*, 315601. [CrossRef] [PubMed]
33. Shubha, J.P.; Adil, S.F.; Khan, M.; Hatshan, M.R.; Khan, A. Facile fabrication of a ZnO/Eu2O3/NiO-based ternary heterostruc-ture nanophotocatalyst and its application for the degradation of methylene Blue. *ACS Omega* **2021**, *6*, 3866–3874. [CrossRef] [PubMed]
34. Lee, S.Y.; Do, H.T.; Kim, J.H. Microplasma-assisted synthesis of TiO2–Au hybrid nanoparticles and their photocatalytic mechanism for degradation of methylene blue dye under ultraviolet and visible light irradiation. *Appl. Surf. Sci.* **2022**, *573*, 151383. [CrossRef]
35. Thuy, N.T.T.; Tung, D.H.; Manh, L.H.; Kim, J.H.; Kudryashov, S.I.; Minh, P.H.; Hien, N.T. Plasma enhanced wet chemical sur-face sctivation of TiO_2 for the synthesis of high-performance photocatalytic Au/TiO_2 nanocomposites. *Appl. Sci.* **2020**, *10*, 3345. [CrossRef]
36. Fatemeh, R.; Patrick, V.; Anton, N.; Rino, M.; Nathalie, D.G. Applications of plasma-liquid systems: A review. *Materials* **2019**, *12*, 2751.
37. Qiang, C.; Junshuai, L.; Yongfeng, L. A review of plasma–liquid interactions for nanomaterial synthesis. *J. Phys. D Appl. Phys.* **2015**, *48*, 424005.
38. Hu, H.; Huang, X.; Deng, C.; Chen, X.; Qian, Y. Hydrothermal synthesis of ZnO nanowires and nanobelts on a large scale. *Mater. Chem. Phys.* **2007**, *106*, 58–62. [CrossRef]
39. Yu, F.; Wang, C.; Li, Y.; Ma, H.; Wang, R.; Liu, Y.; Suzuki, N.; Terashima, C.; Ohtani, B.; Ochiai, T.; et al. Enhanced solar photothermal catalysis over solution plasma activated TiO_2. *Adv. Sci.* **2020**, *7*, 2000204. [CrossRef]
40. Qiang, C.; Toshiro, K.; Rikizo, H. Reductants in Gold Nanoparticle Synthesis Using Gas–Liquid Interfacial Discharge Plasmas. *Appl. Phys. Express* **2012**, *5*, 086201.
41. Li, Y.; Wang, Y.; Lin, J.; Shi, Y.; Zhu, K.; Xing, Y.; Li, X.; Jia, Y.; Zhang, X. Solution-plasma-induced oxygen vacancy enhances MoO_x/Pt electrocatalytic counter electrode for bifacial dye-sensitizes solar cells. *Front. Energy Res.* **2022**, *10*, 924515. [CrossRef]
42. Pitchaimuthu, S.; Honda, K.; Suzuki Shoki Naito, A.; Suzuki, N.; Katsumata, K.; Nakata, K.; Ishida, N.; Kitamura, N.; Idemoto, Y.; Kondo, T.; et al. Solution plasma process-derived defect-induced heterophase anatase/brookite TiO_2 nanocrystals for enhanced gaseous photocatalytic performance. *ACS Omega* **2018**, *3*, 898–905. [CrossRef]
43. Kondeti, V.S.; Gangal, U.; Yatom, S.; Bruggeman, P.J. Ag^+ reduction and silver nanoparticle at the plasma-liquid interface by an RF driven atmospheric pressure plasma jet: Mechanisms and the effect of surfactant. *J. Vac. Sci. Technol. A Vac. Surf. Film.* **2017**, *35*, 061302. [CrossRef]
44. Leonardo, V.; Ignacio, C.; Javier, E.; Nancy, B.; Dinesh, P.S. Ascorbic acid based controlled growth of various Cu and Cu_2O nanostructures. *Mater. Res. Express* **2019**, *6*, 065033.

45. Mariotti, D.; Patel, J.; Švrček, V.; Maguire, P. Plasma-liquid interactions at atmospheric pressure for nanomaterials synthesis and surface engineering. *Plasma Process. Polym.* **2012**, *9*, 1074–1085. [CrossRef]
46. Zheng, K.; Zhang, Z.; Wang, X.; Zhan, R.; Chen, H.; Deng, S.; Xu, N.; Chen, J. Mechanism of photoluminescence quenching in visible and ultraviolet emissions of ZnO nanowires decorated with gold nanoparticles. *Jpn. J. Appl. Phys.* **2019**, *58*, 051005. [CrossRef]
47. Wu, H.-Y.; Jian, W.-J.; Dang, H.-F.; Zhao, X.-F.; Zhang, L.-Z.; Li, J.-H. Hierarchical Ag-ZnO Microspheres with Enhanced Photocatalytic Degradation Activities. *Pol. J. Environ. Stud.* **2017**, *26*, 871–880. [CrossRef]
48. Cui, J.; Ye, L.; Chen Xi Li, J.; Bian Yang Yang, M.; Yang, Q.; Yun, D.; Sun, S. Simultaneously promoting adsorption and charge separation in Z-scheme ZnO/Cu_2O heterojunctions for efficient removal of tetracycline. *Appl. Surf. Sci.* **2023**, *638*, 158046. [CrossRef]
49. Reynolds, D.C.; Look, D.C.; Jogai, B.; Litton, C.W.; Cantwell, G.; Harsch, W.C. Valence-band ordering in ZnO. *Phys. Rev. B* **1999**, *60*, 2340–2344. [CrossRef]
50. Chaudhary, K.; Shaheen, N.; Zulfiqar, S.; Sarwar, M.I.; Suleman, M.; Agboola, P.O.; Shakir, I.; Warsi, M.F. Binary WO_3-ZnO nanostructures supported rGO ternary nanocomposite for visible light driven photocatalytic degradation of methylene blue. *Synth. Met.* **2020**, *269*, 116526. [CrossRef]
51. Tran Thi, V.H.; Pham, T.N.; Pham, T.T.; Le, M.C. Synergistic Adsorption and Photocatalytic Activity under Visible Irradiation Using Ag-ZnO/GO Nanoparticles Derived at Low Temperature. *J. Chem.* **2019**, *2019*, 2979517. [CrossRef]
52. Munawar, T.; Yasmeen, S.; Hussain, F.; Mahmood, K.; Hussain, A.; Asghar, M.; Iqbal, F. Synthesis of novel heterostructured ZnO-CdO-CuO nanocomposite: Characterization and enhanced sunlight driven photocatalytic activity. *Mater. Chem. Phys.* **2020**, *29*, 122983. [CrossRef]
53. Ashiegbu, D.C.; Potgieter, H.J. ZnO-based heterojunction catalysts for the photocatalytic degradation of methyl orange dye. *Heliyon* **2023**, *9*, e20674. [CrossRef] [PubMed]
54. Wu, Z.; Chen, X.; Liu, X.; Yang, X.; Yang, Y. A Ternary Magnetic Recyclable ZnO/Fe_3O_4/g-C_3N_4 Composite Photocatalyst for Efficient Photodegradation of Monoazo Dye. *Nanoscale Res. Lett.* **2019**, *14*, 147. [CrossRef]

inorganics

MDPI

Article

Tunable Electronic and Magnetic Properties of 3*d* Transition Metal Atom-Intercalated Transition Metal Dichalcogenides: A Density Functional Theory Study

Yujie Liu [1], Guang Yang [1], Zhiwen He [1], Yanbiao Wang [2,*], Xianghong Niu [3,*], Sake Wang [4], Yongjun Liu [1] and Xiuyun Zhang [1]

[1] College of Physics Science and Technology, Yangzhou University, Yangzhou 225002, China; 212701228@stu.yzu.edu.cn (Y.L.); yguang@gs.zzu.edu.cn (G.Y.); 220903214@stu.yzu.edu.cn (Z.H.); yjliu@yzu.edu.cn (Y.L.); xyzhang@yzu.edu.cn (X.Z.)
[2] Department of Fundamental Courses, Wuxi Institute of Technology, Wuxi 214121, China
[3] Institute of Advanced Materials (IAM), College of Electronic and Optical Engineering, School of Science, Nanjing University of Posts & Telecommunications, Nanjing 210023, China
[4] College of Science, Jinling Institute of Technology, 99 Hongjing Avenue, Nanjing 211169, China; isaacwang@jit.edu.cn
* Correspondence: wangyb@wxit.edu.cn (Y.W.); xhniu@njupt.edu.cn (X.N.)

Abstract: Currently, intercalation has become an effective way to modify the fundamental properties of two-dimensional (2D) van der Waals (vdW) materials. Using density functional theory, we systematically investigated the structures and electronic and magnetic properties of bilayer transition metal dichalcogenides (TMDs) intercalated with 3*d* TM atoms (TM = Sc–Ni), TM@BL_MS_2 (M = Mo, V). Our results demonstrate that all the studied TM@BL_MS_2s are of high stability, with large binding energies and high diffusion barriers of TM atoms. Interestingly, most TM@BL_MoS_2s and TM@BL_VS_2s are found to be stable ferromagnets. Among them, TM@BL_MoS_2s (TM = Sc, Ti, Fe, Co) are ferromagnetic metals, TM@BL_MoS_2 (TM = V, Cr) and TM@BL_VS_2 (TM = Sc, V) are ferromagnetic half-metals, and the remaining systems are found to be ferromagnetic semiconductors. Exceptions are found for Ni@BL_MoS_2 and Cr@BL_VS_2, which are nonmagnetic semiconductors and ferrimagnetic half-metals, respectively. Further investigations reveal that the electromagnetic properties of TM@BL_MoS_2 are significantly influenced by the concentration of intercalated TM atoms. Our study demonstrates that TM atom intercalation is an effective approach for manipulating the electromagnetic properties of two-dimensional materials, facilitating their potential applications in spintronic devices.

Keywords: intercalation; transition metal; transition metal dichalcogenides; ferromagnetic; first-principles calculations

1. Introduction

Since the discovery of graphene [1,2], various two-dimensional (2D) materials with finite thickness have attracted wide attention due to their versatile properties and potential applications in various fields [3,4]. Particularly, the 2D candidates with rich magnetic properties are regarded as a good platform for the development of spintronic devices [2–6], which facilitates low energy consumption and fast device operation. Experimentally, a few 2D magnetic materials have been produced, including ferromagnetic (FM) transition metal trihalide monolayer (e.g., CrI_3, VI_3) [7,8], $Cr_2Ge_2Te_6$ monolayer [9], Fe_3GeTe_2 monolayer [10], and antiferromagnetic (AFM) $MnPS_3$ monolayer [11–13]. Inspired by the above progress, considerable theoretical efforts have been made on the development of 2D magnetic candidates, such as transition metal (TM) borides [14–16], TM phosphides [17–19], TM disulfides [20–22], etc.

Despite the theoretical and experimental achievements, the members of known 2D FM or antiferromagnetic (AFM) candidates are still limited. Moreover, to advance the development of 2D spintronics devices, two major challenges should be solved: (i) robust magnetic orders with high transition temperature and (ii) the feasibility of fabrication experimentally, for example, many predicted 2D FM or antiferromagnetic (AFM) candidates so far face challenges in preparation. Therefore, finding a feasible way to obtain 2D magnetic materials is still a great challenge. It is found that introducing magnetism to grown non-magnetic 2D materials is an effective way to solve the challenge (ii) mentioned above. Various modification strategies have been developed for nonmagnetic materials in terms of tuning their electronic and magnetic properties, such as atom doping or substitution [23], introducing vacancies [24], external strains [25], etc. Unfortunately, these methods usually bring significant distortion to the structures.

Alternatively, the intercalation of magnetic species into the vdW gap of 2D materials provides a promising way to explore novel 2D magnets without destroying the host lattices [26–30]. Taking 2D transition-metal dichalcogenides (TMDs) as an example, the TM atom-intercalated vdW TMDs with different stoichiometries have been produced by electrochemical intercalations, which has been identified to be an effective way to modulate the electronic and magnetic properties [31,32]. For example, Guzman et al. presented a comprehensive investigation into the structural evolution of CVD-grown $V_{1+x}Se_2$ nanoplates. They found the VSe_2-to-V_5Se_8 phase and V_3Se_4-to-V_5Se_8 structural transformation under 300 °C and 250 °C annealing [33]. Zhao et al. extended the ferromagnetic order in tantalum-intercalated $TaS(Se)_y$, such as Ta_9S_{16}, Ta_7S_{12}, $Ta_{10}S_{16}$, and Ta_8Se_{12} [34]. Besides, FM orders were observed in ultra-thin films of V_5X_8 (X = S, Se) [35] and $V_5Se_8/NbSe_2$ 2D-heterostructures [36]. In addition, a number of different $Cr_{1+\delta}Te_2$ phases were found to have a broad range of magnetic orders and novel magnetic phases [37,38]. Using density functional theory calculation, Kumar et al. found that the late TM-intercalated WSe_2 exhibited substantial magnetic moments and pronounced ferromagnetic order [27]. Guo et al. found that filling either Cr or I atoms into the van der Waals gap of stacked and twisted CrI_3 bilayers can induce the double exchange effect and significantly strengthen the interlayer ferromagnetic coupling [39]. Correspondence to the experimental progress, the theoretical understanding of the electronic and magnetic properties of these intercalated systems is still very limited.

In this work, by using the density functional theory, we systematically studied the structural, electronic, and magnetic properties of 3*d* TM atom-intercalated bilayers MS_2 (M = Mo, V), TM@BL_MS_2 (TM = Sc–Ni). Our results show that all the systems are very stable with large binding energies. Except for Ni@BL_MoS_2 and Cr@BL_VS_2, all the other TM@BL_MoS_2s and TM@BL_VS_2s are ferromagnetic (half-)metals and semiconductors. Ni@BL_MoS_2 is a nonmagnetic semiconductor, and Cr@BL_VS_2 is a ferrimagnetic half-metal. Most importantly, the electronic and magnetic properties of these TM@BL_MoS_2s are largely influenced by the concentration of TM atoms.

2. Results and Discussion

2.1. *Structure and Stability of TM@BL_MS_2 (TM = Sc–Ni, M = Mo, V)*

The space groups for H-MoS_2 and H-VS_2 monolayers are the same, $P\bar{6}m2$, with lattice constants of 3.16 Å and 3.17 Å, respectively. The MoS_2 monolayer is a nonmagnetic semiconductor [40–42], and the H-VS_2 monolayer is predicted to be a FM semiconductor with a magnetic moment of 1.0 μB per V atom [43–47]. Three types of stacking configurations were considered for MoS_2 or VS_2 bilayers: (i) AA-stacking, in which the atoms from the top MS_2 layer sit exactly atop those in the bottom layer [see Figures 1a,f and S1a,f]; (ii) AB-stacking, where both layers rotate by 30 degrees relative to each other, with the M atoms from the top layer sitting atop those in the bottom layer, and the S atoms from one layer located on the hexagonal hole site of the adjacent layer [see Figures 1b,g and S1b,g]; (iii) AB′-stacking, which is similar to AB-stacking but with reversed interlayer configurations of the M/S atoms [see Figures 1c,h and S1c,h]. Our results show that the favored stacking structures for bilayer MoS_2 and VS_2 are the AB-stacking ones, which are 0.60 eV/0.56 eV and 0.55 eV/0.55 eV lower in energy than their AA- and AB′-stacking counterparts, respectively. In the following study, we only focus on the AB-stacked MoS_2 bilayer and VS_2 bilayer, whose interlayer spacing is 3.19 Å and 3.04 Å, respectively. In order to study the most stable configurations of TM@BL_Mo(V)S_2s, two configurations with different intercalation sites for TM atoms are considered: (i) Conf_I: in which the intercalated TM atom sits below the Mo/V atom (A site) in the top layer of Mo(V)S_2 [see Figure 1d,i and Figure S1d,i]; (ii) Conf_II: in which the TM atom is located below the S atom (B site) of the upper layer of Mo(V)S_2 [see Figures 1e,j and S1e,j]. Testing calculations show that the TM@BL_MoS_2s with Conf_I intercalation are the favored ones, which are 1.3 eV and 0.9 eV lower than that with B Site intercalating for Ti@BL_MoS_2 and V@BL_MoS_2, respectively. Therefore, we will only use Conf_I for further exploration, and the optimized structures are shown in Figure 2 and Figure S2 in the Supporting Information (SI). The detailed structural information is summarized in Table 1.

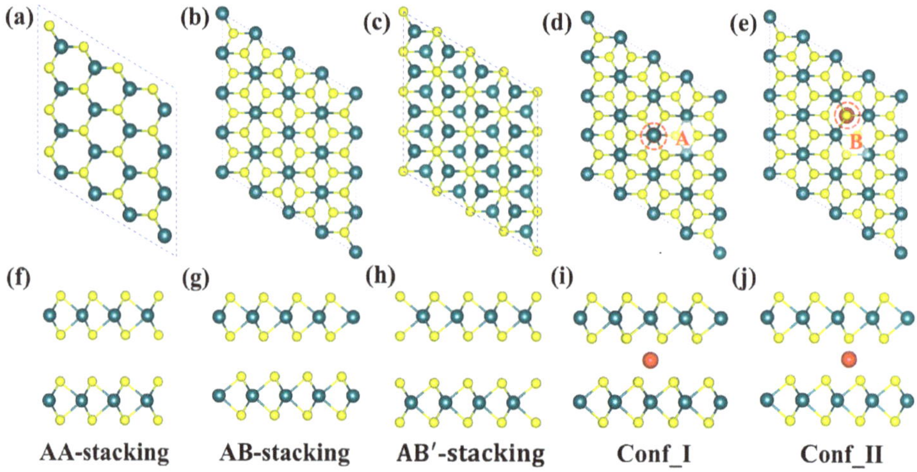

Figure 1. Top and side views of (**a**,**f**) AA-, (**b**,**g**) AB-, and (**c**,**h**) AB′- stacking of bilayer MoS_2. Top and side views of TM@BL_MoS_2 with TM intercalating in the (**d**,**i**) A site and (**e**,**j**) B site.

Except for Ni@BL_MoS_2, all the TM atoms prefer to occupy the center of the vertical interlayer space of both TM@BL_MoS_2s and TM@BL_VS_2s (see Figures 2 and S2), in which the TM atoms are distanced from the two-side MS_2 layer by about 1.09 Å~1.73 Å and 1.50 Å~1.69 Å for TM@BL_MoS_2s and TM@BL_VS_2s, respectively. Compared with free-standing Mo(V)S_2 bilayers, the intercalation of TM atoms enlarges the interlayer spacing to 3.16 Å~3.46 Å and 3.05 Å~3.37 Å in the case of TM@BL_MoS_2s and TM@BL_VS_2s, respectively (see Table 1). To quantitatively characterize the structural stability of these TM@BL_Mo(V)S_2 systems, we calculate the binding energies (E_bs) of the TM atom to the Mo(V)S_2 bilayers based on the following equation:

$$E_b = E_{Total} - E_{TM} - 2E_{MS_2} \quad (1)$$

where E_{Total}, E_{TM}, and E_{MS2} represent the total energy of the TM@BL_MS_2, the isolated transition metal atom, and the free-standing Mo(V)S_2 bilayers, respectively. As shown in Table 1 and Figure 3a,the calculated E_bs are around −7.21~−5.19 eV and −6.69~−4.23 eV for TM@BL_MoS_2s and TM@BL_VS_2s, respectively, which are comparable to those of TM-intercalated graphene/TMD heterostructures [48]. Such large negative E_b values indicate that the intercalation of TM atom in Mo(V)S_2 bilayers is energetically favorable. In the case of TM@BL_MoS_2, the E_b of Ni@BL_MoS_2 system is the lowest (−7.21 eV), and that of Ti@BL_MoS_2 is the highest (−5.19 eV). For the TM@BL_VS_2 system, the E_b of Ti intercalation system is the lowest (−6.69 eV), and that of Cr intercalation system is the highest (−4.23 eV).

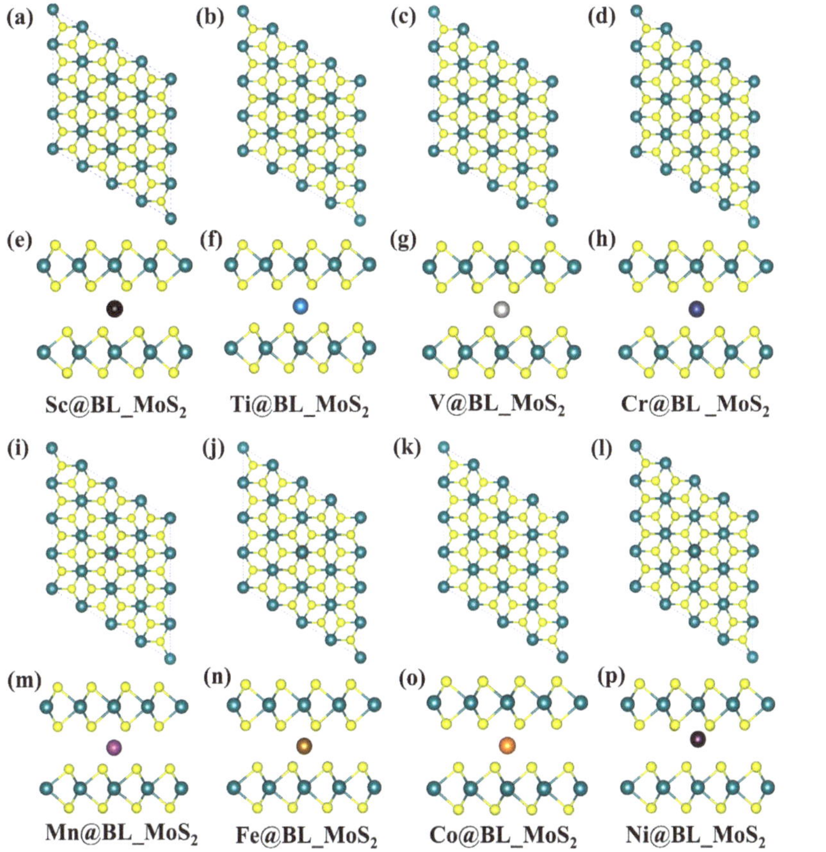

Figure 2. (**a–d**,**i–l**) Top views and (**e–h**,**m–p**) side views of the optimized structures for TM@BL_MoS_2s (TM = Sc–Ni).

Table 1. The interlayer spacing ($d_{S\text{-}S}$), the distance between the TM atoms and the two-side MS_2 layer ($d_{TM\text{-}S}$), magnetic moment (μ) per unit cell, binding energy (E_b), the Bader charges transferred from TM to adjacent MS_2 layers (ΔQ), and ground state (GS) of TM@BL_MS_2 systems.

Metal	$d_{S\text{-}S}$/Å	$d_{TM\text{-}S}$/Å	μ (μB)	E_b (eV)	ΔQ (e^-)	GS
			TM@BL_MoS_2			
Sc	3.46	1.73	2.78	−6.50	1.48	M
Ti	3.34	1.67	2.45	−5.19	1.28	M
V	3.28	1.64	3.44	−6.80	1.10	HM
Cr	3.37	1.69	5.93	−5.66	0.92	HM
Mn	3.37	1.68	5.00	−5.82	0.89	SC
Fe	3.17	1.58	2.29	−5.42	0.73	M
Co	3.19	1.60	2.92	−7.11	0.60	M
Ni	3.16	1.09	0	−7.21	0.32	SC

Table 1. *Cont.*

Metal	$d_{S\text{-}S}$/Å	$d_{TM\text{-}S}$/Å	μ (μB)	E_b (eV)	ΔQ (e^-)	GS
			TM@BL_VS_2			
Sc	3.37	1.69	36.86	−6.67	1.53	HM
Ti	3.31	1.66	37.65	−6.69	1.41	SC
V	3.16	1.58	38.94	−6.51	1.37	HM
Cr	3.03	1.51	33.79	−4.23	1.26	HM
Mn	3.05	1.52	40.62	−4.63	1.18	SC
Fe	3.09	1.55	39.71	−5.22	1.04	SC
Co	3.01	1.50	36.69	−5.78	0.65	SC
Ni	3.05	1.53	37.72	−5.45	0.61	SC

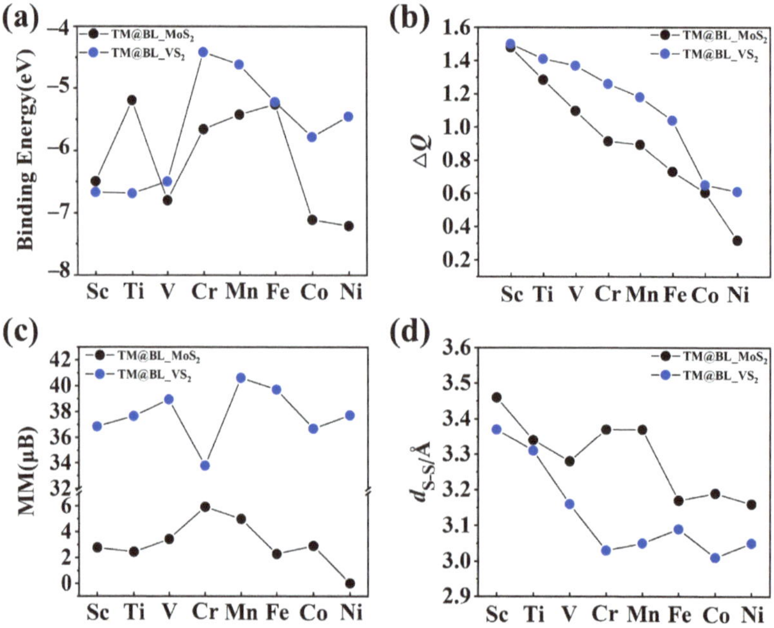

Figure 3. (**a**) The binding energy (E_b) of TM-embedded MS_2 (M = Mo, V) bilayers. (**b**) The number of electrons transferred from TM atoms to adjacent MS_2 layers. (**c**) The magnetic moments (MM) and (**d**) the interlayer distance ($d_{S\text{-}S}$) of TM@BL_MS_2 systems.

In order to clarify the bonding characteristics of these systems, the charge density differences (CDDs) of TM@BL_MS_2s (TM = Ti, V, Cr) are given in Figure S3, which are calculated based on the following equation:

$$\Delta\rho = \rho_{[TM@BL\_MS_2]} - \rho_{[TM]} - \rho_{[MS_2(T)]} - \rho_{[MS_2(L)]} \quad (2)$$

where $\rho_{[TM@BL\_MS2]}$, $\rho_{[TM]}$, $\rho_{[MS2(T)]}$, $\rho_{[MS2(L)]}$ represent the charge density of the whole system, the TM atom, the top MS_2 layer, and the low MS_2 layer, respectively. Clearly, the embedding of TM atoms resulted in charge redistribution in the Mo(V)S_2 sublayers. It was found that the charge densities were reduced at the center of TM atoms and accumulated between the TM-S bonds, indicating the obvious covalent bonding characteristics in the systems. Quantitatively, the Bader charges of the TM atoms were calculated (see Table 1). It can be seen that the amount of charge transfer gradually decreased from Sc to Ni in both TM@BL_MoS_2 and TM@BL_VS_2 systems (see Figure 3b). Moreover, the diffusion of Ti atom in Ti@BL_MoS_2 along the MoS_2 interface was examined (see Figure S4), and the large diffusion barrier (1.4 eV) confirms the difficulty of clustering of Ti atoms.

2.2. Electronic and Magnetic Properties of TM@BL_MS_2 (TM = Sc–Ni, M = Mo, V)

The total magnetic moments of both TM@BL_MoS_2s and TM@BL_VS_2s are summarized in Table 1 and Figure 3c. Interestingly, the intercalation of TM atoms in MoS_2 bilayers introduces magnetism into the systems. Except for Ni@BL_MoS_2, all the other TM@BL_MoS_2s (TM = Sc–Co) exhibit ferromagnetism. Among them, Cr@BL_MoS_2 has the largest magnetic moment (5.93 μB), and Fe@BL_MoS_2 displays the smallest one (2.29 μB) per unit cell (see Table 1). Such induced ferromagnetisms in nonmagnetic MoS_2 bilayers are similar to those of TM-intercalated TMDs [27] and other 2D materials [28,48]. The partial density of states (PDOS) and spin density plots are shown in Figure 4. The magnetic moments of these TM@BL_MoS_2 systems are mainly contributed by the 3*d* electrons from TM atoms. In addition, spin polarization of some Mo atoms close to the TM atoms occurs, as shown in the spin density plots in Figure 4. Taking Cr@BL_MoS_2 as an example, five *d* orbitals (d_{xy}, d_{x2-y2}, d_{xz}, d_{yz}, d_{z2}) from the spin-up channel of the Cr atom contribute 5.0 μB magnetic moment, and one d orbital from the spin-up channel of the two Mo atoms above/below the Cr atom contributes 1.0 μB magnetic moment. As a result, the total magnetic moment of this system is about 6.0 μB. Similar analysis can also be applied to other systems.

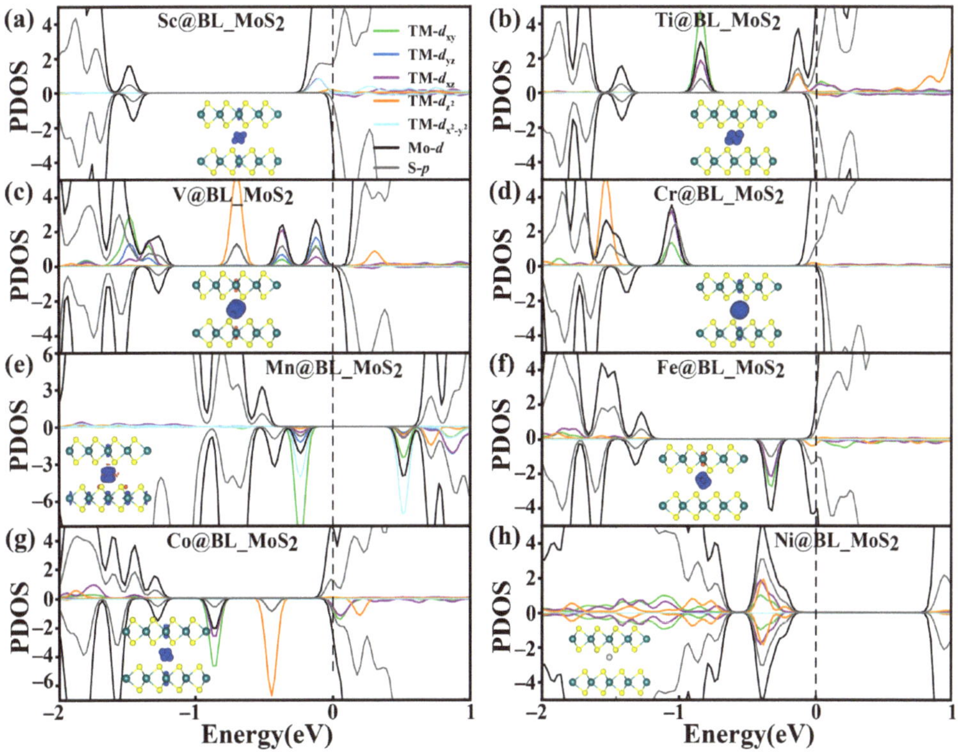

Figure 4. (**a**–**h**) Partial density of states (PDOS) and spin density plots for TM@BL_MoS_2 (TM = Sc–Ni). Inside the PDOS graph is its corresponding spin density graph; blue represents positive spin, and red represents negative spin.

As for TM@BL_VS_2s, most systems are found to be ferromagnetic, with ferromagnetic coupling between V atoms and intercalated TM atoms (see Figure S5). The magnetic moments range from 36.69 μB to 40.62 μB per unit cell. Among them, Co@BL_VS_2 has the smallest magnetic moment (36.69 μB), and Mn@BL_VS_2 has the largest magnetic moment (40.62 μB). Exceptions are found for Sc@BL_VS_2 and Cr@BL_VS_2. For the former, zero magnetic moment is found for the Sc atom, and for the latter, it is ferromagnetic, with the Cr atom displaying opposite magnetic moments to those of the V atoms.

Compared with the semiconducting properties of free-standing MoS_2 monolayers, TM@BL_MoS_2s (TM = Sc, Ti, Fe, Co) are changed to be ferromagnetic metals (see Figure 4 and Figure S6 and Table 1), in which the conduction band bottom passes through the Fermi level. V@BL_MoS_2 and Cr@BL_MoS_2 are ferromagnetic half-metals, in which one spin channel shows conducting behavior, and the contrast spin channel shows semiconducting properties. Besides, Mn@BL_MoS_2 and Ni@BL_MoS_2 are ferromagnetic semiconductor and non-magnetic semiconductor with band gaps of about 0.74 eV and 0.99 eV, respectively, which are significantly reduced compared to those of the MoS_2 monolayer (1.8 eV) [41]. Moreover, the embedded TM atom produces impurity energy bands in the gap of MoS_2, leading to the Fermi level of the energy band shifting upwards moderately. For TM@BL_VS_2s, as shown in Figure S7, the systems with TM = Sc, V, Cr become ferromagnetic or ferrimagnetic half-metals. In the case of TM@BL_VS_2s with TM = Ti, Mn, Fe, Co, Ni, they are all ferromagnetic semiconductors with a band gap of about 0.52, 0.46, 0.53, 0.19, and 0.52 eV, respectively.

2.3. *Effect of TM Intercalation Concentration on the TM BL_MS_2 (TM = V, Cr, Mn, Fe, M = Mo)*

In order to determine the influence of TM intercalation ratios on the structures and electronic properties of these systems, the structures intercalated with two or more TM atoms (TM = V, Cr, Mn, Fe), i.e., 2TM@BL_MoS_2 and TMC@BL_MoS_2, were considered, in which two adjacent TM atoms inserted in the 4 × 4 supercell and one chain intercalated TM atoms inserted in the 4 × 1 supercell were considered. The optimized structures of 2TM@BL_MoS_2s and TMC@BL_MoS_2s are shown in Figure 5. Similar to the TM@BL_MoS_2 systems, the TM atoms from the 2TM@BL_MoS_2 and TMC@BL_MoS_2 systems were firmly stabilized in the middle of the space of the MoS_2 bilayer. Moreover, for the 2TM@BL_MoS_2 systems, the interlayer distances were around 3.44 Å, 3.55 Å, 3.50 Å, and 3.24 Å, and the distances between TM atoms and the two-side MoS_2 layer were 1.72 Å, 1.77 Å, 1.75 Å, and 1.62 Å in the systems with TM = V, Cr, Mn, Fe, respectively (see Table 2). As for TMC@BL_MoS_2s, the interlayer distances were around 3.14 Å, 4.56 Å, 3.43 Å, and 3.54 Å for the structures inserted with chains of V, Cr, Mn, and Fe, respectively, and the distances between TM atoms and adjacent MoS_2 layers were around 1.57 Å, 2.27 Å, 1.71 Å, and 1.77 Å, respectively (see Table 2). The binding energies of these 2TM@BL_MoS_2 and TMC@BL_MoS_2 systems were in the range of -4.21~-3.45 eV and -3.39~-1.95 eV, respectively, in which the systems with Mn atoms exhibited the lowest E_bs, while those with the V atoms showed the largest E_bs. Similar to those TM@BL_MoS_2 systems, the charges transferred from TM to adjacent MoS_2 layers per unit cell were in the range of 0.73~1.13*e* and 0.73~1.09*e* for the 2TM@BL_MoS_2 and TMC@BL_MoS_2 systems, respectively (see Figure S8 and Table 2).

The PDOS, spin density diagram, and band structures of these 2TM@BL_MoS_2s and TMC@BL_MoS_2s are shown in Figure 6 and Figure S9. It was found that all 2TM@BL_MoS_2s and TMC@BL_MoS_2s are FM metals except for 2V@BL_MoS_2, which is FM half-metal. For 2TM@BL_MoS_2s, the magnetic moments were 7.85 μB, 11.89 μB, 10 μB, and 6.14 μB per unit cell for the systems with TM = V, Cr, Mn, Fe, respectively, nearly double those of their TM@BL_MoS_2 counterparts (see Table 2). As for TMC@BL_MoS_2s, the magnetic moments were 3.03 μB, 5.17 μB, 3.61 μB, and 2.44 μB per unit cell for the ones with TM = V, Cr, Mn, Fe, respectively. Similar to TM@BL_MoS_2s, the magnetic moments of the above systems were mainly contributed by the inserted 3*d* TM atoms (see Figure 6). However, the magnetic moments per unit cell of TMC@BL_MoS_2s differed from those of TM@BL_MoS_2s, indicating that the magnetic properties of these systems are sensitive to the intercalated TM concentrations. Moreover, with the intercalated TM atom increasing, the electronic states around the Fermi level increased, leading to the metallic character in them. Specifically, the number of impurity bands from TM atoms in 2TM@BL_MoS_2s increased and almost doubled those in TM@BL_MoS_2s (see Figure S9). In contrast, in TMC@BL_MoS_2s, the electronic bands around the Fermi level were no longer flat and became much more dispersed, indicating that the hybridization between TM_d and MoS_2_p orbitals was largely strengthened. Therefore, it is very effective to control the electronic and magnetic properties of these systems by varying the TM ratios.

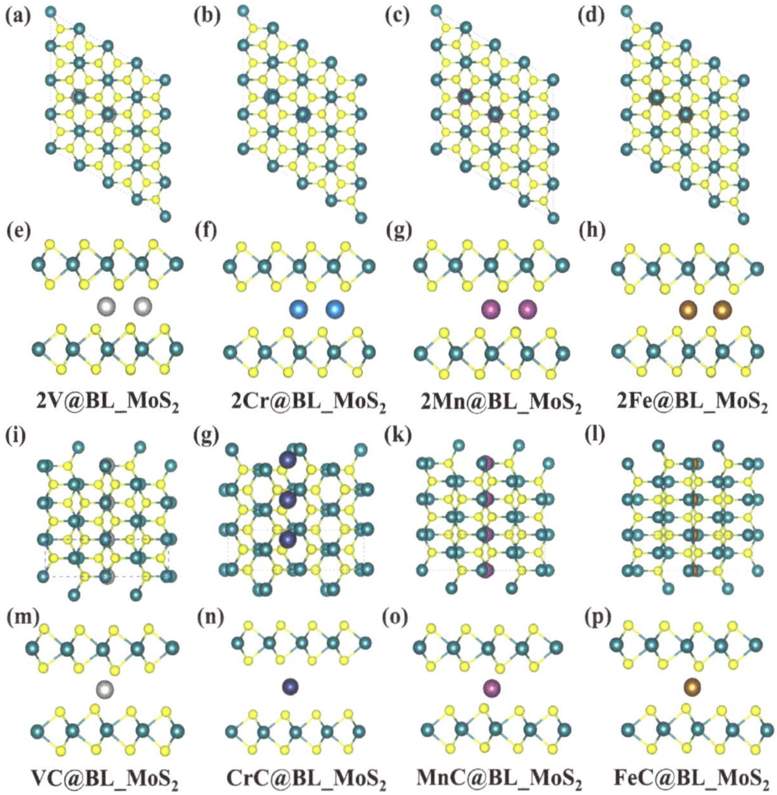

Figure 5. (**a–d**,**i–l**) Top views and (**e–h**,**m–p**) side views of the optimized structure of 2TM@BL_MoS_2s and TMC@BL_MoS_2s (TM = V, Cr, Mn, Fe), respectively.

Table 2. The interlayer spacing ($d_{S\text{-}S}$), the distance between the TM atoms and the two-side MoS_2 layer ($d_{TM\text{-}S}$), magnetic moment (μ) per unit cell, binding energy (E_b) per TM atom, the Bader charges transferred from TM to adjacent MoS_2 layers (ΔQ), and ground states (GS) of 2TM@BL_MoS_2s and TMC@BL_MoS_2s.

Metal	$d_{S\text{-}S}$/Å	$d_{TM\text{-}S}$/Å	μ (μB)	E_b (eV)	ΔQ (e^-)	GS
			2TM@BL_MoS_2			
2V	3.44	1.72	7.85	−4.21	1.13	HM
2Cr	3.55	1.77	11.89	−3.79	0.98	M
2Mn	3.50	1.75	10.00	−3.45	0.95	M
2Fe	3.23	1.62	6.14	−4.18	0.73	M

Table 2. *Cont.*

Metal	$d_{S\text{-}S}$/Å	$d_{TM\text{-}S}$/Å	μ (μB)	E_b (eV)	ΔQ (e^-)	GS
			$TMC@BL\_MoS_2$			
VC	3.14	1.57	3.03	−3.39	1.09	M
CrC	4.55	2.27	5.14	−2.35	0.64	M
MnC	3.43	1.71	3.61	−1.95	0.95	M
FeC	3.54	1.77	2.44	−3.04	0.73	M

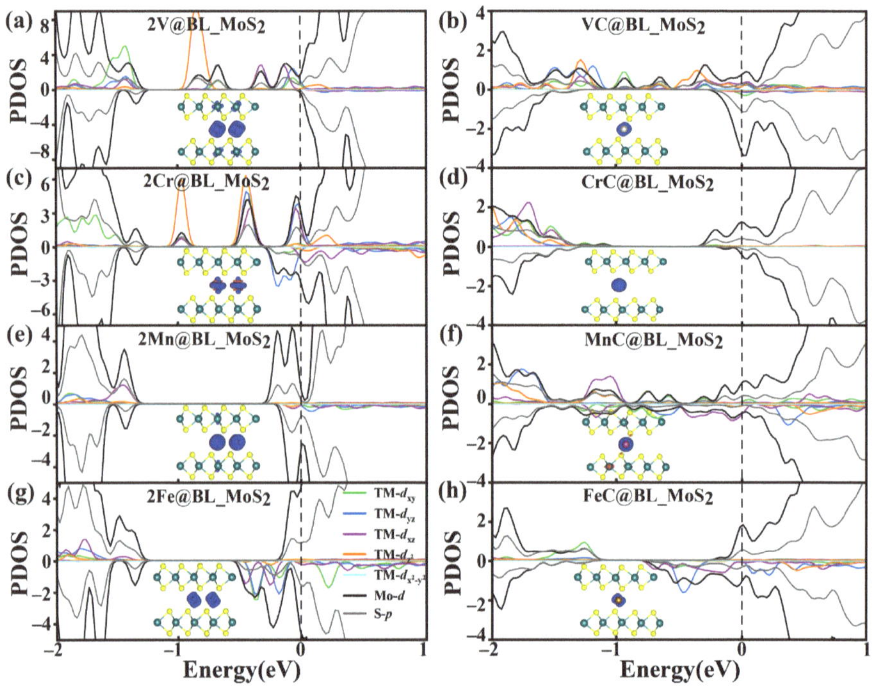

Figure 6. (**a**–**d**) Partial density of states (PDOS) and spin density plots for $2TM@BL\_MoS_2$ (TM = V, Cr, Mn, Fe). (**e**–**h**) Partial density of states (PDOS) and spin density plots for $TMC@BL\_MoS_2$ (TM = V, Cr, Mn, Fe). Inside the PDOS graph is its corresponding spin density graph; blue represents positive spin, and red represents negative spin.

3. Methods

All the calculations were performed using the Vienna ab initio simulation package (VASP) [49–51]. The exchange–correlation potentials were treated by the generalized gradient approximation (GGA) of optimized Perdew–Burke–Ernzerhof (PBE) [52]. The interactions between the ion core and the valence electrons were modeled with projector augmented wave (PAW) potentials [53], and the DFT-D2 method was used to account for van der Waals (vdW) interaction [54]. A vacuum layer of 20 Å was added in the *z* direction to eliminate the interaction of interlayer caused by periodic boundary conditions. In order to consider the Coulomb interaction and exchange interactions on TM electrons, the GGA+U method with U_{eff} = 4.0 eV was adopted according to the previous literature, which has been proven to give very close results for these systems [55–58]. A plane-wave basis set with a kinetic cutoff energy of 500 eV was employed. The 4 × 4 supercells with the lattice constants of $a = b = 12.76$ Å and $a = b = 12.72$ Å for bilayer MoS_2 and VS_2 were applied, respectively. The Brillouin zone (BZ) was sampled using 3 × 3 × 1 and 5 × 5 × 1 gamma-centered Monkhorst–Pack grids for the calculation of structural relaxation and electronic structures, respectively. The criteria for energy and atomic force convergence were set to 10^{-5} eV per unit cell and 0.01 eV $Å^{-1}$, respectively.

4. Conclusions

In summary, the structural, electronic, and magnetic properties of *3d* TM atom (TM = Ti–Ni)-intercalated Mo(V)S_2 bilayers, TM@BL_ Mo(V)S_2, are explored using density functional theory methods. All the studied systems are thermal dynamically stable with large binding energies, ranging from −4.23 to −7.21eV. The intercalation of TM atoms introduces rich electronic and magnetic properties to Mo(V)S_2 bilayers. Except for Ni@BL_MoS_2, which is a nonmagnetic semiconductor, most studied systems are found to display robust magnetic properties. TM@BL_MoS_2s (TM = Sc, Ti, Fe, Co) are ferromagnetic metals, TM@BL_MoS_2 (TM = V, Cr) and TM@BL_VS_2 (TM = Sc, V, Cr) are ferromagnetic or ferrimagnetic half-metals, and the remaining systems are found to be ferromagnetic semiconductors. Furthermore, the electronic and magnetic properties of these TM@BL_MoS_2s are largely influenced by the concentration of TM atoms. Our study proposes that the intercalation of TM atoms is a feasible way to tune the properties of 2D materials and promote their potential applications in electronic and spin electronic devices.

Supplementary Materials: The following supporting information can be downloaded at https://www.mdpi.com/article/10.3390/inorganics12090237/s1; Figure S1: Top and side views of AA-, AB-, and AB′-stacking of bilayer VS_2 and TM@BL_VS_2, with TM intercalating at the A site and B site; Figure S2: Top and side views of the optimized structures for TM@BL_VS_2s; Figure S3: The CDDs plots of TM@BL_MoS_2 (TM = Ti, V, Cr) and TM@BL_VS_2 (TM = Ti, V, Cr); Figure S4: Energy barrier for Ti@BL_MoS_2 diffusion from position A to position B; Figure S5: Partial density of states (PDOS) and spin density plots for TM@BL_VS_2s; Figure S6: Band structures of TM@BL_MoS_2s; Figure S7: Band structures of TM@BL_VS_2s; Figure S8: The CDDs plots of 2TM@BL_MoS_2 (TM = V, Cr, Mn, Fe) and TMC@BL_MoS_2 (TM = V, Cr, Mn, Fe); Figure S9: Band structures of 2TM@BL_MoS_2 (TM = V, Cr, Mn, Fe) and TMC@BL_MoS_2 (TM = V, Cr, Mn, Fe).

Author Contributions: Conceptualization, Y.L. (Yujie Liu) and X.Z.; data curation, Y.L. (Yujie Liu) and G.Y.; formal analysis, Y.L. (Yujie Liu), G.Y., Z.H. and Y.W.; investigation, G.Y., Z.H. and X.Z.; writing—original draft, Y.L. (Yujie Liu) and G.Y.; writing—review and editing, Y.L. (Yujie Liu), Y.W., Y.L. (Yongjun Liu) and X.Z.; supervision, Y.W., X.N., S.W. and X.Z.; project administration, Y.L. (Yongjun Liu), X.N. and S.W. All authors have read and agreed to the published version of the manuscript.

Funding: This research was funded by the Six Talent Peaks Project in Jiangsu Province (XCL-104), the Natural Science Foundation of Jiangsu Province (No. BK20211002), as well as Qinglan Project of Jiangsu Province of China. China Postdoctoral Science Foundation (No. 2022M711691), National Natural Science Foundation of China (No. 22473060).

Data Availability Statement: Data are included within the article and Supplementary Materials.

Acknowledgments: We thank the computational resources at Yangzhou University.

Conflicts of Interest: The authors declare no conflicts of interest.

References

1. Geim, A.K.; Novoselov, K.S. The rise of graphene. *Nat. Mater.* **2007**, *6*, 183–191. [CrossRef] [PubMed]
2. Novoselov, K.S.; Geim, A.K.; Morozov, S.V.; Jiang, D.; Zhang, Y.; Dubonos, S.V.; Grigorieva, I.V.; Firsov, A.A. Electric field effect in atomically thin carbon films. *Science* **2004**, *306*, 666–669. [CrossRef]
3. Wang, Q.H.; Kalantar-Zadeh, K.; Kis, A.; Coleman, J.N.; Strano, M.S. Electronics and optoelectronics of two-dimensional transition metal dichalcogenides. *Nat. Nanotechnol.* **2012**, *7*, 699–712. [CrossRef]
4. Chen, S.; Shi, G.Q. Two-dimensional materials for halide perovskite-based optoelectronic devices. *Adv. Mater.* **2017**, *29*, 1605448. [CrossRef]
5. Geim, A.K.; Grigorieva, I.V. Van der Waals heterostructures. *Nature* **2013**, *499*, 419–425. [CrossRef]
6. Williams, T.J.; Aczel, A.A.; Lumsden, M.D.; Nagler, S.E.; Stone, M.B.; Yan, J.Q.; Mandrus, D. Magnetic correlations in the quasi-two-dimensional semiconducting ferromagnet $CrSiTe_3$. *Phys. Rev. B* **2015**, *92*, 144404. [CrossRef]
7. Jiang, S.W.; Li, L.Z.; Wang, Z.F.; Mak, K.F.; Shan, J. Controlling magnetism in 2D CrI_3 by electrostatic doping. *Nat. Nanotechnol.* **2018**, *13*, 549–553. [CrossRef]

8. Son, S.; Coak, M.J.; Lee, N.; Kim, J.; Kim, T.Y.; Hamidov, H.; Cho, H.; Liu, C.; Jarvis, D.M.; Brown, P.A.C.; et al. Bulk properties of the van der Waals hard ferromagnet VI_3. *Phys. Rev. B* **2019**, *99*, 041402. [CrossRef]
9. Gong, C.; Li, L.; Li, Z.L.; Ji, H.W.; Stern, A.; Xia, Y.; Cao, T.; Bao, W.; Wang, C.Z.; Wang, Y.A.; et al. Discovery of intrinsic ferromagnetism in two-dimensional van der Waals crystals. *Nature* **2017**, *546*, 265–269. [CrossRef]
10. Meng, L.J.; Zhou, Z.; Xu, M.Q.; Yang, S.Q.; Si, K.P.; Liu, L.X.; Wang, X.G.; Jiang, H.N.; Li, B.X.; Qin, P.X.; et al. Anomalous thickness dependence of Curie temperature in air-stable two-dimensional ferromagnetic 1T-$CrTe_2$ grown by chemical vapor deposition. *Nat. Commun.* **2021**, *12*, 809. [CrossRef]
11. Qin, J.; Yang, C.; Yakushi, K.; Nakazawa, Y.; Ichimura, K.; Liu, D. Synthesis and ferromagnetism of a new intercalation compound: $Mn_{0.86}PS_3(bipy)_{0.56}$. *Synth. Met.* **1997**, *85*, 1673–1674. [CrossRef]
12. Chu, H.; Roh, C.J.; Island, J.O.; Li, C.; Lee, S.; Chen, J.J.; Park, J.G.; Young, A.F.; Lee, J.S.; Hsieh, D. Linear magnetoelectric phase in ultrathin $MnPS_3$ probed by optical second harmonic generation. *Phys. Rev. Lett.* **2020**, *124*, 027601. [CrossRef] [PubMed]
13. Chittari, B.L.; Park, Y.; Lee, D.; Han, M.; MacDonald, A.H.; Hwang, E.; Jung, J. Electronic and magnetic properties of single-layer MPX_3 metal phosphorous trichalcogenides. *Phys. Rev. B* **2016**, *94*, 184428. [CrossRef]
14. Li, Q.; Zhou, D.; Zheng, W.; Ma, Y.; Chen, C. Global structural optimization of tungsten borides. *Phys. Rev. Lett.* **2013**, *110*, 136403. [CrossRef]
15. Sharma, A.; Rangra, V.S.; Thakur, A. Synthesis, properties, and applications of MBenes (two-dimensional metal borides) as emerging 2D materials: A review. *J. Mater. Sci.* **2022**, *57*, 12738–12751. [CrossRef]
16. Zhang, B.; Zhou, J.; Sun, Z. MBenes: Progress, challenges and future. *J. Mater. Chem. A* **2022**, *10*, 15865–15880. [CrossRef]
17. Du, M.; Li, D.; Liu, S.; Yan, J. Transition metal phosphides: A wonder catalyst for electrocatalytic hydrogen production. *Chin. Chem. Lett.* **2023**, *34*, 108156. [CrossRef]
18. Zhang, N.; Amorim, I.; Liu, L. Multimetallic transition metal phosphide nanostructures for supercapacitors and electrochemical water splitting. *Nanotechnology* **2022**, *33*, 432004. [CrossRef]
19. Yang, Y.; Zhou, C.; Wang, W.; Xiong, W.; Zeng, G.; Huang, D.; Zhang, C.; Song, B.; Xue, W.; Li, X.; et al. Recent advances in application of transition metal phosphides for photocatalytic hydrogen production. *Chem. Eng. J.* **2021**, *405*, 126547. [CrossRef]
20. Sun, L.; Leong, W.S.; Yang, S.; Chisholm, M.F.; Liang, S.-J.; Ang, L.K.; Tang, Y.; Mao, Y.; Kong, J.; Yang, H.Y. Concurrent synthesis of high-performance monolayer transition metal disulfides. *Adv. Funct. Mater.* **2017**, *27*, 1605896. [CrossRef]
21. Cheng, J.; Jin, Y.; Zhao, J.; Jing, Q.; Gu, B.; Wei, J.; Yi, S.; Li, M.; Nie, W.; Qin, Q.; et al. From VIB- to VB-group transition metal disulfides: Structure engineering modulation for superior electromagnetic wave absorption. *Nano-Micro Lett.* **2023**, *16*, 29. [CrossRef] [PubMed]
22. Liu, H.; Qi, G.; Tang, C.; Chen, M.; Chen, Y.; Shu, Z.; Xiang, H.; Jin, Y.; Wang, S.; Li, H.; et al. Growth of large-area homogeneous monolayer transition-metal disulfides via a molten liquid intermediate process. *ACS Appl. Mater.* **2020**, *12*, 13174–13181. [CrossRef] [PubMed]
23. Zhao, C.Y.; Jin, C.H.; Wu, J.L.; Ji, W. Magnetism in molybdenum disulphide monolayer with sulfur substituted by 3*d* transition metals. *J. Appl. Phys.* **2016**, *120*, 144305. [CrossRef]
24. Lopez-Bezanilla, A.; Lado, J.L. Defect-induced magnetism and Yu-Shiba-Rusinov states in twisted bilayer graphene. *Phys. Rev. Mater.* **2019**, *3*, 084003. [CrossRef]
25. Zhang, X.Y.; Xi, B.; Liu, Y.J.; Yao, X.J.; Wu, X.J. Antiferromagnetic semimetal in Ti-intercalated borophene heterobilayer. *J. Phys. Chem. C* **2020**, *124*, 4709–4716. [CrossRef]
26. Wan, J.Y.; Lacey, S.D.; Dai, J.Q.; Bao, W.Z.; Fuhrer, M.S.; Hu, L.B. Tuning two-dimensional nanomaterials by intercalation: Materials, properties and applications. *Chem. Soc. Rev.* **2016**, *45*, 6742–6765. [CrossRef]
27. Kumar, P.; Skomski, R.; Pushpa, R. Magnetically ordered transition-metal-intercalated WSe_2. *ACS Omega* **2017**, *2*, 7985–7990. [CrossRef]
28. Zhang, X.Y.; Xu, W.X.; Dai, J.P.; Liu, Y.J. Role of embedded 3d transition metal atoms on the electronic and magnetic properties of defective bilayer graphene. *Carbon* **2017**, *118*, 376–383. [CrossRef]
29. Chen, Z.; Leng, K.; Zhao, X.; Malkhandi, S.; Tang, W.; Tian, B.; Dong, L.; Zheng, L.; Lin, M.; Yeo, B.S.; et al. Interface confined hydrogen evolution reaction in zero valent metal nanoparticles-intercalated molybdenum disulfide. *Nat. Commun.* **2017**, *8*, 14548. [CrossRef]
30. Zhang, M.; Zhang, C.; You, Y.; Xie, H.; Chi, H.; Sun, Y.; Liu, W.; Su, X.; Yan, Y.; Tang, X.; et al. Electron density optimization and the anisotropic thermoelectric properties of Ti self-intercalated $Ti_{1+x}S_2$ compounds. *ACS Appl. Mater.* **2018**, *10*, 32344–32354. [CrossRef]
31. Li, F.; Hu, R.; Huang, Z.; Luo, S.; Qiao, H.; Zhong, J.; Qi, X. Properties tuning and applications for two dimension materials in electrochemical intercalation process. *Appl. Mater. Today* **2024**, *36*, 102069. [CrossRef]
32. Rajapakse, M.; Karki, B.; Abu, U.O.; Pishgar, S.; Musa, M.R.K.; Riyadh, S.M.S.; Yu, M.; Sumanasekera, G.; Jasinski, J.B. Intercalation as a versatile tool for fabrication, property tuning, and phase transitions in 2D materials. *NPJ 2D Mater. Appl.* **2021**, *5*, 30. [CrossRef]
33. Guzman, R.; Liu, H.; Bian, C.; Bao, L.; Shen, C.-M.; Gao, H.-J.; Zhou, W. Temperature-induced structural evolution and magnetism in self-intercalated $V_{1+x}Se_2$ nanoplates. *Adv. Funct. Mater.* **2024**, 2401304. [CrossRef]
34. Zhao, X.X.; Song, P.; Wang, C.C.; Riis-Jensen, A.C.; Fu, W.; Deng, Y.; Wan, D.Y.; Kang, L.X.; Ning, S.C.; Dan, J.D.; et al. Engineering covalently bonded 2D layered materials by self-intercalation. *Nature* **2020**, *581*, 171–177. [CrossRef]

35. Niu, J.; Yan, B.; Ji, Q.; Liu, Z.; Li, M.; Gao, P.; Zhang, Y.; Yu, D.; Wu, X. Anomalous hall effect and magnetic orderings in nanothick V_5S_8. *Phys. Rev. B* **2017**, *96*, 075402. [CrossRef]
36. Matsuoka, H.; Barnes, S.E.; Ieda, J.I.; Maekawa, S.; Bahramy, M.S.; Saika, B.K.; Takeda, Y.; Wadati, H.; Wang, Y.; Yoshida, S.; et al. Spin–orbit-induced ising ferromagnetism at a van der Waals interface. *Nano Lett.* **2021**, *21*, 1807–1814. [CrossRef] [PubMed]
37. Fujisawa, Y.; Pardo-Almanza, M.; Garland, J.; Yamagami, K.; Zhu, X.; Chen, X.; Araki, K.; Takeda, T.; Kobayashi, M.; Takeda, Y.; et al. Tailoring magnetism in self-intercalated $Cr_{1+\delta}Te_2$ epitaxial films. *Phys. Rev. Mater.* **2020**, *4*, 114001. [CrossRef]
38. Saha, R.; Meyerheim, H.L.; Göbel, B.; Hazra, B.K.; Deniz, H.; Mohseni, K.; Antonov, V.; Ernst, A.; Knyazev, D.; Bedoya-Pinto, A.; et al. Observation of Neel-type skyrmions in acentric self-intercalated $Cr_{1+\delta}Te_2$. *Nat. Commun.* **2022**, *13*, 3965. [CrossRef]
39. Guo, Y.; Liu, N.; Zhao, Y.; Jiang, X.; Zhou, S.; Zhao, J. Enhanced ferromagnetism of CrI_3 bilayer by self-intercalation. *Chin. Phys. Lett.* **2020**, *37*, 107506. [CrossRef]
40. Tan, F.; Li, J.; Fang, X.; Guan, L. The optical properties of few-layer MoS_2 by DFT calculations. *Phys. E* **2024**, *155*, 115813. [CrossRef]
41. Mak, K.F.; Lee, C.; Hone, J.; Shan, J.; Heinz, T.F. Atomically thin MoS_2: A new direct-gap semiconductor. *Phys. Rev. Lett.* **2010**, *105*, 136805. [CrossRef] [PubMed]
42. Szary, M.J. Adsorption of ethylene oxide on doped monolayers of MoS_2: A DFT study. *Mater. Sci. Eng. B* **2021**, *265*, 115009. [CrossRef]
43. Dai, J.-Q.; Yuan, J.; Ke, C. Synergistic magnetic proximity and ferroelectric field effect on a 2H-VS_2 monolayer by ferromagnetic termination of a $BiFeO_3$(0001) surface. *J. Mater. Chem. C* **2022**, *10*, 1498–1510. [CrossRef]
44. Kan, M.; Wang, B.; Lee, Y.H.; Sun, Q. A density functional theory study of the tunable structure, magnetism and metal-insulator phase transition in VS_2 monolayers induced by in-plane biaxial strain. *Nano Res.* **2015**, *8*, 1348–1356. [CrossRef]
45. Karthick, K.; Bijoy, T.K.; Sivakumaran, A.; Mansoor Basha, A.B.; Murugan, P.; Kundu, S. Enhancing hydrogen evolution reaction activities of 2H-phase VS_2 layers with palladium nanoparticles. *Inorg. Chem.* **2020**, *59*, 10197–10207. [CrossRef] [PubMed]
46. Cui, Y.; Fan, W.; Liu, X.; Ren, J.; Gao, Y. Electronic conductivity of two-dimensional VS_2 monolayers: A first principles study. *Comp. Mater. Sci.* **2021**, *200*, 110767. [CrossRef]
47. Liu, X.; Pyatakov, A.P.; Ren, W. Magnetoelectric coupling in multiferroic bilayer VS_2. *Phys. Rev. Lett.* **2020**, *125*, 247601. [CrossRef] [PubMed]
48. Zhang, X.; Sun, Y.; Gao, W.; Lin, Y.; Zhao, X.; Wang, Q.; Yao, X.; He, M.; Ye, X.; Liu, Y. Sizable bandgaps of graphene in 3d transition metal intercalated defective graphene/WSe_2 heterostructures. *RSC Adv.* **2019**, *9*, 18157–18164. [CrossRef] [PubMed]
49. Kresse, G.; Furthmüller, J. Efficient iterative schemes for ab initio total-energy calculations using a plane-wave basis set. *Phys. Rev. B* **1996**, *54*, 11169–11186. [CrossRef]
50. Kresse, G.; Hafner, J. Ab initio molecular dynamics for liquid metals. *Phys. Rev. B* **1993**, *47*, 558–561. [CrossRef]
51. Kresse, G.; Hafner, J. Ab initio molecular-dynamics simulation of the liquid-metal—Amorphous-semiconductor transition in germanium. *Phys. Rev. B* **1994**, *49*, 14251–14269. [CrossRef]
52. Perdew, J.P.; Burke, K.; Ernzerhof, M. Generalized gradient approximation made simple. *Phys. Rev. Lett.* **1996**, *77*, 3865–3868. [CrossRef] [PubMed]
53. Blöchl, P.E. Projector augmented-wave method. *Phys. Rev. B* **1994**, *50*, 17953–17979. [CrossRef] [PubMed]
54. Grimme, S. Semiempirical GGA-type density functional constructed with a long-range dispersion correction. *J. Comput. Chem.* **2006**, *27*, 1787–1799. [CrossRef]
55. Anasori, B.; Shi, C.; Moon, E.J.; Xie, Y.; Voigt, C.A.; Kent, P.R.C.; May, S.J.; Billinge, S.J.L.; Barsoum, M.W.; Gogotsi, Y. Control of electronic properties of 2D carbides (MXenes) by manipulating their transition metal layers. *Nanoscale Horiz.* **2016**, *1*, 227–234. [CrossRef]
56. Zhang, X.; Wei, J.; Li, R.; Zhang, C.; Zhang, H.; Han, P.; Fan, C. DFT + U predictions: Structural stability, electronic and optical properties, oxidation activity of BiOCl photocatalysts with 3d transition metals doping. *J. Mater. Sci.* **2018**, *53*, 4494–4506. [CrossRef]
57. Liu, C.; Fu, B.; Yin, H.; Zhang, G.; Dong, C. Strain-tunable magnetism and nodal loops in monolayer MnB. *Appl. Phys. Lett.* **2020**, *117*, 103101. [CrossRef]
58. Islam, M.F.; Canali, C.M.; Pertsova, A.; Balatsky, A.; Mahatha, S.K.; Carbone, C.; Barla, A.; Kokh, K.A.; Tereshchenko, O.E.; Jiménez, E.; et al. Systematics of electronic and magnetic properties in the transition metal doped Sb_2Te_3 quantum anomalous hall platform. *Phys. Rev. B* **2018**, *97*, 155429. [CrossRef]

inorganics

Review

Synthesis and Study of Correlated Phase Transitions of CrN Nanoparticles

Khan Alam [1,2]

[1] Department of Physics, King Fahd University of Petroleum and Minerals, Dhahran 31261, Saudi Arabia; khan.alam@kfupm.edu.sa

[2] Interdisciplinary Research Center for Sustainable Energy Systems, King Fahd University of Petroleum and Minerals, Dhahran 31261, Saudi Arabia

Abstract: Chromium nitride is an important transition metal nitride for studying fundamental properties and for advanced technological applications. It is considered a model system for exploring structural, electronic, and magnetic transitions. These transitions occur at 275 ± 10 K and appear to be coupled; however, many discrepant studies on these transitions can be found in the published literature. The underlying reasons for these controversies are suspected to be the CrN nanoparticles preparation methods, strains, impurities, stoichiometry, nanoparticle size, characterization methods, and ambient conditions for characterizing them. This article is focused on the review of the nanoparticle synthesis methods and the use of these nanoparticles for studying structural, electronic, and magnetic transitions. The focus is mainly on the experimental methods, while theoretical simulations are briefly reviewed at the end of the article.

Keywords: chromium nitride; phase change material; magnetostructural transition; electric transition; magnetic transition; crystallography

1. Introduction

Transition metal nitrides (TMNs) exhibit interesting structural, electronic, magnetic, and noble-metal-like characteristics [1–6]. These unique characteristics of TMNs arise from the combination of the d-orbitals of transition metals and the p-orbital of nitrogen; the d-orbital contributes to both magnetic and electronic properties, whereas the p-orbital contributes mainly to the electronic properties of materials. TMNs are found at room temperature and normal pressure in face-centered cubic (W_2N) [7], hexagonal close-packed (ε-Ti_2N) [8], and hexagonal crystal (BN) [9] structures. The crystal structures of some TMNs transform into other crystal structures with temperature [10] or pressure [11,12] variations. These crystal structures can be understood as metal lattices in which nitrogen occupies the interstitial sites. The nature of the bond between a metal and nitrogen is mostly covalent with an overtone of the ionic bonding nature. The bond between the metal and nitrogen leads to the expansion of the metal lattice but causes constriction in the d-bonds between the metals [13]. Therefore, the d-orbital constriction and high density of states at the Fermi level produce noble-metal-like properties in electrolysis, making them ideal candidates for electrochemical conversion and energy storage applications [14–20]. Their chemical inertness enables TMNs to function across a broad pH range, increasing their versatility for use in various electrolytes. Due to these characteristics, they could solve the persistent issues of capacitor and battery electrode erosion or collapse while reacting with lithium, zinc, or other electrolytes [21–26]. Many TMNs such as molybdenum nitride (MoN), vanadium nitride (VN), titanium nitride (TiN), niobium nitride (NbN), cobalt nitride (CoN), nickel nitride (NiN), Boron nitride (BN), iron nitride (FeN), chromium nitride (CrN), etc. have been successfully synthesized and studied as electrode materials for water splitting, electrolysis, hydrogen production, supercapacitors, and batteries [27–34].

Among TMNs, CrN is considered an important model system both for understanding the fundamental science behind the interesting properties of TMNs [35,36] and a candidate for advanced technological applications. Some of the interesting applications that CrN is studied for include photovoltaics, thermoelectrics [37], supercapacitor electrodes [30,38,39], battery electrode [40–43], dilute magnetic semiconductors [44–46], and spintronics [47–49]. In a recent study published by Liu et al. [50], they used a novel method for suppressing Cr_2N phases and enhancing the thermoelectric performance of CrN to a maximum power factor of 1002 $\mu W \cdot m^{-1} \cdot K^{-2}$, a thermal conductivity of 4.78 $W \cdot m^{-1} \cdot K^{-1}$, and a figure of merit (zT) value of up to 0.20 at 973 K. A similar study by Yuan et al. [51] reported a power factor of 1230 $\mu W \cdot m^{-1} \cdot K^{-2}$, a thermal conductivity of 6.8 $W \cdot m^{-1} \cdot K^{-1}$, and a zT value of up to 0.18 at 973 K. These interesting studies indicate that CrN is a potential candidate for thermoelectric applications. Furthermore, Gubert et al. [52] reported a mixture of dichromium nitride (Cr_2N) and CrN nanoparticles by using pulsed laser irradiation on a chromium target submerged in liquid nitrogen. The Cr ions removed from the target reacted with the readily available nitrogen and formed cylindrically shaped, chemically stable nanoparticles. The particles exhibited surface plasmon resonance, which can be used for plasmonics. Another interesting study reported by Das et al. [53] showed the Li-cycling performance using CrN nanoparticles as an active material in a battery's anode. They achieved a first-cycle reversible capacity of up to 635 $\pm$ 10 mA·h/g, which stabilized at around 500 $\pm$ 10 mA·h/g up to 80 cycles, which is greater than the graphite specific capacity of 372 mA·h/g [54]. Shuang et al. [55] reported a Soret-effect-induced phase transition in CrN. Upon the application of repeated electrical pulses, they observed a nonvolatile and reversible crystalline-to-crystalline phase transition. This transition was facilitated by nitrogen diffusion from the hotter region to a relatively colder region, thereby stabilizing CrN_2 in the hexagonal phase. Their resistive switching device offers a programming window of 10^5, low operating energy of about 100 pJ, and a fast speed of about 30 ns. A study by Guo et al. [52] investigated the stability of lattice nitrogen and deactivation mechanisms in cubic CrN nanoparticles during the electrochemical nitrogen reduction reaction. Their findings reveal that cubic CrN nanoparticles undergo degradation under operational conditions due to the loss of lattice nitrogen and metal dissolution caused by ammonia poisoning from surface-accumulated ammonia. These observations are crucial for understanding the long-term performance and durability of CrN nanoparticles as electrocatalysts in nitrogen reduction reactions.

Materials with a face-centered cubic crystal structure can have magnetization in any of five types of spin arrangements: type I, type II, type III, type IA, and type IV [56]. The magnetic structure of CrN is type IV, where two ferromagnetically coupled atomic layers along [110] are antiferromagnetically coupled to the next two adjacent layers. Thus, the net magnetic moment of all four atomic layers is zero [56]. The magnetization in CrN is mostly contributed by Cr, as Cr is a transition metal with a partially filled 3d-orbital, and nitrogen has an unfilled 2p-orbital. In a paramagnetic state, the net magnetization is zero because the unpaired spins are in a high energy state; they move randomly with no correlation to each other. The number of spins pointed in any arbitrary direction is equal to the number of spins pointed in the opposite direction. By decreasing the energy of the spins by lowering the temperature of the paramagnet, at a characteristic temperature, the spins can form an order. When the low-temperature state is antiferromagnetic, its magnetic susceptibility produces a cusp at the transition in the susceptibility-versus-temperature plot. If the low-temperature state is ferromagnetic, then the susceptibility diverges at the transition temperature. By cooling down a paramagnet below the Néel temperature, it transitions to an antiferromagnetic state, where the spins are in a low energy state and have a long-range correlation.

The magnetic and structural transitions from a paramagnetic (cubic) to an antiferromagnetic (orthorhombic) phase in chromium nitride (CrN) nanoparticles were reported by Corliss et al. in 1960 [10]. Mrozinska et al. [57] and Filippetti et al. [58] studied the magnetic transition using their analytical and computational models, respectively. However, many researchers found contradictory results to the initial observations. The proposed reasons included the crystallinity of CrN, epitaxial constraints for thin films grown on crystalline substrates [59,60], nitrogen vacancies and chromium-rich phases such as Cr_2N [61], and lattice strains [62]. Similarly, Jin et al. suggested epitaxial-dependent phase transition in CrN thin films [63]. Their films grown by pulsed laser deposition on (010), (001), and (110) planes of $NdGaO_3$ showed electronic transitions from semiconductor to semiconductor, metallic to semiconductor, and metallic to semiconductor, respectively [63].

This controversy is also evident in the high-pressure study of CrN nanoparticles. For example, Rivadulla et al. observed a 25% drop in the bulk modulus of CrN at high pressures, accompanied by a structural transition from cubic to orthorhombic at 1 GPa [64], whereas Yan et al. [65] observed the transition at 5 GPa. The calculations of Alling et al. [66] did not show the 25% reduction in the bulk modulus associated with the structural transition.

This article provides a comprehensive review of various methods for synthesizing nanoparticles, with a particular emphasis on their application in studying phase transitions. The primary focus is on experimental techniques, exploring how different synthesis methods can influence the structural, electronic, and magnetic phase transitions. The synthesis methods can induce different levels of stress, size variations, nitrogen vacancies, defects, and impurities. These transitions appear to be coupled, with the onset of one transition triggering the other. The transitions have been reported over a wide temperature range from 240 to 285 K, but most studies report them at 275 ± 10 K. In addition to experimental methods, the article briefly reviews theoretical simulations towards the end.

2. Chromium Nitride Nanoparticles

Corliss et al. [10] prepared chromium nitride nanoparticles by passing ammonia (NH_3) over chromium nanoparticles at 1100 °C, followed by grinding and nitriding to minimize nitrogen-deficient phases. They studied the structural transition using variable-temperature X-ray diffraction and the magnetic transition using variable-temperature neutron diffraction. They were the first group to report that CrN is paramagnetic at room temperature and undergoes a magnetic transition to an antiferromagnetic phase around 290 K, consistent with later reports [3,67,68]. At the transition temperature, their neutron diffraction data revealed a change in pattern associated with the ordered magnetic state (antiferromagnetic phase). Concurrently, the variable-temperature X-ray diffraction showed that the crystal structure changes from FCC to orthorhombic. CrN is paramagnetic at room temperature with a face-centered cubic (FCC) crystal structure and two-atom basis (rock salt) crystal structure (i.e., the nitrogen occupies interstitial lattice sites in the FCC lattice structure). Its lattice parameters are a = b = c = 4.15 Å and $\alpha = \beta = \gamma = 90°$ [68–70]. Figure 1a,b shows the 2D and 3D crystal structure of CrN at room temperature, in which the large ball (yellow color) represents the Cr atom and small ball (dark green color) represents the N atom. Corliss et al. revealed that at the Neel temperature, it makes a structural transition to an orthorhombic crystal structure with two atom bases upon cooling, and its lattice parameters are $a' = \sqrt{(a_x + b_x)^2 + (a_y + b_y)^2}$, $b' = \frac{1}{2}\sqrt{(a_x - b_x)^2 + (a_y - b_y)^2}$, and $c' = c$, $\alpha' = 88°, \beta' = \gamma' = 90°$ [10]. Figure 1c,d show the 2D and 3D crystal structure of CrN at low temperatures. The model shows that the antiferromagnetic phase with the orthorhombic crystal structure consists of alternating double magnetic sheets parallel to a [110] direction, known as a type IV antiferromagnet. This magnetic transition induces stress in the crystal structure, as two adjacent parallel spins repel each other and two adjacent antiparallel spins attract each other [71–73]. The stresses thus produced are released by shear deformation, which results in $\alpha' = 88°$ instead of 90° [58,74].

Nasr-Eddine et al. studied CrN nanoparticles using variable temperature X-ray diffraction (VT-XRD) and neutron diffraction [75]. The sample preparation method was not provided. Their VT-XRD and neutron diffraction data were acquired over a temperature range from 5 K to 360 K. The VT-XRD data revealed a first-order transition from a rock salt crystal structure to an orthorhombic crystal structure at 285 K, which is in agreement with the results of Corliss et al. [10]. The neutron diffraction showed a transition from a paramagnetic to an antiferromagnetic state at $T_N = 285$ K upon cooling, with a hysteresis of about 5 K during the transition between heating and cooling cycles.

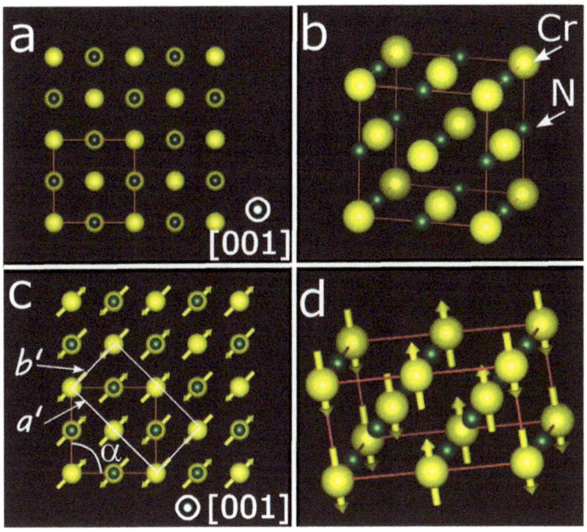

Figure 1. (**a**) A view of the fcc lattice of CrN along the [001] direction. (**b**) A 3D model representation of the fcc lattice. (**c**) A 2D view along [001] of the orthorhombic model for CrN at a low temperature. A shear distortion of the cubic crystal structure leading to $\alpha \approx 88°$. The double ferromagnetic layers alternate along [110] and spins choose an orthorhombic unit cell arrangement. (**d**) A 3D view of the orthorhombic unit cell.

A study by Browne et al. [76] showed structural, electronic, and magnetic transitions between 278 and 287 K. Their samples were prepared by enclosing Cr powder in a tube where high-purity nitrogen gas was passed at 1 atm pressure and 1123 K for about 100 h. During the preparation process, sample oxidation was prevented by inserting titanium pieces at the entrance port of the tube. The structural transition was determined by measuring the length of the specimen loaded in a silica dilatometer as a function of temperature. In the dilatometer, a differential transformer measured the relative change in the length of the outer diameter of the silica tube, in which the specimen was placed, and a spring-loaded silica rod held in contact with the specimen. The resistivity versus temperature data showed a transition from one metallic phase to another metallic phase between 284 and 287 K, which is different from what Browne et al. reported. The magnetic transition was determined by studying magnetic susceptibility as a function of temperature. Using the dilatometer, they observed a volume contraction of −0.59% at 284 K, which corresponds to the reported difference between the volumes of the cubic and orthorhombic crystals of CrN [10,75]. The data were recorded by a modified Sucksmith ring balance over a temperature range from 85 to 500 K for both cooling and warming parts of a cycle in an applied magnetic field of up to 20 kG. Magnetic susceptibility in these samples showed a sharp, first-order phase transition at 287 K. The marginal temperature difference between the two observed transitions indicates that one transition triggers the other.

Interesting thermometric properties were reported in polycrystalline CrN-powered samples [3,77] as well as in a thin film form [78]. Quintela et al. synthesized CrN by ammonolysis of Cr_3S_4 at 800 °C for 10 h, with an iteration of grinding. They produced samples with 0.3 GPa pressure and sintered them for high-density and transport measurements [77]. The electrical resistivity of their stoichiometric CrN samples at room temperature was measured to be 30×10^{-3} Ω·cm. The sample showed electronic transition from one semiconducting phase to another at 286 K. This observation is different from the report by

Wang et al. [79] of metallic (room temperature) and metallic (low temperature) phase transitions, which they observed in nanoparticles produced by a high-pressure synthesis method of Na_2CrO_4 (99.5% purity) and hexagonal-BN (99.9% purity). It also contradicts the semiconductor (room temperature) to metallic (low temperature) transition observed in CrN thin films [70,80]. Quintela et al. [77] measured the activation energy to be 75 meV, which matches the value reported by Constantine et al. [70].

The sample shows a kink at the transition temperature in Seebeck versus temperature, thermoelectric power factor versus temperature, and thermal conductivity versus temperature data [78]. The thermal conductivity versus temperature data of $Cr_{1-x}V_xN$ are shown in Figure 2. Regarding the change in the heat conduction mechanism due to the latent heat involvement in the phase transition, the thermal conductivity shows a cusp at 286 K. In the scope of this paper, the interest is in the $x = 0$ curve, which represents the stoichiometric CrN sample. The other vanadium doped CrN samples also show the cusp, which indicates that the structural transition strongly alters the heat transport mechanism even in $Cr_{1-x}V_xN$ samples. The broad maximum observed at around 150 K, which is related to phonons at a low-temperature propagation of phonons, is limited by the grain size. A more detailed discussion of the phonon peaks is provided by Ebad-Allah et al. [81]. Vanadium incorporation in CrN reduces the thermal conductivity at around 150 K, but the cusp temperature remains almost unchanged.

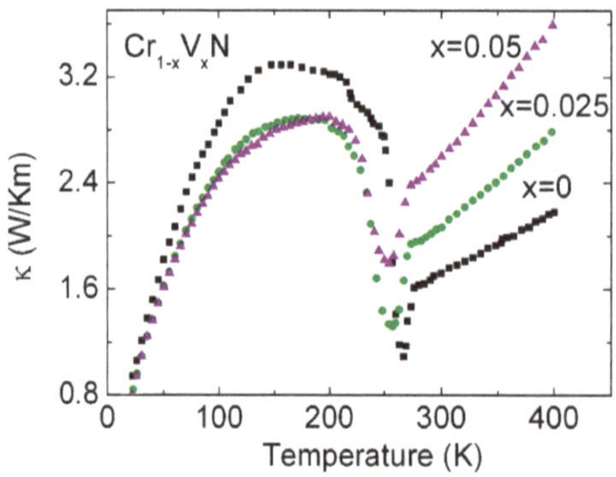

Figure 2. Thermal conductivity of $Cr_{1-x}V_xN$. The dip around 286 K corresponds to the temperature of the orthorhombic-to-cubic structural transition [77].

The polycrystalline CrN samples prepared by ammonolysis of Cr_3S_4 at 800 °C for 10 h was studied by Ebad-Allah et al. [81]. They studied optical transmittance as a function of temperature in a frequency range of 100 to 20,000 cm^{-1}. They mixed the polycrystalline CrN powered with CsI to prepare pellets for the study. Figure 3 shows that the major changes in the absorbance spectra due to the phonon mode are between 550 cm^{-1} and 670 cm^{-1}. The phonon mode is excited at around 270 K; the peak first broadens and then splits during further cooling of the samples. The change is particularly noticeable at 630 cm^{-1}: the absorbance increases by cooling the sample and decreases by warming the sample. The onset of the new phonon mode at 270 K matches is related to the lowering symmetry of crystal structure from cubic to orthorhombic. The measure bandgap at room temperature from the data is between 100 and 150 meV [59,70,77,78,82]. They also studied optical absorption as a function of applied pressure on the sample at room temperature. The samples show an onset of phonon mode related to the lowering in crystal symmetry at a 0.6 GPa pressure. It confirms the structural transition from a cubic to an orthorhombic crystal structure. The pressure induces a metallic phase at 0.6 GPa as the optical reflectance increases with an increase in the pressure. The data reveal that at room temperature and atmospheric pressure, CrN is a charge-transfer insulator than a Mott insulator [81,83]. A report by Yan et al. [65], which focused on a pressure-driven phase transition in CrN nanoparticles, contradicts the reports by Ebad-Allah et al. [81] and Rivadullah et al. [64]

on the structural phase transitions. Yan et al. [65] observed a cubic-to-orthorhombic phase transition at ∼5 GPa, which is about 8 and 5 times larger than what Ebad-Allah et al. [81] and Rivadullah et al. [64] reported.

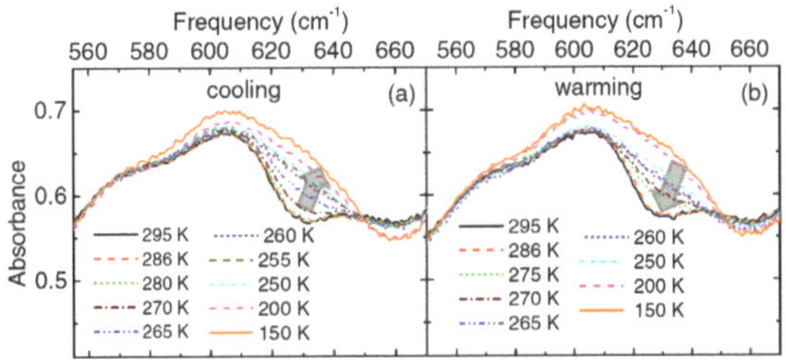

Figure 3. Absorbance of a CrN pellet as a function of temperature for (**a**) cooling and (**b**) warming parts of cycles. The arrows highlight changes in the data acquired while cooling and warming the samples [81].

A study on the correlated paramagnetic cubic-to-antiferromagnetic metal transition published by Bhobe et al. investigated the magnetostructural transition in polycrystalline CrN samples [84]. Their CrN powdered samples were prepared by heating 3N pure $CrCl_3$ in the continuous flow of 5N NH_3 at a 1173 K temperature for 20 h. For further characterization, it was pressed into a pellet and sintered at 1323 K for 30 h under N_2 gas. The resistivity-versus-temperature data showed $d(ln(\sigma))/d(ln(T)) \rightarrow 0$ as $T \rightarrow 0$, which indicates that the sample was metallic below T_N [70,80,81]. Resonant photoemission spectroscopy (Res-PES) revealed a bandgap in the Cr 3d-orbital density of states near the Fermi level, and bulk-sensitive laser photoemission spectroscopy showed a Fermi edge associated with the antiferromagnetic metal phase below T_N. Moreover, Bhobe et al. determined the Coulomb energy term U = 4.5 eV. It is consistent with the value used by Alam et al. [68] for their computational modeling of local density approximation with Hubbard correction (LDA + U)-based calculations [85].

Zieschang et al. prepared CrN nanoparticles by a chemical reaction of 9.9 mg Na and $CrCl_3$ with 50 mL liquid ammonia at 195 K [74]. The solution was kept at a low temperature for 1.5 h. Intermittently, the flask was taken out for stirring for 1 min after a duration of 20 min. Once the reaction was completed, the mixture was allowed to warm to room temperature. After evaporating the ammonia, the powders were further dried in vacuum and annealed at 873 K. The powders were then washed with methanol. They studied the structural and magnetic transitions within their samples with variable temperature X-ray diffraction and SQUID magnetometer, respectively. Their as-prepared CrN nanoparticles were amorphous and did not produce any peaks in XRD spectra. The crystal quality was improved after annealing and washing the samples. According to their report, they observed only CrN peaks in the XRD data and Cr-rich phases such as Cr_2N. They studied two samples, one annealed at 773 K and another at 873 K. The sample annealed at a higher temperature had a larger grain size and better crystal quality.

The VT-XRD data of the nanoparticles annealed at 873 K and Rietveld refinement simulated data are shown in Figure 4. The data shown in Figure 4a were acquired at 298 K and show a cubic crystal structure of the space group Fm$\underline{3}$m; and the data in Figure 4b were acquired at 193 K and show an orthorhombic crystal structure of the space group Pnma. These indicate that the samples went through a structural transition between 298 K and 193 K. The main difference is that the 222 peak ($2\theta = 34.49°$ produced by the cubic symmetry was split into three peaks—022, 402, and 122—related to the orthorhombic symmetry. Scanning the 022, 402, and 122 peaks in steps of 20 K while warming the sample from 193 K to 298 K revealed the transition temperature between 253 and 273 K.

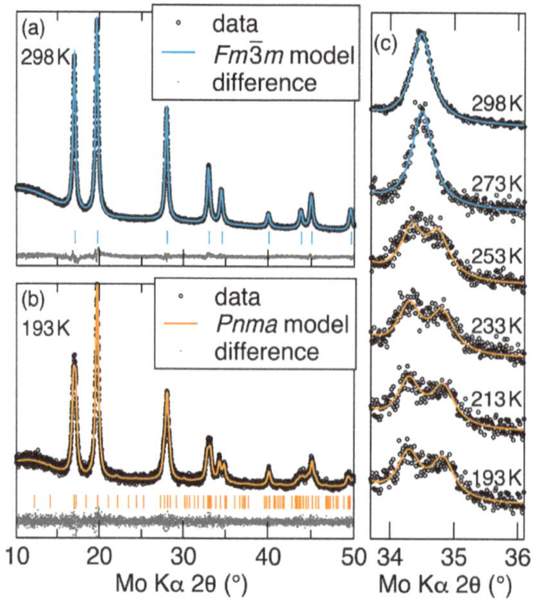

Figure 4. Variable temperature X-ray diffraction spectra of CrN nanoparticles (**a**) at 298 K and (**b**) at 193 K. The theoretical data are the Rietveld refinement of (**a**) cubic CrN (Fm3̲m) and (**b**) orthorhombic CrN (Pnma). (**c**) The 222 peak related to the cubic crystal structure is split into three peaks, which are the 022, 402, and 122 peaks, related to the orthorhombic crystal structure [74].

Furthermore, Zieschang et al. studied transition in heat capacity and magnetic transition as a function of temperature by differential scanning calorimetry and magnetic susceptibility by SQUID as shown in Figure 5. In Figure 5a, the data recorded for the cooling and warming cycles show a transition in the heat capacity at around 259 K, which indicates that the transition is reversible and reproducible. Similarly, the magnetic susceptibility-versus-temperature data in Figure 5b reveal magnetic phase transition from the paramagnetic to the antiferromagnetic phase between 248 and 269 K, which matches the structural and heat capacity transition. The observation of all these transitions in CrN particles at about the same temperature strengthens the earlier theoretical and experimental claims that these transitions are coupled [58,68].

Jankovsky et al. [86] synthesized CrN nanoparticles by ammonolysis of anhydrous $CrCl_3$. The $CrCl_3$ was first dried in thionylchloride under reflux for 36 h in a control ambient in a glove box. The ammonolysis of anhydrous $CrCl_3$ was conducted in a continuous flow of 100 cm^3/min NH_3 and 50 cm^3/min N_2 at 1073 K for 72 h. After cooling the sample to room temperature, one part of the powder sample was sintered at 1073 K under the same flow of NH_3 and N_2, and other parts were processed with spark plasma sintering at 1023 K, 1073 K, and 1123 K. All samples were kept under an 80 MPa pressure during sintering. The samples showed a peak in the heat capacity-versus-temperature data at 291 K and exhibited a magnetic transition from the paramagnetic state to the antiferromagnetic state at around the same temperature. All the samples prepared by different methods as described above showed metallic behavior above and below the Neel temperature, which agrees with Gui et al. [87]. The common part of synthesizing the nanoparticles was the use of high pressure either during the synthesis step or during the sintering process. Further investigation is needed to check whether the pressure induces any permanent changes in the nanoparticles.

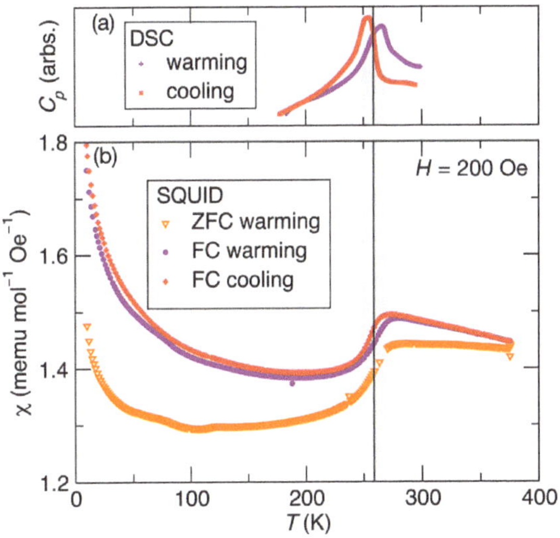

Figure 5. (**a**) Heat capacity versus temperature and (**b**) magnetic susceptibility versus temperature of CrN nanoparticles. The heat capacity and susceptibility data were recorded for cooling as well as for warming cycle [74].

CrN nanoparticles were synthesized by a chemical reaction of 1 g of nonahydrated chromium nitrate [$Cr(NO_3)_39H_2O$] with 6 g urea (NH_2CONH_2) at 623 K for 2 h by Singh et al. [88]. Ammonia [NH_3] was used for the nitridation reaction of the prepared precursor in a combustion quartz tube furnace. The ammonia was flowing at 180–200 mL/min at a temperature of 1073 K for 6 h. The nanoparticles thus prepared were 22.92 nm in diameter. They used a superconducting quantum interference device (SQUID) magnetometer for studying magnetization as a function of temperature. Both field cold and zero field cold magnetization data of the samples were obtained. The data revealed that the magnetization of the sample increased by warming up the samples from the liquid helium temperature to 265 K, which is a typical antiferromagnetic behavior. At 265 K, the magnetization quickly saturated and showed no appreciable increase when the temperature was further increased beyond 275 K. The Neel temperature of 265 K is slightly lower than most of the reported values in the literature [68,69,76], which could be attributed to the unintentional oxygen or carbon doping of their samples as both elements exist in their precursor or due to the finite size of the nanoparticles [89,90]. In a paper published in 2021, Jin et al. demonstrated that by decreasing the size of the thin film, the Neel temperature decreased following $T_N = 14t^{0.46}$, where t represents the thickness of the film [91].

A meticulous study of the finite-size effect on the magnetic and structural transitions in CrN nanoparticles was reported by Wang et al. [92]. They synthesized CrN nanoparticles by mixing anhydrous $CrCl_3$ of 99.5% purity and $NaNH_2$ of 99.5% purity in a 1:4 ratio. The mixture was compressed into a cylindrical shape 10 mm in diameter and height in argon ambient. The samples were then encapsulated into a molybdenum capsule and loaded into a high-pressure cell. For controlling the nanoparticles' size, the cell was first exposed to a different target pressure between 1 and 5 GPa, followed by heating to a different temperature between 573 and 1773 K for 20 min, before quenching and decompressing the cell to atmospheric pressure. The nanoparticles thus prepared were washed with distilled water. The average diameter of the nanoparticles synthesized under different conditions was in a range of 4 nm to 2750 nm. The lattice constant determined by XRD is provided in Figure 6a. The lattice sharply increases from 4.086 Å to 4.418 Å for nanoparticle diameters of 4 nm and 100 nm, respectively, and it shows a plateau for all nanoparticle diameters larger than 100 nm, which matches the reported values in [68,69,80,93]. It shows that smaller particles have a larger stress compared to larger ones. Figure 6b shows heat capacity versus reduced temperature $\varepsilon = |1 - \frac{T}{T_N}|$, where T is the samples' temperature. Heat capacity manifests discontinuity across T_N and shows a peak at 274.41 K for the bulk samples, which is related to the spin reordering in the magnetic phase transition from the paramagnetic to the antiferromagnetic phase. The critical behaviors of heat capacity can be explored by fitting the data by a power law of $C/R \sim \varepsilon^{-\alpha}$, with α values of 1.39 and 1.28 for the paramagnetic and antiferromagnetic phases, which are larger than any conventional universality classes with α values of less than 0.25, which leads to a conclusion that the spins were reordered by the lattice due to its spin–lattice coupling [58,94]. In Figure 6c, the temperature derivative of dc susceptibility ($d\chi/dT$) as a function of temperature shows peaks associated with the magnetic transition in both the cooling and warming data. The peaks can be observed in particles of size 550 nm to 9 nm, but below 9 nm, the transition vanishes due to a small size effect [92]. The Neel temperature as a function of the particle size is illustrated in Figure 6d, where T_N for 9 nm particles is about 256 K and for the 1000 nm particle, it is 285 K, which matches the values reported by [68,70].

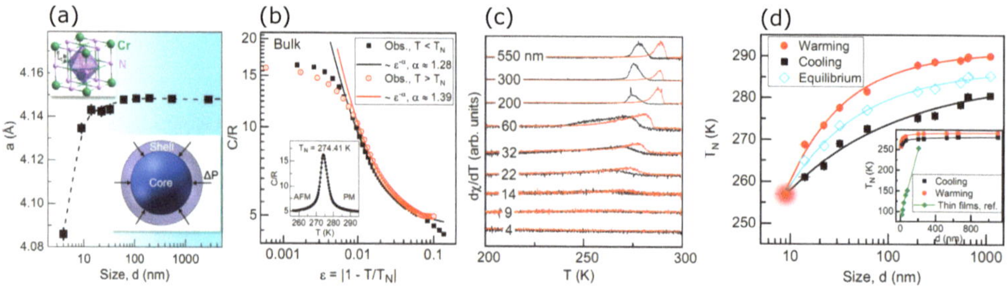

Figure 6. (**a**) Lattice parameter of CrN versus the grain size. It shows cubic CrN, and the model shows the core and shell of the CrN nanoparticle with surface compression ΔP. (**b**) Specific heat (C/R) as a function of temperature is shown for the bulk sample. The inset shows the specific heat against temperature in a linear scale. (**c**) Temperature derivative of susceptibility of CrN nanoparticles recorded for cooling (black colored data) and warming (red colored data) processes. (**d**) T_N versus particle size and its linear plot provided in the inset. The inset also shows the thickness-dependent T_N Ref. [92].

Stoichiometric CrN nanoparticles synthesized by Wang et al. [95] using $NaCrO_2$ and hexagonal-BN precursors at 5 GPa and heated to 1573 K for 20 min. The solid-state ion-exchange reaction produced CrN along with $NaBO_2$; the byproduct $NaBO_2$ can be removed by washing the product with distilled water. The nanoparticles' preparation method was reported in another study by the same group [79]. They studied the the structural transition by XRD as a function of pressure and temperature and observed cubic to orthorhombic at 5.46 GP while keeping the sample at 315 K. The pressure value at which the transition occurred is in agreement with 5 GPa as reported by Yan et al. [65], but it is more than 5 times higher than Revidulla et al.'s [64] value of ~1 GPa and about 9 times higher than the value reported by Ebad-Allah et al. [81]. The electrical transition studied with resistivity-versus-temperature data occurs at 240 K, which is much smaller than most of the reported values [68,70]. The low transition temperature could be either due to the small size of their nanoparticles [92], impurities [80], or strain in the nanoparticles [91]. They [95] determined the bulk modulus of CrN cubic and orthorhombic crystal structures to be 257 GPa and 262 GPa, which indicates no huge reduction in the bulk modulus of CrN with phase change. The determined value of bulk modulus supports Alling et al. [66] and contradicts Revidulla et al.'s [64] reports.

Improper multiferroic-like transition in CrN nanoparticles was reported by Gui et al. [87]. The nanoparticles were prepared by chemical reaction at a high pressure as reported in [79,95]. Their neutron powder diffraction experiment revealed a structural transition from cubic to orthorhombic upon cooling. However, their electrical resistivity-versus-temperature data showed metal-to-metal transition around 240 K, which contradicts other reported observations of semiconducting to metallic [70], semiconducting to semiconducting [77], and impurities [80] and epitaxial constraints [63]-dependent transitions. The low-temperature metallic state appears only in a tetragonal/cubic crystal structure [68]; however, their neutron diffraction shows the prevailing orthorhombic structure at low temperatures. They used DFT calculations with a Hubbard correction of 0.5 eV to match the observed metallicity and magnetic moment in the orthographic phase; however, 0.5 eV is a much smaller value of U determined both experimentally [84] as well as adopted in theoretical models [48,68,96,97].

Gui et al. meticulously explained the lattice modulation produced by the frustrated chain of Cr spins in the orthorhombic phase [87]. Accompanied by the magnetostructural transition, CrN nanoparticles also showed an antiferroelectric-like transition, which is driven by the frustrated Cr t_{2g} spins in the antiferromagnetic order. Metallicity, on the other hand, originates from the Cr e_g and N p orbitals [87]. The magnetic and lattice modulations buckle up the lattice along the c-axis, which moves the negative and positive charge centers in CrN_6 octahedra in opposite directions to electric dipole. These results agree with the reported buckled CrN (001)surface [68] observed with low-temperature high-energy electron diffraction. Alam et al. reported the doubling of the structural periodicity on the CrN (001) surface, which appeared at 273 K in an epitaxial CrN film grown on a MgO (001) substrate [68].

Table 1 shows the experimental results reviewed in this article. The sample synthesis methods, transition temperature, and structural, electronic, and magnetic transitions are summarized in the table. The temperature above the transition temperature is labeled as HT and that below the phase transition temperature is labeled as LT.

Table 1. CrN nanoparticles' preparation methods and structural, electronic, and magnetic transitions.

Reference	Preparation Method	T_N (K)	Structural Transition	Electronic Transition	Magnetic Transition
Corliss et al. [10]	Cr + NH_3 at 1100 °C	273	FCC (HT) to orthorhombic (LT)		Paramagnetic (HT) to antiferromagnetic (LT)
Nasr-Eddine et al. [75]		285	FCC (HT) to orthorhombic (LT)		Paramagnetic (HT) to antiferromagnetic (LT)
Browne et al. [76]	Cr + N_2 at 950 °C for 100 h	~285	FCC (HT) to orthorhombic (LT)	metallic (HT) to metallic (LT)	Paramagnetic (HT) to antiferromagnetic (LT)
Quintela et al. [77]	Cr_3S_4 + NH_3 at 800 °C for 10 h	286		semiconducting (HT) to metallic (LT)	
Ebad-Allah et al. [81]	Cr_3S_4 + NH_3 at 800 °C for 10 h	270	FCC (HT) to orthorhombic (LT) (also shows transtion at RT under 0.6 GPa)		
Yan et al. [65]	Na_2CrO_4 + h-BN at 5 GPa at 1573 K for 20 min [98]		FCC (HT) to orthorhombic (LT) at ~5 GPa		
Bhobe et al. [84]	$CrCl_3$ + NH_3 at 1173 K for 20 h	286		semiconducting (HT) to metallic (LT)	Paramagnetic (HT) to antiferromagnetic (LT)
Zieschang et al. [74]	Na + $CrCl_3$ + NH_3 at 195 K for 1.5 h	248–273	FCC (HT) to orthorhombic (LT)		Paramagnetic (HT) to antiferromagnetic (LT)

Table 1. *Cont.*

Reference	Preparation Method	T_N (K)	Structural Transition	Electronic Transition	Magnetic Transition
Jankovsky et al. [86]	$CrCl_3$ + NH_3 at 1073 K for 72 h	291		metallic (HT) to metallic (LT)	Paramagnetic (HT) to antiferromagnetic (LT)
Singh et al. [88]	$[Cr(NO_3)_39H_2O]$ + NH_2CONH_2 at 623 K for 2 h $\Longrightarrow$ NH_3 at temperature of 1073 K for 6 h	265		metallic (HT) to metallic (LT)	Paramagnetic (HT) to antiferromagnetic (LT)
Wang et al. [92]	$CrCl_3$ + $NaNH_2$ compressed at 1–5 GPa heated to 573–1773 K for 20 min then quenching	256–285 (size-dependent)			Paramagnetic (HT) to antiferromagnetic (LT)
Gui et al. [87]	Na_2CrO_4 + h-BN at 3.5–5 GPa at 1573 K for 20 min [79]	273	FCC (HT) to orthorhombic (LT)	metallic (HT) to metallic (LT)	Paramagnetic (HT) to antiferromagnetic (LT)

3. Chromium Nitride Computational Models

Although analytical and computational research is not the focus of this review, I provide a brief overview of the research carried out on CrN. Many groups have applied various theories to understand the structural and magnetic transitions in CrN [57,83,93,94,96,99–101]. The analytical methods include applying renormalization group theory and the Landau–Ginsburg–Wilson models to check critical behavior around the transition temperature [57], and other similar models [75]. Computational methods often use density functional theory (DFT) in combination with the projector augmented wave (PAW) method, typically applied in the Vienna ab-initio simulation package (VASP). The local spin density approximation (LDA) [102], the generalized gradient approximation (GGA) [103], and the LDA combined with a Hubbard Coulomb term (LDA + U) [104,105] are commonly used to understand electron exchange-correlation effects. The Hubbard term is used to understand the Cr 3d-orbital interactions [106] and the opening of the bandgap. Tawinan et al. [107] studied CrN using self-consistent GW methods. They started with the local spin density approximation with U and then removed U to let the GW energy define the bandgap. They found a bandgap of about 1 eV, which is in agreement with Botana et al. [108], who determined the bandgap using LDA + U.

Following the renormalization group theory method, Mrozinska et al. [57] constructed a 12-component order parameter system, which led to the formation of a relatively simple Landau–Ginsburg–Wilson Hamiltonian. They noted that the order parameter in CrN transforms with respect to the 12-dimensional irreducible representation of the Fm3m space group. They proved that the RG transformations have no stable fixed points, indicating that the transition should be first-order.

4. Conclusions and Prospects

Various methods for synthesizing CrN nanoparticles and their influence on structural, electronic, and magnetic phase transitions have been reviewed. It has been observed that the synthesis methods can induce different levels of stress, size variations, nitrogen vacancies, changes in stoichiometry, defects, and impurities. For instance, particle size and strain can alter the transition temperature or even eliminate the transition. Characterization methods and the ambient conditions for studying properties also affect the physical properties of the materials. It is observed that the onset of one transition is often concurrent with another, indicating that these transitions are correlated. Transitions have been reported over a wide temperature range, though most studies report them at 275 ± 10 K. In addition to experimental methods, the article briefly reviews theoretical simulations towards the end. This comprehensive review serves as a detailed reference for researchers and aims to stimulate further scientific interest in understanding the physical properties of CrN nanoparticles and their use in technological applications.

Despite the use of various synthesis methods, nanoparticles are seldom free from defects, stress, and impurities. Oxygen is one of the main impurities in CrN nanoparticles, as Cr reacts more readily with oxygen than with nitrogen. Many nanoparticles have been prepared either in air or in rough vacuum environments. The oxidation issue can be observed more easily when studying the CrN bandgap [109], which is extremely sensitive to oxygen doping. Many nanoparticle preparation methods require NH_3, which is hazardous, expensive, and not environmentally friendly. Some methods can induce defects, such as encapsulating Cr with a nitrogen precursor and then quenching the nanoparticles from high temperature. The nanoparticles thus prepared will be polycrystalline and will have a range of strains. Although nanoparticle preparation by chemical reactions is well-established, it is not as clean and controlled as thin film preparations in ultrahigh vacuum. To converge all these divergent results, nanoparticles need to be prepared in a controlled environment in an ultrahigh vacuum chamber to ensure their quality and reproducibility.

To the best of my knowledge, no study has directly compared the impact of stress versus stoichiometry on the phase transitions of the films, or determined which one has a more dominant effect. However, by comparing different studies, one can conclude that the stoichiometry of the film plays a more dominant role than the stress within the film.

Another deficiency is that not all phase transitions have been studied with the same nanoparticles, making it difficult to understand the delicate correlation between different transitions. Comprehensive studies of all transitions using high-quality nanoparticles are needed.

It is computationally demanding to study phase transitions as a function of temperature or pressure variation; however, it is desirable and would provide the most accurate guide to experimental studies. Most models use only a fixed temperature and pressure and allow a system to relax from a predefined crystal structure. Such studies can sometimes be misleading as simulated models may become stuck in a local minimum rather than reaching to a global minimum. In the GGA+U or LDA+U methods, one has to define the value of U from other sources or by trial and error. The most complete study would begin from a cubic system of CrN and gradually cool down below the Neel temperature, where CrN adopts an antiferromagnetic order and an electronic phase with a certain crystal structure.

Combining experimental studies with theoretical modeling to understand the physical properties under different conditions and using the acquired knowledge to bridge the gap between science and engineering would be productive. Such collaboration would be beneficial for exploring CrN for phase change memory, temperature sensors, electrical sensors, and other advanced applications.

Funding: This work was supported by the Deanship of Research Oversight and Coordination of King Fahd University of Petroleum and Minerals, Dhahran, Saudi Arabia, under grant No. ISP23218.

Data Availability Statement: No new data were created or analyzed in this study. Data sharing is not applicable to this article.

Conflicts of Interest: The author declare no conflict of interest. The funders had no role in the design of the study; in the collection, analyses, or interpretation of data; in the writing of the manuscript; or in the decision to publish the results.

References

1. Cheng, Z.; Qi, W.; Pang, C.H.; Thomas, T.; Wu, T.; Liu, S.; Yang, M. Recent advances in transition metal nitride-based materials for photocatalytic applications. *Adv. Funct. Mater.* **2021**, *31*, 2100553. [CrossRef]
2. Young, A.F.; Sanloup, C.; Gregoryanz, E.; Scandolo, S.; Hemley, R.J.; Mao, H.K. Synthesis of novel transition metal nitrides IrN_2 and OsN_2. *Phys. Rev. Lett.* **2006**, *96*, 155501. [CrossRef]

3. Eklund, P.; Kerdsongpanya, S.; Alling, B. Transition-metal-nitride-based thin films as novel energy harvesting materials. *J. Mater. Chem. C* **2016**, *4*, 3905–3914. [CrossRef] [PubMed]
4. Qi, W.; Zhou, Y.; Liu, S.; Liu, H.; Hui, L.S.; Turak, A.; Wang, J.; Yang, M. Oxidized impurity in transition metal nitride for improving the hydrogen evolution efficiency of transition metal nitride-based catalyst. *Appl. Mater. Today* **2020**, *18*, 100476. [CrossRef]
5. Juneja, S.; Shishodia, M.S. Surface plasmon amplification in refractory transition metal nitrides based nanoparticle dimers. *Opt. Commun.* **2019**, *433*, 89–96. [CrossRef]
6. Sharma, K.; Arora, A.; Tripathi, S.K. Review of supercapacitors: Materials and devices. *J. Energy Storage* **2019**, *21*, 801–825.
7. Smyrnova, K.; Sahul, M.; Haršáni, M.; Čaplovič, L.; Beresnev, V.; Čaplovičová, M.; Kusy, M.; Pogrebnjak, A. Effect of bias voltage on the structural properties of WN/NbN nanolayer coatings deposited by cathodic-arc evaporation. *J. Phys. Conf. Ser.* **2024**, *2712*, 012014. [CrossRef]
8. Weinberger, C.R.; Yu, X.X.; Yu, H.; Thompson, G.B. Ab initio investigations of the phase stability in group IVB and VB transition metal nitrides. *Comput. Mater. Sci.* **2017**, *138*, 333–345. [CrossRef]
9. Biswas, A.; Alvarez, G.A.; Tripathi, M.; Lee, J.; Pieshkov, T.S.; Li, C.; Gao, B.; Puthirath, A.B.; Zhang, X.; Gray, T.; et al. Cubic and hexagonal boron nitride phases and phase boundaries. *J. Mater. Chem. C* **2024**, *12*, 3053–3062. [CrossRef]
10. Corliss, L.; Elliott, N.; Hastings, J. Antiferromagnetic structure of CrN. *Phys. Rev.* **1960**, *117*, 929. [CrossRef]
11. Srivastava, A.; Chauhan, M.; Singh, R. Pressure induced phase transitions in transition metal nitrides: Ab initio study. *Phys. Status Solidi B* **2011**, *248*, 2793–2800. [CrossRef]
12. Ojha, P.; Aynyas, M.; Sanyal, S.P. Pressure-induced structural phase transformation and elastic properties of transition metal mononitrides. *J. Phys. Chem. Solids* **2007**, *68*, 148–152. [CrossRef]
13. Wang, H.; Li, J.; Li, K.; Lin, Y.; Chen, J.; Gao, L.; Nicolosi, V.; Xiao, X.; Lee, J.M. Transition metal nitrides for electrochemical energy applications. *Chem. Soc. Rev.* **2021**, *50*, 1354–1390. [CrossRef] [PubMed]
14. Tokura, Y.; Nagaosa, N. Orbital physics in transition-metal oxides. *Science* **2000**, *288*, 462–468. [CrossRef] [PubMed]
15. Börgel, J.; Campbell, M.G.; Ritter, T. Transition metal d-orbital splitting diagrams: An updated educational resource for square planar transition metal complexes. *J. Chem. Educ.* **2016**, *93*, 118–121. [CrossRef]
16. Li, Y.H.; Hung, T.H.; Chen, C.W. A first-principles study of nitrogen-and boron-assisted platinum adsorption on carbon nanotubes. *Carbon* **2009**, *47*, 850–855. [CrossRef]
17. Mananghaya, M.R.; Santos, G.N.; Yu, D. Nitrogen substitution and vacancy mediated scandium metal adsorption on carbon nanotubes. *Adsorption* **2017**, *23*, 789–797. [CrossRef]
18. Ma, Y.; Foster, A.S.; Krasheninnikov, A.; Nieminen, R.M. Nitrogen in graphite and carbon nanotubes: Magnetism and mobility. *Phys. Rev. B* **2005**, *72*, 205416. [CrossRef]
19. Lin, C.; Liu, X.; Qu, J.; Feng, X.; Seh, Z.W.; Wang, T.; Zhang, Q. Strain-controlled single Cr-embedded nitrogen-doped graphene achieves efficient nitrogen reduction. *Mater. Adv.* **2021**, *2*, 5704–5711. [CrossRef]
20. Feng, H.; Ma, J.; Hu, Z. Nitrogen-doped carbon nanotubes functionalized by transition metal atoms: A density functional study. *J. Mater. Chem.* **2010**, *20*, 1702–1708. [CrossRef]
21. Mateti, S.; Sultana, I.; Chen, Y.; Kota, M.; Rahman, M.M. Boron Nitride-Based Nanomaterials: Synthesis and Application in Rechargeable Batteries. *Batteries* **2023**, *9*, 344. [CrossRef]
22. Thomas, S.A.; Pallavolu, M.R.; Khan, M.E.; Cherusseri, J. Graphitic carbon nitride (g-C3N4): Futuristic material for rechargeable batteries. *J. Energy Storage* **2023**, *68*, 107673. [CrossRef]
23. Xiong, T.; Li, J.; Roy, J.C.; Koroma, M.; Zhu, Z.; Yang, H.; Zhang, L.; Ouyang, T.; Balogun, M.S.; Al-Mamun, M. Hetero-interfacial nickel nitride/vanadium oxynitride porous nanosheets as trifunctional electrodes for HER, OER and sodium ion batteries. *J. Energy Chem.* **2023**, *81*, 71–81. [CrossRef]
24. Jin, C.; Huang, Y.; Li, L.; Wei, G.; Li, H.; Shang, Q.; Ju, Z.; Lu, G.; Zheng, J.; Sheng, O.; et al. A corrosion inhibiting layer to tackle the irreversible lithium loss in lithium metal batteries. *Nat. Commun.* **2023**, *14*, 8269. [CrossRef] [PubMed]
25. Wang, Y.Y.; Zhang, X.Q.; Zhou, M.Y.; Huang, J.Q. Mechanism, quantitative characterization, and inhibition of corrosion in lithium batteries. *Nano Res. Energy* **2023**, *2*, e9120046. [CrossRef]
26. Ren, L.; Hu, Z.; Peng, C.; Zhang, L.; Wang, N.; Wang, F.; Xia, Y.; Zhang, S.; Hu, E.; Luo, J. Suppressing metal corrosion through identification of optimal crystallographic plane for Zn batteries. *Proc. Natl. Acad. Sci. USA* **2024**, *121*, e2309981121. [CrossRef]
27. Zhou, X.; Chen, H.; Shu, D.; He, C.; Nan, J. Study on the electrochemical behavior of vanadium nitride as a promising supercapacitor material. *J. Phys. Chem. Solids* **2009**, *70*, 495–500. [CrossRef]
28. Xie, Y.; Wang, Y.; Du, H. Electrochemical capacitance performance of titanium nitride nanoarray. *Mater. Sci. Eng. B* **2013**, *178*, 1443–1451. [CrossRef]
29. Shen, H.; Wei, B.; Zhang, D.; Qi, Z.; Wang, Z. Magnetron sputtered NbN thin film electrodes for supercapacitors. *Mater. Lett.* **2018**, *229*, 17–20. [CrossRef]
30. Arif, M.; Sanger, A.; Singh, A. Sputter deposited chromium nitride thin electrodes for supercapacitor applications. *Mater. Lett.* **2018**, *220*, 213–217. [CrossRef]
31. Ting, Y.J.B.; Lian, K.; Kherani, N. Fabrication of titanium nitride and molybdenum nitride for supercapacitor electrode application. *ECS Trans.* **2011**, *35*, 133. [CrossRef]

32. Chen, Z.; Song, Y.; Cai, J.; Zheng, X.; Han, D.; Wu, Y.; Zang, Y.; Niu, S.; Liu, Y.; Zhu, J.; et al. Tailoring the d-band centers enables Co4N nanosheets to be highly active for hydrogen evolution catalysis. *Angew. Chem. Int. Ed.* **2018**, *57*, 5076–5080. [CrossRef] [PubMed]
33. Li, D.; Xing, Y.; Yang, R.; Wen, T.; Jiang, D.; Shi, W.; Yuan, S. Holey cobalt–iron nitride nanosheet arrays as high-performance bifunctional electrocatalysts for overall water splitting. *ACS Appl. Mater. Interfaces* **2020**, *12*, 29253–29263. [CrossRef] [PubMed]
34. Mahadik, S.; Surendran, S.; Kim, J.Y.; Janani, G.; Lee, D.K.; Kim, T.H.; Kim, J.K.; Sim, U. Syntheses and electronic structure engineering of transition metal nitrides for supercapacitor applications. *J. Mater. Chem. A* **2022**, *10*, 14655–14673. [CrossRef]
35. Zhang, X.; Gall, D. CrN electronic structure and vibrational modes: An optical analysis. *Phys. Rev. B* **2010**, *82*, 045116. [CrossRef]
36. Zhou, Y.; Guo, W.; Li, T. A review on transition metal nitrides as electrode materials for supercapacitors. *Ceram. Int.* **2019**, *45*, 21062–21076. [CrossRef]
37. Gharavi, M.A.; Kerdsongpanya, S.; Schmidt, S.; Eriksson, F.; Nong, N.; Lu, J.; Balke, B.; Fournier, D.; Belliard, L.; Le Febvrier, A.; et al. Microstructure and thermoelectric properties of CrN and CrN/Cr2N thin films. *J. Phys. D Appl. Phys.* **2018**, *51*, 355302. [CrossRef]
38. Wei, B.; Liang, H.; Zhang, D.; Wu, Z.; Qi, Z.; Wang, Z. CrN thin films prepared by reactive DC magnetron sputtering for symmetric supercapacitors. *J. Mater. Chem. A* **2017**, *5*, 2844–2851. [CrossRef]
39. Gao, Z.; Wan, Z.; Wu, Z.; Huang, X.; Li, H.; Zhang, T.F.; Mayrhofer, P.H.; Wang, Q. Synthesis and electrochemical properties of nanoporous CrN thin film electrodes for supercapacitor applications. *Mater. Des.* **2021**, *209*, 109949. [CrossRef]
40. Sun, Q.; Fu, Z.W. An anode material of CrN for lithium-ion batteries. *Electrochem. Solid-State Lett.* **2007**, *10*, A189. [CrossRef]
41. Zhang, L.; Liang, P.; Shu, H.B.; Man, X.L.; Du, X.Q.; Chao, D.L.; Liu, Z.G.; Sun, Y.P.; Wan, H.Z.; Wang, H. Design rules of heteroatom-doped graphene to achieve high performance lithium–sulfur batteries: Both strong anchoring and catalysing based on first principles calculation. *J. Colloid Interface Sci.* **2018**, *529*, 426–431. [CrossRef] [PubMed]
42. Heckmann, A.; Krott, M.; Streipert, B.; Uhlenbruck, S.; Winter, M.; Placke, T. Suppression of Aluminum Current Collector Dissolution by Protective Ceramic Coatings for Better High-Voltage Battery Performance. *ChemPhysChem* **2017**, *18*, 156–163. [CrossRef] [PubMed]
43. Sun, Q.; Fu, Z.W. $Cr_{1-x}Fe_xN$ ($0 \leqq x \leqq 1$) Ternary Transition-Metal Nitrides as Anode Materials for Lithium-Ion Batteries. *Electrochem. Solid-State Lett.* **2008**, *11*, A233. [CrossRef]
44. Gu, L.; Wu, S.Y.; Liu, H.; Singh, R.; Newman, N.; Smith, D.J. Characterization of Al (Cr) N and Ga (Cr) N dilute magnetic semiconductors. *J. Magn. Magn. Mater.* **2005**, *290*, 1395–1397. [CrossRef]
45. Ney, A.; Rajaram, R.; Parkin, S.; Kammermeier, T.; Dhar, S. Magnetic properties of epitaxial CrN films. *Appl. Phys. Lett.* **2006**, *89*, 112504. [CrossRef]
46. Alsaad, A. Magnetic and structural properties of Cr-based diluted magnetic semiconductors and alloys. *Phys. B Condens. Matter* **2010**, *405*, 951–954. [CrossRef]
47. Song, Y.; Zhang, Y.F.; Pan, J.; Du, S. Ultra-low Young's modulus and high super-exchange interactions in monolayer CrN: A promising candidate for flexible spintronic applications. *Chin. Phys. B* **2021**, *30*, 047105. [CrossRef]
48. Ponce-Pérez, R.; Cocoletzi, G.H.; Takeuchi, N. Antiferromagnetic coupling in the initial stages of the MnN epitaxial growth on the CrN (001) surface. *Appl. Surf. Sci.* **2022**, *573*, 151451.
49. Feng, X.; Bao, K.; Huang, Y.; Ma, S.; Tao, Q.; Zhu, P.; Cui, T. Complete ligand reinforcing the structure of cubic-CrN. *J. Alloys Compd.* **2019**, *783*, 232–236. [CrossRef]
50. Liu, L.; He, Z.; Peng, J.; Guo, D.; Xu, Z.; Wang, C. Enhancing thermoelectric performance of CrN ceramics by optimizing sintering temperature. *J. Eur. Ceram. Soc.* **2024**, *44*, 7660–7667. [CrossRef]
51. Yuan, M.; Wan, X.; Meng, Q.; Lu, X.; Sun, L.; Wang, W.; Jiang, P.; Bao, X. Suppression of secondary phase in CrN matrix to boost the high-temperature thermoelectric performance. *Mater. Today Phys.* **2021**, *19*, 100420. [CrossRef]
52. Guo, W.; Liang, Z.; Tang, Y.; Cai, K.; Qiu, T.; Wu, Y.; Zhang, K.; Gao, S.; Zou, R. Understanding the lattice nitrogen stability and deactivation pathways of cubic CrN nanoparticles in the electrochemical nitrogen reduction reaction. *J. Mater. Chem. A* **2021**, *9*, 8568–8575. [CrossRef]
53. Das, B.; Reddy, M.; Rao, G.S.; Chowdari, B. Synthesis and Li-storage behavior of CrN nanoparticles. *RSC Adv.* **2012**, *2*, 9022–9028. [CrossRef]
54. Zhao, L.; Ding, B.; Qin, X.Y.; Wang, Z.; Lv, W.; He, Y.B.; Yang, Q.H.; Kang, F. Revisiting the roles of natural graphite in ongoing lithium-ion batteries. *Adv. Mater.* **2022**, *34*, 2106704. [CrossRef] [PubMed]
55. Shuang, Y.; Mori, S.; Yamamoto, T.; Hatayama, S.; Saito, Y.; Fons, P.J.; Song, Y.H.; Hong, J.P.; Ando, D.; Sutou, Y. Soret-Effect Induced Phase-Change in a Chromium Nitride Semiconductor Film. *ACS Nano* **2024**, *18*, 21135–21143. [CrossRef] [PubMed]
56. Cox, D. Neutron-diffraction determination of magnetic structures. *IEEE Trans. Magn.* **1972**, *8*, 161–182. [CrossRef]
57. Mrozińska, A.; Przystawa, J.; Sòlyom, J. First-order antiferromagnetic transition in CrN. *Phys. Rev. B* **1979**, *19*, 331. [CrossRef]
58. Filippetti, A.; Hill, N.A. Magnetic stress as a driving force of structural distortions: The case of CrN. *Phys. Rev. Lett.* **2000**, *85*, 5166. [CrossRef]
59. Gall, D.; Shin, C.; Haasch, R.; Petrov, I.; Greene, J. Band gap in epitaxial NaCl-structure CrN (001) layers. *J. Appl. Phys.* **2002**, *91*, 5882. [CrossRef]
60. Zhang, X.; Chawla, J.; Deng, R.; Gall, D. Epitaxial suppression of the metal-insulator transition in CrN. *Phys. Rev. B* **2011**, *84*, 073101. [CrossRef]

61. Biswas, B.; Chakraborty, S.; Joseph, A.; Acharya, S.; Pillai, A.I.K.; Narayana, C.; Bhatia, V.; Garbrecht, M.; Saha, B. Secondary phase limited metal-insulator phase transition in chromium nitride thin films. *Acta Mater.* **2022**, *227*, 117737. [CrossRef]
62. Inumaru, K.; Koyama, K.; Imo-Oka, N.; Yamanaka, S. Controlling the structural transition at the Néel point of CrN epitaxial thin films using epitaxial growth. *Phys. Rev. B* **2007**, *75*, 054416. [CrossRef]
63. Jin, Q.; Zhao, J.; Roldan, M.A.; Qi, W.; Lin, S.; Chen, S.; Hong, H.; Fan, Y.; Rong, D.; Guo, H.; et al. Anisotropic electronic phase transition in CrN epitaxial thin films. *Appl. Phys. Lett.* **2022**, *120*, 073103. [CrossRef]
64. Rivadulla, F.; Bañobre-López, M.; Quintela, C.X.; Piñeiro, A.; Pardo, V.; Baldomir, D.; López-Quintela, M.A.; Rivas, J.; Ramos, C.A.; Salva, H.; et al. Reduction of the bulk modulus at high pressure in CrN. *Nat. Mater.* **2009**, *8*, 947–951. [CrossRef]
65. Yan, M.; Zhou, X.; Cheng, H.; Wang, L.; Zhang, J.; Yu, X.; He, D.; Mei, J.W.; Zhao, Y.; Wang, S. Compressibility and thermoelasticity of CrN. *High Press. Res.* **2020**, *40*, 423–433. [CrossRef]
66. Alling, B.; Marten, T.; Abrikosov, I.A. Questionable collapse of the bulk modulus in CrN. *Nat. Mater.* **2010**, *9*, 283–284. [CrossRef]
67. Alam, K.; Ponce-Pérez, R.; Sun, K.; Foley, A.; Takeuchi, N.; Smith, A.R. Study of the structure, structural transition, interface model, and magnetic moments of CrN grown on MgO (001) by molecular beam epitaxy. *J. Vac. Sci. Technol. A* **2023**, *41*, 053411. [CrossRef]
68. Alam, K.; Disseler, S.M.; Ratcliff, W.D.; Borchers, J.A.; Ponce-Pérez, R.; Cocoletzi, G.H.; Takeuchi, N.; Foley, A.; Richard, A.; Ingram, D.C.; et al. Structural and magnetic phase transitions in chromium nitride thin films grown by rf nitrogen plasma molecular beam epitaxy. *Phys. Rev. B* **2017**, *96*, 104433. [CrossRef]
69. Alam, K.; Meng, K.Y.; Ponce-Pérez, R.; Cocoletzi, G.H.; Takeuchi, N.; Foley, A.; Yang, F.; Smith, A.R. Exchange bias and exchange spring effects in Fe/CrN bilayers. *J. Phys. D Appl. Phys.* **2020**, *53*, 125001. [CrossRef]
70. Constantin, C.; Haider, M.B.; Ingram, D.; Smith, A.R. Metal/semiconductor phase transition in chromium nitride (001) grown by rf-plasma-assisted molecular-beam epitaxy. *Appl. Phys. Lett.* **2004**, *85*, 6371–6373. [CrossRef]
71. Singh, A.; Tešanović, Z. Collective excitations in a doped antiferromagnet. *Phys. Rev. B* **1990**, *41*, 614. [CrossRef] [PubMed]
72. Darradi, R.; Derzhko, O.; Zinke, R.; Schulenburg, J.; Krüger, S.; Richter, J. Ground state phases of the spin-1/2 J 1–J 2 Heisenberg antiferromagnet on the square lattice: A high-order coupled cluster treatment. *Phys. Rev. B* **2008**, *78*, 214415. [CrossRef]
73. Völkel, A.; Mertens, F.; Bishop, A.; Wysin, G. Motion of vortex pairs in the ferromagnetic and antiferromagnetic anisotropic Heisenberg model. *Phys. Rev. B* **1991**, *43*, 5992. [CrossRef] [PubMed]
74. Zieschang, A.M.; Bocarsly, J.D.; Dürrschnabel, M.; Kleebe, H.J.; Seshadri, R.; Albert, B. Low-temperature synthesis and magnetostructural transition in antiferromagnetic, refractory nanoparticles: Chromium nitride, CrN. *Chem. Mater.* **2018**, *30*, 1610–1616. [CrossRef]
75. Nasr-Eddine, M.; Bertaut, E. Etude de la transition de premier ordre dans CrN. *Solid State Commun.* **1971**, *9*, 717–723. [CrossRef]
76. Browne, J.; Liddell, P.; Street, R.; Mills, T. An investigation of the antiferromagnetic transition of CrN. *Phys. Status Solidi A* **1970**, *1*, 715–723. [CrossRef]
77. Quintela, C.; Rivadulla, F.; Rivas, J. Thermoelectric properties of stoichiometric and hole-doped CrN. *Appl. Phys. Lett.* **2009**, *94*, 152103. [CrossRef]
78. Quintela, C.X.; Podkaminer, J.P.; Luckyanova, M.N.; Paudel, T.R.; Thies, E.L.; Hillsberry, D.A.; Tenne, D.A.; Tsymbal, E.Y.; Chen, G.; Eom, C.B.; et al. Epitaxial CrN thin films with high thermoelectric figure of merit. *Adv. Mater.* **2015**, *27*, 3032–3037. [CrossRef]
79. Wang, S.; Yu, X.; Zhang, J.; Wang, L.; Leinenweber, K.; He, D.; Zhao, Y. Synthesis, hardness, and electronic properties of stoichiometric VN and CrN. *Cryst. Growth Des.* **2016**, *16*, 351–358. [CrossRef]
80. Alam, K.; Haider, M.B.; Al-Kuhaili, M.F.; Ziq, K.A.; Haq, B.U. Electronic phase transition in CrN thin films grown by reactive RF magnetron sputtering. *Ceram. Int.* **2022**, *48*, 17352–17358. [CrossRef]
81. Ebad-Allah, J.; Kugelmann, B.; Rivadulla, F.; Kuntscher, C.A. Infrared study of the magnetostructural phase transition in correlated CrN. *Phys. Rev. B* **2016**, *94*, 195118. [CrossRef]
82. Herle, P.S.; Hegde, M.; Vasathacharya, N.; Philip, S.; Rao, M.R.; Sripathi, T. Synthesis of TiN, VN, and CrN from ammonolysis of TiS2, VS2, and Cr2S3. *J. Solid State Chem.* **1997**, *134*, 120–127. [CrossRef]
83. Herwadkar, A.; Lambrecht, W.R. Electronic structure of CrN: A borderline Mott insulator. *Phys. Rev. B* **2009**, *79*, 035125. [CrossRef]
84. Bhobe, P.; Chainani, A.; Taguchi, M.; Takeuchi, T.; Eguchi, R.; Matsunami, M.; Ishizaka, K.; Takata, Y.; Oura, M.; Senba, Y.; et al. Evidence for a correlated insulator to antiferromagnetic metal transition in CrN. *Phys. Rev. Lett.* **2010**, *104*, 236404. [CrossRef]
85. Himmetoglu, B.; Floris, A.; De Gironcoli, S.; Cococcioni, M. Hubbard-corrected DFT energy functionals: The LDA+ U description of correlated systems. *Int. J. Quantum Chem.* **2014**, *114*, 14–49. [CrossRef]
86. Jankovský, O.; Sedmidubský, D.; Huber, Š.; Šimek, P.; Sofer, Z. Synthesis, magnetic and transport properties of oxygen-free CrN ceramics. *J. Eur. Ceram. Soc.* **2014**, *34*, 4131–4136. [CrossRef]
87. Gui, Z.; Gu, C.; Cheng, H.; Zhu, J.; Yu, X.; Guo, E.J.; Wu, L.; Mei, J.; Sheng, J.; Zhang, J.; et al. Improper multiferroiclike transition in a metal. *Phys. Rev. B* **2022**, *105*, L180101. [CrossRef]
88. Singh, D.; Tamrakar, S.; Shrivas, K.; Dewangan, K. Nitridation of Cr–urea complex into nanocrystalline CrN and its antiferromagnetic magnetostructural transition study. *New J. Chem.* **2022**, *46*, 20879–20885. [CrossRef]
89. Lang, X.; Zheng, W.; Jiang, Q. Size and interface effects on ferromagnetic and antiferromagnetic transition temperatures. *Phys. Rev. B* **2006**, *73*, 224444. [CrossRef]
90. Fisher, M.E.; Barber, M.N. Scaling theory for finite-size effects in the critical region. *Phys. Rev. Lett.* **1972**, *28*, 1516. [CrossRef]

91. Jin, Q.; Cheng, H.; Wang, Z.; Zhang, Q.; Lin, S.; Roldan, M.A.; Zhao, J.; Wang, J.O.; Chen, S.; He, M.; et al. Strain-Mediated High Conductivity in Ultrathin Antiferromagnetic Metallic Nitrides. *Adv. Mater.* **2021**, *33*, 2005920. [CrossRef] [PubMed]
92. Wang, L.; Xu, W.; Zhou, X.; Gu, C.; Cheng, H.; Chen, J.; Wu, L.; Zhu, J.; Zhao, Y.; Guo, E.J.; et al. Anomalous finite-size effect on the magnetostructural transition in CrN. *Phys. Rev. B* **2023**, *107*, 174112. [CrossRef]
93. Alam, K.; Ponce-Pérez, R.; Sun, K.; Foley, A.; Takeuchi, N.; Smith, A.R. Investigating the magnetic and atomic interface configuration for a model Fe/CrN bilayer system. *J. Vac. Sci. Technol. A* **2021**, *39*, 063209. [CrossRef]
94. Filippetti, A.; Pickett, W.; Klein, B. Competition between magnetic and structural transitions in CrN. *Phys. Rev. B* **1999**, *59*, 7043. [CrossRef]
95. Wang, S.; Yu, X.; Zhang, J.; Chen, M.; Zhu, J.; Wang, L.; He, D.; Lin, Z.; Zhang, R.; Leinenweber, K.; et al. Experimental invalidation of phase-transition-induced elastic softening in CrN. *Phys. Rev. B* **2012**, *86*, 064111. [CrossRef]
96. Ponce-Pérez, R.; Alam, K.; Cocoletzi, G.H.; Takeuchi, N.; Smith, A.R. Structural, electronic, and magnetic properties of the CrN (001) surface: First-principles studies. *Appl. Surf. Sci.* **2018**, *454*, 350–357. [CrossRef]
97. Hernández, J.C.M.; Ponce-Pérez, R.; Cocoletzi, G.H.; Takeuchi, N.; Hoat, D. Tuning the half-metallicity in reconstructed CrN (111) surfaces. *Surf. Interfaces* **2022**, *35*, 102420. [CrossRef]
98. Chen, M.; Wang, S.; Zhang, J.; He, D.; Zhao, Y. Synthesis of Stoichiometric and Bulk CrN through a Solid-State Ion-Exchange Reaction. *Chem. Eur. J.* **2012**, *18*, 15459–15463. [CrossRef]
99. Rojas, T.; Ulloa, S.E. Strain fields and electronic structure of antiferromagnetic CrN. *Phys. Rev. B* **2017**, *96*, 125203. [CrossRef]
100. Tenelanda-Osorio, L.I.; Vélez, M.E. First principles study of the thermodynamic, mechanical and electronic properties of crystalline phases of Chromium Nitrides. *J. Phys. Chem. Solids* **2021**, *148*, 109692. [CrossRef]
101. Haq, B.U.; Alam, K.; Haider, M.B.; Alsharari, A.M.; Ullah, S.; Kim, S.H. Structural, electronic, magnetic, and optical properties of exfoliated chromium nitride ultrathin films. *Phys. E Low Dimens. Syst. Nanostruct.* **2023**, *150*, 115697. [CrossRef]
102. Ceperley, D.M.; Alder, B.J. Ground state of the electron gas by a stochastic method. *Phys. Rev. Lett.* **1980**, *45*, 566. [CrossRef]
103. Perdew, J.P.; Burke, K.; Ernzerhof, M. Generalized gradient approximation made simple. *Phys. Rev. Lett.* **1996**, *77*, 3865. [CrossRef] [PubMed]
104. Anisimov, V.I.; Zaanen, J.; Andersen, O.K. Band theory and Mott insulators: Hubbard U instead of Stoner I. *Phys. Rev. B* **1991**, *44*, 943. [CrossRef]
105. Dudarev, S.L.; Botton, G.A.; Savrasov, S.Y.; Humphreys, C.; Sutton, A.P. Electron-energy-loss spectra and the structural stability of nickel oxide: An LSDA+ U study. *Phys. Rev. B* **1998**, *57*, 1505. [CrossRef]
106. Alling, B.; Marten, T.; Abrikosov, I. Effect of magnetic disorder and strong electron correlations on the thermodynamics of CrN. *Phys. Rev. B* **2010**, *82*, 184430. [CrossRef]
107. Cheiwchanchamnangij, T.; Lambrecht, W.R. Quasiparticle self-consistent G W band structure of CrN. *Phys. Rev. B* **2020**, *101*, 085103. [CrossRef]
108. Botana, A.S.; Pardo, V.; Baldomir, D.; Blaha, P. Conducting states caused by a surface electric dipole in CrN (001) very thin films. *Phys. Rev. B* **2013**, *87*, 075114. [CrossRef]
109. Kamoru, W.S.; Haider, M.B.; Haq, B.U.; Aleithan, S.H.; Alsharari, A.M.; Ullah, S.; Alam, K. Structural, electronic, and optical properties of chromium oxynitride thin films grown by RF magnetron sputtering. *Res. Phys.* **2024**, *57*, 107387.

inorganics

MDPI

Article

Effect of Deposition Temperature and Thermal Annealing on the Properties of Sputtered NiO_x/Si Heterojunction Photodiodes

Roumen Nedev [1], David Mateos-Anzaldo [1,*], Eddue Osuna-Escalante [1], Oscar Perez-Landeros [1], Mario Curiel-Alvarez [1], Esteban Osorio-Urquizo [1], Jhonathan Castillo-Saenz [1], Javier Lopez-Medina [2], Benjamin Valdez-Salas [1] and Nicola Nedev [1,*]

1 Instituto de Ingeniería, Universidad Autónoma de Baja California, Blvd. Benito Juárez s/n, Mexicali C.P. 21280, Mexico; roumen.nedev@uabc.edu.mx (R.N.); eddue.osuna@uabc.edu.mx (E.O.-E.); operez55@uabc.edu.mx (O.P.-L.); mcuriel@uabc.edu.mx (M.C.-A.); esteban.osorio@uabc.edu.mx (E.O.-U.); jhonathan.saenz@uabc.edu.mx (J.C.-S.); benval@uabc.edu.mx (B.V.-S.)

2 Conahcyt Centro de Nanociencias y Nanotecnología, Universidad Nacional Autónoma de México, Ensenada C.P. 22800, Mexico; javierlo21@ens.cnyn.unam.mx

* Correspondence: dmateos@uabc.edu.mx (D.M.-A.); nicolan@uabc.edu.mx (N.N.)

Abstract: NiO_x is a p-type semiconductor with excellent stability, which makes it interesting for a wide range of applications. Broadband photodetectors with high responsivity (R) were fabricated by depositing r.f.-sputtered NiO_x layers on n-Si at room temperature (RT), 50 °C and 100 °C. In self-powered mode the RT diodes have R between 0.95 and 0.39 A/W for wavelengths between 365 and 635 nm, while at a reverse bias of −4 V, the responsivity increases to values between 22 A/W and 10.7 A/W for wavelengths in the same range. The increase of the deposition temperature leads to a decrease of R but also to a smaller reverse dark current. Thus, the 100 °C photodiodes might be more appropriate for applications where high responsivity is required, because of their smaller power consumption compared to the RT diodes. In addition, it was found that the increase of the deposition temperature leads to an increase of the diodes' series resistance and the resistivity of NiO_x. The effect of Rapid Thermal Annealing (RTA) on the properties of the photodiodes was studied. Annealing at 550 °C for 6 min leads to much higher responsivity compared to R of diodes with as-deposited NiO_x. However, a disadvantage of the annealed diode is that the reverse current depends on the amplitude and polarity of previously applied bias voltage. The higher responsivity of the RTA photodiodes makes them useful as light sensors.

Keywords: NiO_x-nSi broadband photodetectors; high responsivity; r.f. sputtering; effect of RTA treatment; light sensors

1. Introduction

NiO_x is a p-type semiconductor with excellent chemical stability and high transmittance in the visible part of the electromagnetic spectrum [1]. Its bandgap varies between 3.4 and 4.3 eV depending on the deposition technique and conditions. Due to its electrical and optical properties, NiO_x is considered an attractive material for various applications, such as photovoltaic devices [2], electrochromic devices [3], gas sensors [4] and ultraviolet and broadband photodetectors [5–8]. Various physical and chemical deposition techniques have been used to prepare NiO_x films. Among the most commonly used methods are sputtering [9], sol–gel spin-coating [10–12], electron beam evaporation [7] and chemical vapor deposition [13]. The structural, optical and electrical properties of the NiO_x layers depend strongly on the deposition technique [14–16]. The magnetron r.f. sputtering is widely used because of the following advantages: high film quality, low density of defects, high adhesion, low cost and high deposition rate. NiO_x films have been deposited by d.c. and r.f. sputtering. Metallic Ni in an Ar/O_2 gas atmosphere has been used in the d.c. sputtering process. The r.f. sputtering of an NiO target has been carried out in Ar/O_2 and pure Ar atmospheres [14,17–21]. Despite recent advances, some challenges remain regarding the optimization of the optical and electrical properties of r.f.-sputtered NiO_x layers, especially for application in photodetectors. Variations of process parameters, such as substrate temperature, deposition power, base and working pressures, affect the deposited film characteristics.

Wang et al. reported the effect of varying the deposition temperature during reactive sputtering with a fixed r.f. power of 110 W [17]. Jamal et al. studied the effect of the variation of the growth temperature in the range between room temperature (RT) and 400 °C on the properties of r.f.-sputtered NiO thin films, demonstrating that increasing the substrate temperature to 100 °C results in suitable properties for photovoltaic applications [18]. Ahmed et al. reported the effect of r.f. power used in the deposition and showed that the best structural, morphological and optical properties of NiO were obtained at sputtering power of 200 W [19]. Elmassi et al. investigated the variation of r.f. power on structural properties of NiO thin films obtained in an O_2/Ar atmosphere, finding a relationship between power increment and crystallite size [20]. Abdur et al. investigated the impact of open-air post-annealing on the optoelectrical and structural properties of NiO films, reporting that films deposited at 100 °C showed favorable crystalline quality with the highest observed carrier mobility [21].

However, the results for the effect of the deposition temperature and thermal annealing processes on the properties of broadband photodetectors are limited. In this work, we study the properties of NiO_x layers deposited by r.f sputtering from an NiO target at a fixed r.f. power of 60 W and temperatures between RT and 100 °C. In addition, the effect of post-deposition annealing of NiO_x films on the optical and electrical properties of the metal/NiO_x/nSi heterojunction diodes was studied.

2. Results

Figure 1a shows the experimental dependencies of the ellipsometric parameters Ψ and Δ, and the fitted curves using the Cauchy dispersion model for an NiO_x layer deposited at 100 °C and subjected to Rapid Thermal Annealing (RTA). Similar fitting results were obtained for the as-deposited layers prepared at RT, 50 °C and 100 °C, and for the RTA layers deposited at the other two temperatures, RT and 50 °C. Figure 1b,c show the refractive index of the as-deposited and annealed NiO_x films. From results in Figure 1b,c, it may be concluded that the RTA process does not lead to an important change of the optical constants. The layer thicknesses, deposition rate, the refractive index (n) and the extinction coefficient (k) at 632.8 nm of the as-deposited layers are shown in Table 1. The smaller thickness of the films deposited at 100 °C may be explained by assuming that during the deposition process at 100 °C, the substrate provides sufficient thermal energy to weakly bonded atoms to leave the surface. The obtained values for the optical constants n and k are close to ones reported in the literature for NiO_x [19].

Figure 2a shows transmission spectra of a corning glass (CG) substrate and CG coated with NiO_x layers deposited at RT, 50 °C and 100 °C. The transparency of CG was ≥80% in the range of 350–1000 nm. The transmittance of the samples with an NiO_x layer varies between ~60 and 80% in the visible—near IR range (400–1000 nm). Using the Tauc plot method, the band gaps E_g of the films were obtained for the three deposition temperatures, 3.79, 3.82 and 3.91 eV, respectively (Figure 2b). These results are in agreement with previously published values for E_g of NiO [7,22–24].

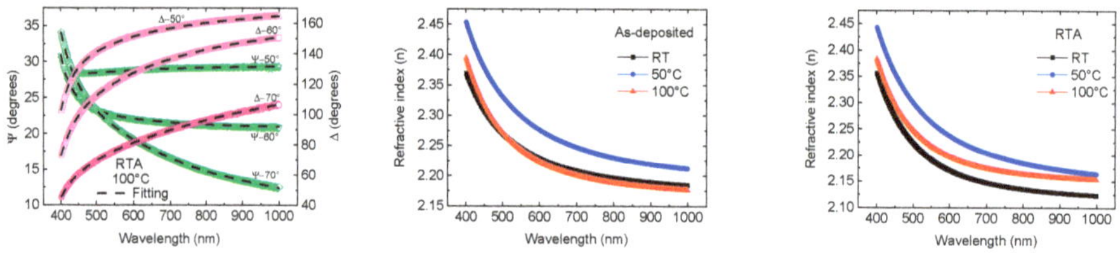

(a) (b) (c)

Figure 1. (**a**) Experimental Ψ and Δ dependences and the fitted curves for a layer deposited at 100 °C and subjected to RTA. Refractive index for: (**b**) As-deposited layers; (**c**) RTA annealed layers.

Table 1. Thickness, deposition rate and optical constants n and k at 632.8 nm for layers deposited at the three temperatures.

Temperature (°C)	Thickness (nm)	Deposition Rate (nm/min)	n	k
RT	41.5	6.9	2.25	0.03
50	48.9	8.2	2.29	0.044
100	36.0	6.0	2.25	0.01

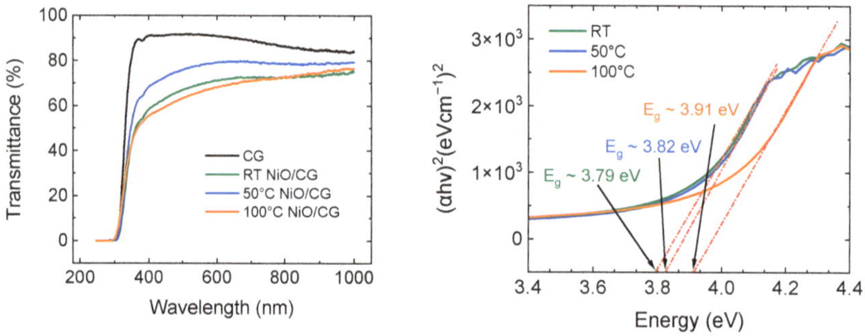

(a) (b)

Figure 2. (**a**) Transmittance of CG substrate and CG coated with NiO_x layers deposited at RT, 50 °C and 100 °C; (**b**) band gaps of the films determined by the Tauc plot method.

Atomic force microscopy (AFM) was used to characterize the surface morphology of as-deposited and annealed layers on an Si substrate (Figure 3). Smooth surfaces were obtained for films deposited at RT, 50 °C and 100 °C with root mean square (RMS) values of 0.6, 0.49 and 0.46 nm, respectively. The slight decrease of RMS with the temperature may be explained by assuming that the thermal energy provided by the substrate leads to some rearrangement of the surface atoms, which results in smoother surface. After the RTA, the values of RMS for layers deposited at RT, 50 °C and 100 °C are 0.72, 0.42 and 0.62 nm. Interestingly, the annealing process leads to a slight increase of the roughness of the RT and 100 °C films, but to a small decrease of RMS of the 50 °C layer.

A schematic diagram of the structure of the studied photodiodes and the experimental setup used in the electrical characterization is shown in Figure 4. The applied bias was defined relative to the back contact of the Si substrate; therefore, reverse bias means negative voltage applied on the top metal electrode.

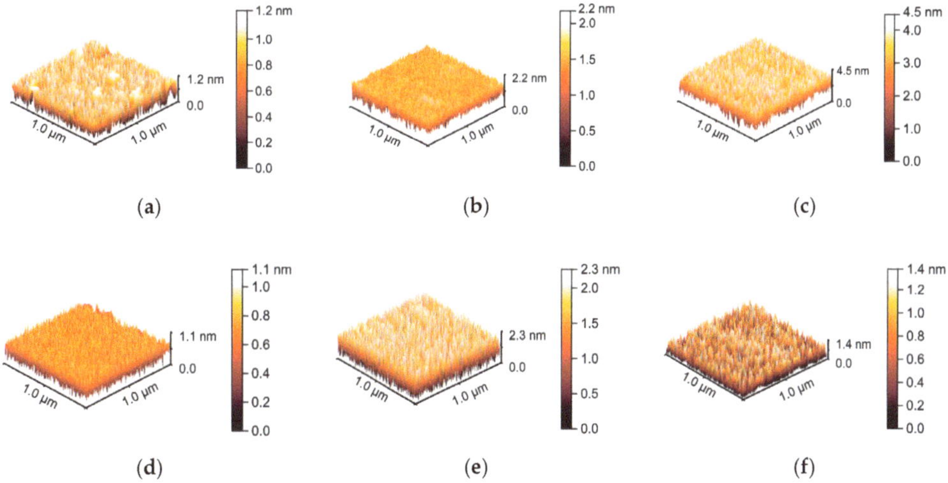

Figure 3. AFM results of as-deposited NiO_x layers obtained at: (**a**) RT, (**b**) 50 °C and (**c**) 100 °C and of RTA annealed layers deposited at: (**d**) RT, (**e**) 50 °C and (**f**) 100 °C.

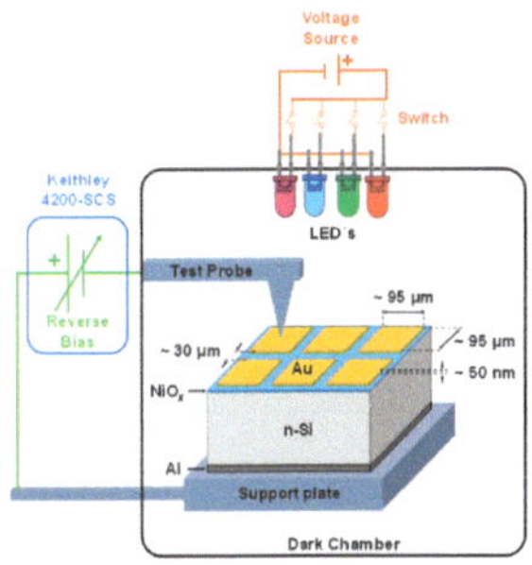

Figure 4. Schematic diagram of the structure of the studied devices with definition of the bias voltage polarity.

Current–voltage (*I-V*) characteristics of photodiodes with NiO_x layers deposited at RT, 50 °C and 100 °C were measured by ramping the bias voltage in both directions (Figure 5). No hysteresis was observed in these measurements as well as after various consecutive measurements, indicating low density of deep traps in the NiO_x layer and in the NiO_x/Si interfacial region. Typical rectifying dependences were measured with reverse saturation dark currents of ~3×10^{-7}, 7×10^{-7} and 1×10^{-7} *A* for the three deposition temperatures. Under forward bias, the current increases exponentially to voltages of ~1 V for the RT diode and to ~0.7 V for the other two diodes, and then the characteristics change their shape because of the diodes' series resistance. All diodes show photoresponses under reverse bias voltages (V_r < 0 V) and in self-powered mode (V = 0 V). The short-circuit currents (I_{SC}) under illumination with red, green, blue and UV light for the diodes with NiO_x deposited at the three temperatures are given in Table 2. The open-circuit voltage (V_{OC}) depends slightly on the light intensity and varies between 0.2 V and 0.35 V depending on the light intensity and the NiO_x deposition temperature (Figure 5a–c). The photoresponses of the diodes with the NiO_x layer deposited at 100 °C are higher than those of the diodes obtained at 50 °C and are close to those of the diodes obtained at room temperature. However, an advantage of the 100 °C photodiodes is their smaller reverse dark current compared to the RT diodes.

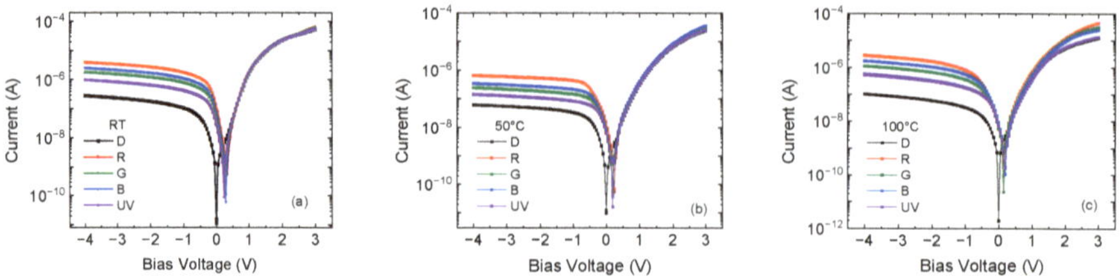

Figure 5. *I-V* characteristics of photodiodes with NiO_x layers deposited at: (**a**) RT, (**b**) 50 °C and (**c**) 100 °C measured in the dark and under illumination with red, green, blue and UV light.

Table 2. Short-circuit currents under illumination with red, green, blue and UV light for the diodes with NiO_x deposited at the three temperatures.

Light Color	I_{SC} (A)		
	RT	50 °C	100 °C
Red	1.30×10^{-7}	1.50×10^{-8}	7.85×10^{-9}
Green	5.70×10^{-8}	6.70×10^{-9}	9.15×10^{-9}
Blue	7.97×10^{-8}	8.50×10^{-9}	5.25×10^{-9}
UV	2.87×10^{-8}	6.40×10^{-9}	6.85×10^{-9}

Figure 6 shows a schematic band diagram of a p-NiO_x/n-Si photodiode under short-circuit conditions. Under illumination with each of the four light-emitting diodes (LEDs) used in this study, the light passes through the NiO_x layer since the photons' energy is smaller than the NiO_x bandgap. Then, the photons are absorbed in Si, generating electron–hole pairs. The pairs generated in the depletion region and within a hole diffusion length from the transition region are separated by the internal electric field, resulting in I_{sc} flowing through the diode. At reverse bias, the space charge region and the electric field at the heterojunction increase, leading to an increase of the photocurrent (Figure 5).

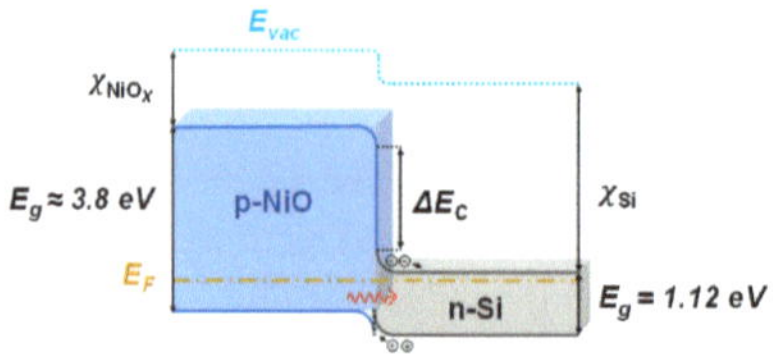

Figure 6. Schematic energy band diagram of a p-NiO_x/n-Si diode under short-circuit conditions, $V = 0$ V.

I-V characteristics of diodes with NiO_x layers deposited at the three temperatures and subjected to the RTA process are shown in Figure 7. The thermal annealing leads to several important changes in the *I-V* dependences. Thus, the dark current and the photocurrents depend on the direction in which the bias voltage changes, leading to hysteresis at both the forward and reverse bias ramps. For forward biases, the current measured with bias increasing from 0 V towards positive voltages is smaller than the current measured in the opposite direction, most clearly seen in Figure 7b. The hysteresis is more pronounced under reverse biases and increases with the NiO_x deposition temperature. The reverse bias voltages at which the current reaches saturation are close to or greater than −10 V (Figure 7a–c). The photocurrents at these large negative voltages are greater than the photocurrents of the diodes with as-deposited NiO_x layers. The highest values were obtained for the 50 °C, RTA diode (Figure 7b). However, the *I-V* hysteresis of the annealed photodiodes can be a disadvantage in their practical application. The inset in Figure 7b shows a zoomed view of the current in the region of −0.5–0.5 V of a characteristic that was measured between 5 V and −10 V and then in the opposite direction. It can be seen that when ramping the bias from positive voltages towards 0 V, the characteristic crosses the *x*-axis at a positive voltage of 0.3 V, indicating increase of the negative charge trapped in the structure. In contrast, when the bias voltage varies from −10 V towards 0 V, the characteristic crosses the *x*-axis at −0.3 V, indicating increase of the positive trapped charge. Therefore, the observed hysteresis is more likely due to change of the charge state of shallow traps at the NiO_x/n-Si heterojunction interface and in the interfacial region with the bias voltage. As a result, no open-circuit voltage and short-circuit current were measured. Since no shift was observed of the *I-V* curves of diodes with as-deposited NiO_x layers, it may be concluded that the RTA process has created defects, which can exchange carriers with Si, and as a result change their charge state.

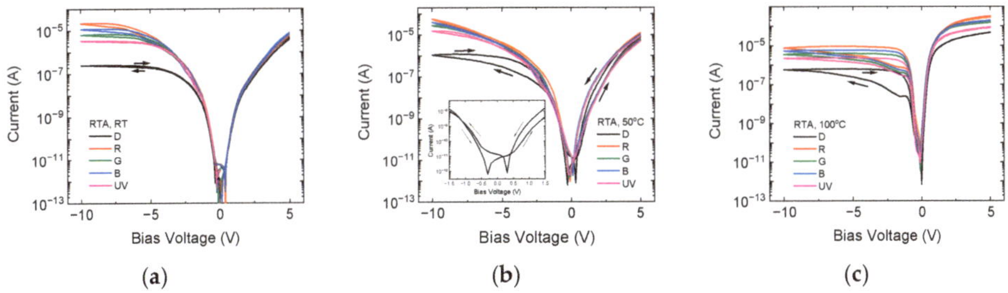

Figure 7. *I-V* characteristics of photodiodes with NiO_x layers deposited at: (**a**) RT, (**b**) 50 °C and (**c**) 100 °C and subjected to RTA annealing. The dependences were measured in the dark and under illumination with red, green, blue and UV light. The inset in (**b**) shows a zoomed view of the current in the region of −0.5–0.5 V of a characteristic measured between 5 V and −10 V and then in the opposite direction.

Capacitance–voltage (*C-V*) measurements were carried out in order to study in more detail the effect of consecutive bias voltage ramps on the charge state of the diodes. The results in Figure 8a show that ramping the voltage in both directions does not lead to hysteresis in the *C-V* curves of diodes with NiO_x layers deposited at the three temperatures, indicating that no charge was trapped during these measurements. In contrast, the *C-V* characteristics of the diodes subjected to RTA when measured under reverse bias sweep from 0 V to −10 V and back to 0 V show a displacement of the *C-V* curves, corresponding to increase of the trapped positive charge in the diodes (Figure 8b). In accordance with the *I-V* measurements, the diodes with NiO_x layers deposited at 50 °C and 100 °C show higher hysteresis in dark.

Time-dependent photoresponses of all photodiodes upon red, green, blue and UV light illumination was measured. The on/off switching curves (Figure 9) are characterized by the rise (τ_r) and fall (τ_f) times, defined as the time needed for the photocurrent to increase from 10% to 90% of its maximum value or to drop from 90 to 10%, respectively. The τ_r and τ_f values for the diodes with as-deposited layers are less than 0.1 s (Figure 9a–c). As could be expected from the *I-V* results obtained at reverse bias ramps, the RTA process can lead to an increase of the switching times. Thus, the τ_r values for the annealed samples deposited at RT, 50 °C and 100 °C are 9.5, 1.7 and 8.6 s, respectively. However, the τ_f times for the annealed diodes are <0.1 s. Another important observation is that at the voltage pulses used o $V_r = -8$ V, the RTA diodes with an NiO_x layer deposited at 50 °C reach faster saturation of the photocurrent (Figure 9e) compared to the other two diodes (Figure 9d,f). As it was mentioned above, the variation of the current at applied bias under light illumination could be related to change of the charge state of shallow traps at the NiO_x/Si interface and in the interfacial region.

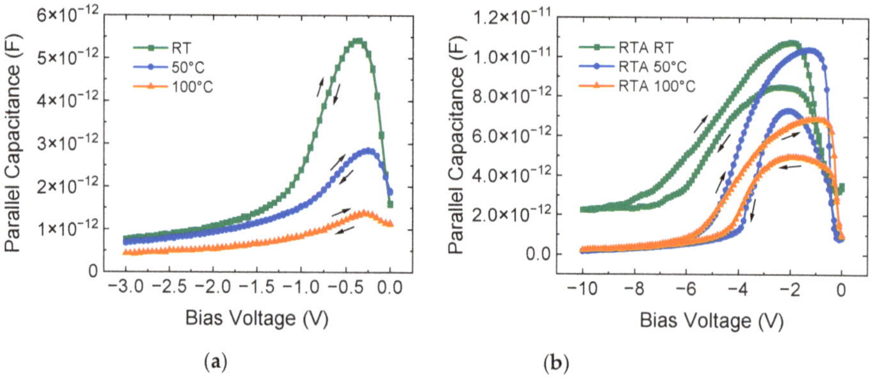

Figure 8. *C-V* characteristics measured in the dark under reverse bias of: (**a**) diodes with as-deposited NiO_x layers; (**b**) diodes with RTA annealed layers.

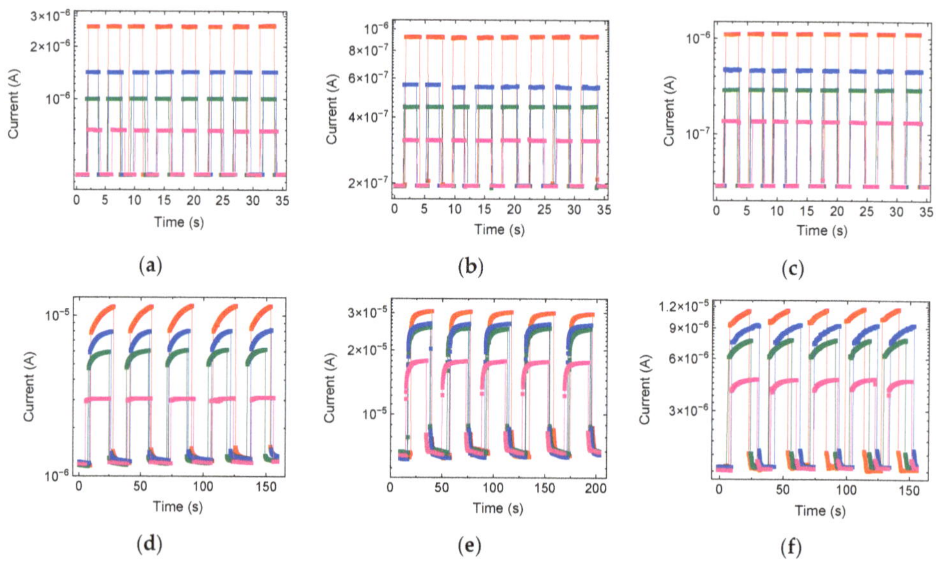

Figure 9. Time-dependent photoresponses of diodes with as-deposited NiO_x layers obtained at: (**a**) RT, (**b**) 50 °C, and (**c**) 100 °C, and of diodes with RTA annealed layers deposited at: (**d**) RT, (**e**) 50 °C, and (**f**) 100 °C. The switching curves of the diodes with as-deposited layers were measured at voltage pulses of −3 V, while the RTA diodes were measured at pulses of V_r = −8 V. The red, green, blue and violet color lines correspond to illumination with red, green, blue and UV LED, respectively.

3. Discussion

The series resistance (R_s) reduces the voltage applied on the heterojunction and, thus, affects the *I-V* characteristic. R_s can be evaluated from the slope of the I_f/g vs. I_f characteristic [25], where I_f is the forward current in the region after the exponential increase of I and g is the conductance dI_f/dV_f. In this region R_s leads to change of the shape of the *I-V* curve. The dependence of I/g on I may be derived by differentiating the *I-V* equation

$$I = I_0\left(e^{\frac{q(V-IR_S)}{nkT}} - 1\right),$$

where n is the ideality factor, which gives

$$\frac{dI}{dV} = I\frac{q}{nkT}\left(1 - R_s\frac{dI}{dV}\right),$$

or

$$g = I\frac{q}{nkT}(1 - R_s g),$$

By simple algebraic transformations, a linear dependence of I/g on the forward current I is obtained

$$\frac{I}{g} = \frac{nkT}{q} + R_s I$$

Figure 10a displays the I/g vs. I dependencies for no annealed diodes with NiO_x layers deposited at RT, 50 °C and 100 °C. The slope of the straight lines increases with the deposition temperature, indicating that R_s also increases. Since the resistivity of the n-type region of the heterojunction, n-Si, is constant, it may be concluded that the increase of the deposition temperature leads to an increase of the resistivity of the NiO_x layers. A similar effect of increased resistivity was observed at a constant deposition temperature with an increase of the r.f. power [26,27].

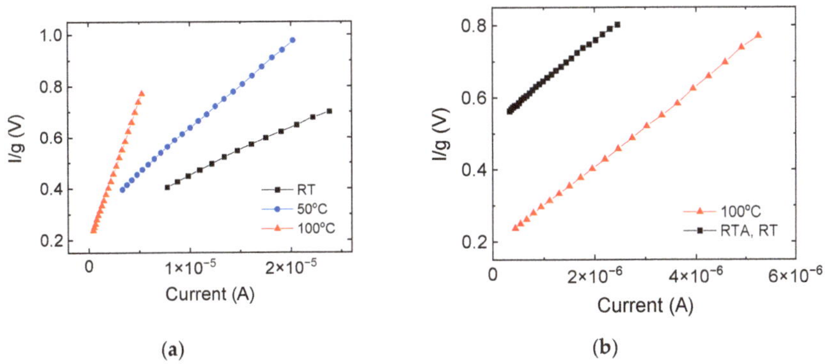

Figure 10. (**a**) I/g vs. I dependencies for diodes with as-deposited NiO_x layers prepared at RT, 50 °C and 100 °C. Higher line slope corresponds to higher R_s; (**b**) comparison of the I/g vs. I dependences for a diode with as-deposited NiO_x prepared at 100 °C (red line) and a RTA annealed diode with NiO_x layer deposited at RT (black line). Note the difference in x-axis scale.

The dependence of the forward current on the voltage sweep direction makes it impossible to evaluate the series resistances of the diodes after RTA. An approximate estimation was obtained for the RTA diode with an NiO_x layer deposited at RT (Figure 7a) by using the I-V characteristic measured from 0 V towards positive voltages. The obtained result is compared with the I/g vs. I dependence for a no annealed diode with an NiO_x layer deposited at 100 °C, the one with the highest R_s (Figure 10b). The line slopes in Figure 10b indicate that the RTA process leads to comparable or even higher values of R_s than those of the 100 °C diodes.

The responsivity (R) of the fabricated photodiodes was calculated from [28]

$$R = \frac{I_{ph}}{P_\lambda \cdot A}$$

where $I_{ph} = I_{light} - I_{dark}$ is the photocurrent, P_λ is the optical power intensity at a specific wavelength λ and A is the area of the top contact. Here, I_{light} and I_{dark} are the currents measured under light illumination and in dark conditions. High responsivity was determined for the no annealed photodiodes. The results are shown in Figure 11a and in Table 3. The responsivity of the diodes with an NiO_x layer deposited at RT is higher than R of the other diodes in self-powered mode, as well as at reverse bias of −4 V. For applications where a self-biased mode of operation is required, the RT diodes are the most appropriate. However, for some applications where higher responsivity is required, the 100 °C photodiodes may be the better option because of their smaller dark current at −4 V and R, which is close to that of the RT diode. The responsivity of the diodes subjected to RTA under $V_r = -10$ V is shown in Figure 11b. Very high values for R were determined under bias voltage of −10 V, much higher than the ones reported in the literature for broadband photodetectors. While the R of the diode RTA annealed at 50 °C is the highest one, this diode can be less appropriate because of its higher dark current and lack of saturation of the photocurrent, even at reverse bias voltage of −10 V. Although the RTA annealed diodes may not be appropriate for applications where precise measurement of the optical power is required, they may be useful as light sensors because of their high responsivity.

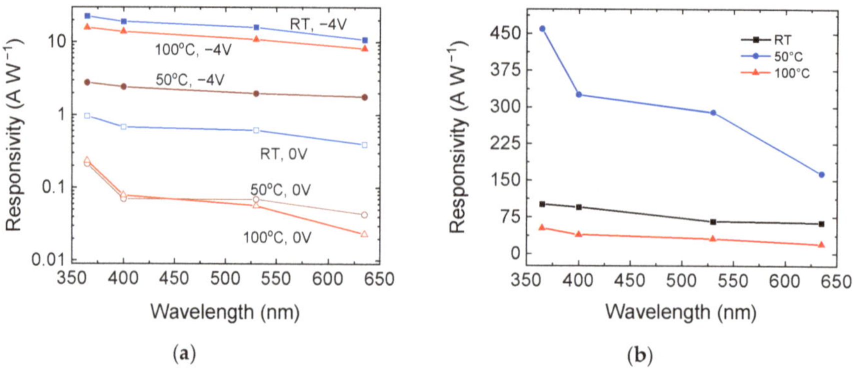

Figure 11. Responsivity of photodiodes with: (**a**) as-deposited NiO_x layers; (**b**) RTA annealed NiO_x layers.

Table 3. Responsivity of photodiodes with as-deposited NiO_x layers.

Wavelength (nm)	*R* (A/W), Self-Powered Mode			*R* (A/W), $V_r = -4$ V		
	RT	**50 °C**	**100 °C**	**RT**	**50 °C**	**100 °C**
635	0.39	0.045	0.023	10.7	1.77	8.1
530	0.61	0.072	0.056	15.8	1.96	10.8
400	0.67	0.072	0.077	18.8	2.4	13.7
365	0.95	0.21	0.23	22.4	2.74	15.6

The obtained results for R at 0 V and at −4 V for photodiodes with as-deposited NiO_x layers are compared with results reported in the literature in Table 4. The determined responsivity of the photodiodes deposited at RT and 100 °C are among the highest reported and are comparable with results published by other authors. Ma et. al. reported a responsivity of 15.8 A/W at −5 V, which is slightly higher than our result for 100 °C diodes, however the maximum bias voltage used here was −4 V.

Table 4. Comparison of NiO-based photodiode parameters reported in literature. P_{RF} stands for r.f. power.

Device	*R* (A/W)	*D* (Jones)	I_d (*A*)	On/Off Ratio	τ_r, τ_f (s)	Year [Ref.]
p-NiO/nSi	4.8×10^{-3} (at 318 nm)	-	-	-	-	2015 [29]
p-NiO/nSi	1.73, 1.81, 2.07 (at 5 V, 365, 625, 850 nm)	-	-	31, 35, 41 (at 5 V, 365, 625, 850 nm)	$3.49/7.68 \times 10^{-3}$	2022 [30]
NiO/TiO_2	0.074 (at 0 V, 380 nm)	-	10^{-9}	922	$33.6/92.6 \times 10^{-3}$	2022 [31]
p-NiO/n-Si	0.95, 0.67, 0.61, 0.39 (at 0 V, 365, 400, 530, 635 nm, P_{RF} = 60 *W*) 22.4, 18.8, 15.8, 10.7 (at −4 V, 365, 400, 530, 635 nm)	5.5×10^{10} (at 0 V, 365 nm)	8.5×10^{-12} (at 0 V) 2.8×10^{-7} (at −4 V)	3.4×10^{3} (at 0 V, 365 nm, P_{RF} = 60 W)	<0.1	2023 [27]
p-NiO/n-GaN	3.11×10^{-3} (at 0 V)	8.69×10^{9}	-	-	$2/47 \times 10^{-3}$	2024 [32]
NiO/Ga_2O_3	5.08×10^{-3} (at −10 V, 254 nm)	3.19×10^{11}	2.76×10^{-12}	802.53	$62/67 \times 10^{-3}$	2024 [9]
NiO/Al_2O_3/n-Si	15.8 (at −5 V, 365 nm)	1.14×10^{14}	0.279×10^{-6}	-	$0.08/0.184 \times 10^{-3}$	2024 [33]
p-NiO/n-Si (100 °C)	15.6 (at −4 V, 365 nm) 0.23 (at 0 V, 365 nm)	8.41×10^{10} (at −4 V, 365 nm) 2.875×10^{11} (at 0 V, 365 nm)	1.02×10^{-7} 1.9×10^{-12}	68.23×10^{-3} 3.66×10^{-3}	<0.1 -	This work

Another relevant parameter is the specific detectivity (D^*), which can be defined as the sensibility of a photodiode to detect weak light signals. D^* involves the R at a given wavelength of the excitation light, the elementary charge (q) and the dark current density (J_D) and is given by

$$D^* = \frac{R_\lambda}{\sqrt{2qJ_D}}$$

The values obtained for the detectivity are close to the best values reported in the literature for NiO_x-based photodetectors (Table 4).

4. Materials and Methods

NiO_x thin films were deposited on n-type (100) Si wafers with resistivity of 4–10 Ω·cm by r.f. magnetron sputtering at 25 °C (RT), 50 °C and 100 °C. The deposition time was 6 min for the three temperatures. Prior to deposition, the substrates were cleaned in an ultrasonic bath for 5 min with acetone, isopropyl alcohol and deionized water and dried with high-purity nitrogen. NiO target (99.9%) with dimensions of 50.8 × 3.175 mm at a working distance of 7 cm was used for deposition. The sputtering gas was ultra-high-purity (UHP) argon (99.9999%). The chamber base pressure was 1×10^{-5} Torr, the working pressure was 5 mTorr and the Ar flow was 18.9 sccm. Before the deposition, the target was pre-sputtered for 20 min. To study the effect of thermal annealing, half of the samples were subjected to the RTA process in an UHP N_2 atmosphere. The annealing was performed in a horizontal tube furnace (Zhengzhou CY Scientific Instrument Co., Zhengzhou, China) equipped with a sliding rail. The samples were placed in a quartz tube under constant N_2 flow of 100 mL/min in a position outside of the furnace. The furnace was heated at a rate of 10 °C/min to the desired temperature and when it was reached, the furnace was moved to the position of the samples. The samples were annealed at 550 °C for 6 min. Following the annealing process, the furnace was returned to its initial position and the samples were allowed to cool to room temperature under the same N_2 environment to ensure controlled conditions. For electrical characterization, semitransparent Au electrodes (~50 nm of thickness) with areas of 95 × 95 μm^2 spaced by 30 μm were thermally evaporated through a shadow mask as top contacts. Aluminum with a thickness of ~300 nm was used as back contact.

The thickness and the optical constants, refractive index and extinction coefficient, were determined by ellipsometric measurements using a Variable Angle Spectroscopic Ellipsometer J.A. Woollam M—2000U, (J.A. Woollam Co., Lincoln, NE, USA). The measurements were carried out at three angles of incidence in the range of 50–70°. The experimental results for the ellipsometric angles Ψ and Δ were fitted in the 400–1000 nm range using the Cauchy dispersion model. The surface morphologies of NiO_x layers deposited on Si were characterized using AFM instrument NX10 (Park Systems Co., Suwon, Republic of Korea) Park Systems, Republic of Korea. The AFM was equipped with an anti-acoustic box and an active vibration isolation table to prevent noises and vibrations that can affect the measurements. The measurements were carried out with a PPP-NCHR probe tip (Park Systems Co., Suwon, Republic of Korea) with a force constant k = 42 N/m and a resonance frequency of 330 kHz using the non-contact mode. The scanned area was 1 × 1 μm^2. Electrical characterization was carried out using a Keithley 4200-SCS Semiconductor Characterization System, (Tektroniks Inc., Beaverton,OR, USA). *I-V* dependencies were measured in the dark and under illumination with red, green, blue and UV light-emitting diodes (LEDs). The LEDs' optical power intensities are given in Table 5. *C-V* characteristics were measured in the dark at a frequency of 100 kHz using a test signal with an amplitude of 30 mV. The bias sweep delay time and hold time for the *I-V* and *C-V* measurements were 0.5 s and 1 s, respectively, while the voltage steps used were 50 or 100 mV.

Table 5. Optical power of the light-emitting diodes used in this study [27].

Color	Photon Wavelength/Energy (nm/eV)	Optical Power Intensity (mW/cm^2)
Red	635/1.95	3.71
Green	530/2.34	1.03
Blue	400/3.10	1.31
UV	365/3.40	0.335

5. Conclusions

Photodetectors with high responsivity in self-powered mode and at a reverse bias of −4 V were fabricated by depositing r.f.-sputtered NiO_x layers on n-Si at RT, 50 °C and 100 °C. The increase of the deposition temperature leads to a decrease of R at $V = 0$ V and $V_r = -4$ V, but at the same time to a smaller dark current at reverse biases. In addition, higher deposition temperature leads to an increase of the diodes' series resistance and of the NiO_x resistivity. The diodes fabricated after RTA annealing show much higher R compared to those with as-deposited layers. A disadvantage of these diodes is the dependence of their dark current and photocurrents on the direction of the bias voltage change. As a result, the annealed diodes may be inappropriate for applications where exact determination of the optical power is required, but because of the very high responsivity, they might be useful as light sensors.

Author Contributions: Conceptualization, N.N., D.M.-A. and R.N.; methodology, D.M.-A., R.N., B.V.-S. and J.L.-M.; software, R.N., E.O.-E. and D.M.-A.; validation, D.M.-A. and R.N.; formal analysis, N.N.; investigation, E.O.-E. and E.O.-U.; resources, B.V.-S., N.N., M.C.-A. and D.M.-A.; data curation, R.N. and E.O.-E.; writing—original draft preparation, R.N., E.O.-E. and D.M.-A.; writing—review and editing, O.P.-L., M.C.-A., J.C.-S., B.V.-S. and N.N.; visualization, N.N., D.M.-A. and R.N.; supervision,

B.V.-S.; project administration, N.N. and M.C.-A. All authors have read and agreed to the published version of the manuscript.

Funding: This research received no external funding

Data Availability Statement: The data presented in this study are available on request from the corresponding author.

Acknowledgments: E. Osuna and E. Osorio, gratefully acknowledges the National Council for Science and Technology (CONACYT), Mexico for a M. Eng. and Ph.D. fellowships. We acknowledge the help of Ivan Cardoza-Navarro and Jorge Salvador-Carlos in sample preparation and analysis.

Conflicts of Interest: The authors declare no conflicts of interest.

References

1. Napari, M.; Huq, T.N.; Hoye, R.L.Z.; MacManus-Driscoll, J.L. Nickel oxide thin films grown by chemical deposition techniques: Potential and challenges in next-generation rigid and flexible device applications. *InfoMat* **2021**, *3*, 536–576. [CrossRef]
2. Aivalioti, C.; Manidakis, E.G.; Pelekanos, N.T.; Androulidaki, M.; Tsagaraki, K.; Viskadourakis, Z.; Spanakis, E.; Aperathitis, E. Niobium-doped NiO as p-type nanostructured layer for transparent photovoltaics. *Thin Solid Films* **2023**, *778*, 139910. [CrossRef]
3. Wang, F.; Zhang, W.; Jia, J.; Chen, Y.; Chen, Z.; Wang, Z.; Zhang, L.; Ma, H. Improvement of electrochromic properties of NiO film doped with ZnO prepared by magnetron sputtering. *J. Mater. Sci. Mater. Electron.* **2024**, *35*, 455. [CrossRef]
4. Li, C.; Choi, P.G.; Kim, K.; Masuda, Y. High performance acetone gas sensor based on ultrathin porous NiO nanosheet. *Sens. Actuators B Chem.* **2022**, *367*, 132143. [CrossRef]
5. Peng, Y.; Jiang, D.; Zhao, M.; Duan, Y.; Wei, H.; Li, H.; Liang, Q.; Wang, S. High-performance UV–visible photodetectors based on ZnO/perovskite heterostructures. *J. Alloys Compd.* **2023**, *965*, 171372. [CrossRef]
6. Chen, L.; Wang, Y.; Zhang, J.; Chen, L.; Zhai, J.; Song, J. Inorganic Semiconductor-Based Flexible UV Photodetector Arrays Achieved by Specific Flip-Chip Bonding. *ACS Appl. Mater. Interfaces* **2024**, *16*, 51089–51096. [CrossRef]
7. Castillo-Saenz, J.R.; Nedev, N.; Valdez-Salas, B.; Bernechea, M.; Martínez-Guerra, E.; Mendivil-Palma, I.; Curiel-Alvarez, M.; Mateos, D.; Perez-Landeros, O. Effect of oxidation temperature on the properties of NiO_x layers for application in optical sensors. *Thin Solid Films* **2021**, *734*, 138849. [CrossRef]
8. Ahmed, A.A.; Hashim, M.R.; Qahtan, T.F.; Rashid, M. Preparation and characteristics study of self-powered and fast response *p*-NiO/n-Si heterojunction photodetector. *Ceram. Int.* **2022**, *48*, 20078–20089. [CrossRef]
9. Wang, J.; Li, Q.; Mi, W.; Wang, D.; Xu, M.; Xiao, L.; Zhang, X.; Luan, C.; Zhao, J. Fast response self-powered solar-blind UV photodetector based on NiO/Ga_2O_3 p-n junction. *Mater. Sci. Semicond. Process.* **2025**, *186*, 109084. [CrossRef]
10. Hwang, J.-D.; Cheng, Y.-A. Enhancing the Performance of NiO-Based Metal-Semiconductor-Metal Ultraviolet Photodetectors Using a MgO Capping Layer. *IEEE Sens. J.* **2021**, *21*, 27400–27404. [CrossRef]
11. Caglar, M.; Sever, K.; Aktas, S.; Demiroglu, A. Improving the electrical performance of NiO based photodiode fabricated by sol-gel process with Al doping. *Sens. Actuators A Phys.* **2023**, *350*, 114099. [CrossRef]
12. Pandit, B.; Parida, B.; Jang, H.-S.; Heo, K. Self-Powered Broadband Photodetector Based on NiO/Si Heterojunction Incorporating Graphene Transparent Conducting Layer. *Nanomaterials* **2024**, *14*, 551. [CrossRef] [PubMed]
13. Wilson, R.L.; Macdonald, T.J.; Lin, C.-T.; Xu, S.; Taylor, A.; Knapp, C.E.; Guldin, S.; McLachlan, M.A.; Carmalt, C.J.; Blackman, C.S. Chemical vapour deposition (CVD) of nickel oxide using the novel nickel dialkylaminoalkoxide precursor $[Ni(dmamp')_2]$ (dmamp′ = 2-dimethylamino-2-methyl-1-propanolate). *RSC Adv.* **2021**, *11*, 22199–22205. [CrossRef] [PubMed]
14. Aydin, E.; Troughton, J.; Bastiani, M.D.; Ugur, E.; Sajjad, M.; Alzahrani, A.; Neophytou, M.; Schwingenschlögl, U.; Laquai, F.; Baran, D.; et al. Room-Temperature-Sputtered Nanocrystalline Nickel Oxide as Hole Transport Layer for p–i–n Perovskite Solar Cells. *ACS Appl. Energy Mater.* **2018**, *1*, 6227–6233. [CrossRef]
15. Meng, X.; Gao, X.; Du, Y. Influence of O_2/Ar flow ratio on the structure, optical and electrical properties of DC reactively magnetron-sputtered face-centered cubic NiO films at room temperature. *Phys. Rev. B Condens. Matter* **2020**, *579*, 411877. [CrossRef]
16. Timoshnev, S.; Kazakin, A.; Shubina, K.; Andreeva, V.; Fedorenko, E.; Koroleva, A.; Zhizhin, E.; Koval, O.; Kurinnaya, A.; Shalin, A.; et al. Annealing Temperature Effect on the Physical Properties of NiO Thin Films Grown by DC Magnetron Sputtering. *Adv. Mater. Interfaces* **2024**, *11*, 2300815. [CrossRef]
17. Wang, H.; Zhao, Y.; Li, X.; Zhen, Z.; Li, H.; Wang, J.; Tang, M. Effect of Sputtering Temperature on Structure and Optical Properties of NiO Films Fabricated by Magnetron Sputtering. *J. Electron. Mater.* **2017**, *46*, 4052–4056. [CrossRef]
18. Jamal, M.S.; Shahahmadi, S.A.; Chelvanathan, P.; Alharbi, H.F.; Karim, M.R.; Dar, M.A.; Luqman, M.; Alharthi, N.H.; Al-Harthi, Y.S.; Aminuzzaman, M.; et al. Effects of growth temperature on the photovoltaic properties of RF sputtered undoped NiO thin films. *Results Phys.* **2019**, *14*, 102360. [CrossRef]

19. Ahmed, A.A.; Hashim, M.R.; Rashid, M. Effect of RF power on structural, morphological and optical properties of NiO thin films. In Proceedings of the AIP Conference, Kelantan, Malaysia, 6 February 2019.
20. Elmassi, S.; Narjis, A.; Nkhaili, L.; Elkissani, E.; Amiri, L.; Drissi, S.; Abali, A.; Bousseta, M.; Outzourhit, A. Effect of annealing on structural, optical and electrical properties of nickel oxide thin films synthesized by the reactive radio frequency sputtering. *Phys. B Condens. Matter* **2022**, *639*, 413980. [CrossRef]
21. Abdur, R.; Choudhury, S.; Bashar, M.S.; Hossain, M.R.; Quddus, M.S.; Akhtar, U.S.; Shaikh, M.A.A.; Hossain, M. Enhancing perovskite solar cell performance: Investigating the impact of post-annealing on the optoelectrical and structural properties of RF-sputtered NiO films via SCAPS-1D device modeling. *Sol. Energy* **2024**, *271*, 112443. [CrossRef]
22. Koushik, D.; Jošt, M.; Dučinskas, A.; Burgess, C.; Zardetto, V.; Weijtens, C.; Verheijen, M.A.; Kessels, W.M.M.; Albrecht, S.; Creatore, M. Plasma-assisted atomic layer deposition of nickel oxide as hole transport layer for hybrid perovskite solar cells. *J. Mater. Chem. C* **2019**, *7*, 12532–12543. [CrossRef]
23. Abzieher, T.; Moghadamzadeh, S.; Schackmar, F.; Eggers, H.; Sutterlüti, F.; Farooq, A.; Kojda, D.; Habicht, K.; Schmager, R.; Mertens, A.; et al. Electronbeam-evaporated nickel oxide hole transport layers for perovskite-based photovoltaics. *Adv. Energy Mater.* **2019**, *9*, 1802995. [CrossRef]
24. Hasan, M.R.; Xie, T.; Barron, S.C.; Liu, G.; Nguyen, N.V.; Motayed, A.; Rao, M.V.; Debnath, R. Self-powered p-NiO/n-ZnO heterojunction ultraviolet photodetectors fabricated on plastic substrates. *APL Mater.* **2015**, *3*, 10610. [CrossRef] [PubMed]
25. Schroder, D.K. *Semiconductor Material and Device Characterization*, 3rd ed.; John Wiley & Sons: Hoboken, NJ, USA, 2006; p. 187.
26. Grilli, M.L.; Aydogan, S.; Yilmaz, M. A study on non-stoichiometric p-NiO_x/n-Si heterojunction diode fabricated by RF sputtering: Determination of diode parameters. *Superlattices Microstruct.* **2016**, *100*, 924–933. [CrossRef]
27. Mateos-Anzaldo, D.; Nedev, R.; Perez-Landeros, O.; Curiel-Alvarez, M.; Castillo-Saenz, J.; Arias-Leon, A.; Valdez-Salas, B.; Silva-Vidaurri, L.; Martinez-Guerra, E.; Osorio-Urquizo, E.; et al. High-performance broadband photodetectors based on sputtered NiO_x/n-Si heterojunction diodes. *Opt. Mater.* **2023**, *145*, 114422. [CrossRef]
28. Chen, X.; Ren, F.; Gu, S.; Ye, J. Review of gallium-oxide-based solar-blind ultraviolet photodetectors. *Photonics Res.* **2019**, *7*, 381–415. [CrossRef]
29. Hammadi, O.A.; Khalaf, M.K.; Kadhim, F.J. Fabrication of UV Photodetector from Nickel Oxide Nanoparticles Deposited on Silicon Substrate by Closed-Field Unbalanced Dual Magnetron Sputtering Techniques. *Opt. Quantum Electron.* **2015**, *47*, 3805–3813. [CrossRef]
30. Ahmed, A.A.; Hashim, M.R.; Qahtan, T.F.; Rashid, M. Multi—Wavelength Photodetectors Based on Porous Spin-Coated and Compact RF-Sputtered NiO Films Grown over Si Substrate: Effect of Surface Morphology. *Optik* **2022**, *255*, 168694. [CrossRef]
31. Xu, J.; Cao, R.; Shi, S.; Li, L.; Zhu, K.; Su, Y. Self-Powered Ultraviolet Photodetectors Based on Match like Quasi One-Dimensional N-TiO_2/P-NiO Core-Shell Heterojunction Arrays with NiO Layer Sputtered at Different Power. *J. Alloys Compd.* **2022**, *928*, 167126. [CrossRef]
32. Enns, Y.; Timoshnev, S.; Kazakin, A.; Shubina, K.; Uvarov, A.; Vorobyev, A.; Nikitina, E.; Mizerov, A.; Andreeva, V.; Fedorenko, E.; et al. Characterization of the P-NiO/N-GaN Heterojunction and Development of Ultraviolet Photodiode. *Mater. Sci. Semicond. Process.* **2024**, *181*, 108624. [CrossRef]
33. Ma, X.; Tang, L.; Jia, M.; Zhang, Y.; Zuo, W.; Cai, Y.; Li, R.; Yang, L.; Teng, K.S. Ultrahigh Performance UV Photodetector by Inserting an Al_2O_3 Nanolayer in NiO/N-Si. *Adv. Electron. Mater.* **2024**, *10*, 2300909. [CrossRef]

inorganics

MDPI

Review

Recent Advances in Aluminum Nitride (AlN) Growth by Magnetron Sputtering Techniques and Its Applications

Nabeel Ahmad Khan Jadoon, Vaigunthan Puvanenthiram, Mayada Ahmed Hassan Mosa, Ashutosh Sharma and Kaiying Wang *

Department of Microsystems, University of South-Eastern Norway, 3184 Horten, Norway; nabeeljadoon21@gmail.com (N.A.K.J.); vaigunthan.puvanenthiram@gmail.com (V.P.); mayada9357@gmail.com (M.A.H.M.); ashutosh.iiitk@gmail.com (A.S.)

* Correspondence: kaiying.wang@usn.no

Abstract: This review explores the processes involved in enhancing AlN film quality through various magnetron sputtering techniques, crucial for optimizing performance and expanding their application scope. It presents recent advancements in growing AlN thin films via magnetron sputtering, elucidating the mechanisms of AlN growth and navigating the complexities of thin-film fabrication. Emphasis is placed on different sputtering methods such as DC, RF, pulsed DC, and high-power impulse DC, highlighting how tailored sputtering conditions enhance film characteristics in each method. Additionally, the review discusses recent research findings showcasing the dynamic potential of these techniques. The practical applications of AlN thin films, including wave resonators, energy harvesting devices, and thermal management solutions, are outlined, demonstrating their relevance in addressing real-world engineering challenges.

Keywords: aluminium nitride (AlN); magnetron sputtering; thin-film growth; sputtering techniques; application-dependent growth

1. Introduction

Aluminum nitride (AlN)-based thin films have gained significant attention due to their unique physicochemical characteristics [1], including high piezoelectricity [2–4], substantial surface acoustic velocity [5–8], electromechanical coupling [9], chemical stability, and broad transparency across various spectral ranges [10–12]. In addition to AlN, other materials such as GaN and InN are significant within the III-nitride family. GaN is particularly well-known for its excellent thermal stability, and ability to emit light across both the visible and ultraviolet spectrum [13]. While GaN is typically grown using techniques like molecular beam epitaxy (MBE) or metal–organic chemical vapor deposition (MOCVD), sputtering presents a more cost-effective alternative, making it an attractive option for certain applications [14]. Furqan et al. have demonstrated that the thin-film growth of GaN with magnetron sputtering creates growth along the (002) plane, as well as demonstrating the effects of growth parameters on the quality of films [15].

Another material, like indium nitride (InN), characterized by Faiza et al., showed excellent growth in the c-plane and related to film hexagonal material properties like electrical resistivity, which is decreased when increasing the power in the process of DC magnetron sputtering [16]. Zinc oxide (ZnO), a hexagonal material known for its high Seebeck coefficient and transparent thermoelectric performance, is suitable for thermoelectric and optoelectronic applications [17]. Among various deposition methods available for synthesizing ZnO films, magnetron sputtering stands out for its ability to maintain precise control over growth conditions, leading to the facilitation of the production of c-axis-oriented ZnO films with reproducible and controllable properties [18]. Although several options can be explored, the focus of this review is to explore AlN for magnetron sputtering.

Achieving the optimal performance of AlN films heavily depends on the methods used for their deposition. The deposition techniques play a critical role in producing AlN films with the desired quality and preferred orientations, which directly impact their effectiveness and functionality in specific technological contexts. The deposition methods include molecular beam epitaxy [19,20], magnetron sputtering [21], and chemical vapor deposition [22,23]. Among these techniques, magnetron sputtering stands out due to its superior parameter control, low-temperature operation, and compatibility with semiconductor technology. For example, studies [24–26] have shown that magnetron sputtering allows for precise control over film thickness and composition, which is crucial for applications requiring high precision.

The utilization of magnetron sputtering has showcased significant progress in maximizing the yield and quality of AlN films. For instance, Barth [24] highlighted the potential of scandium doping in AlN films by pulse magnetron sputtering to enhance energy harvesting capabilities, while Jiang [27] proposed a deposition using middle-frequency magnetron sputtering yielding highly c-axis-oriented AlN films on titanium alloy substrate, particularly suited for piezoelectric MEMS and SAW devices [25]. Additionally, Uchiyama [28] achieved uniform and smooth AlN films on sapphire substrates by RF magnetron sputtering, and Ke et al. [29] enhanced the deposition rate of highly c-axis-oriented AlN films using unbalanced magnetron sputtering. This method involves the use of an asymmetrical magnetic field configuration to increase plasma density near the target, resulting in denser grains and reduced surface roughness. Unlike conventional magnetron sputtering, unbalanced magnetron sputtering enhances ion bombardment and increases plasma density at the substrate surface, thereby resulting in improved film properties such as higher density and better crystalline quality, which are critical for high-performance electronic and optical devices.

While magnetron sputtering is acknowledged as an effective technique for AlN growth, it presents several challenges, such as predicting thin-film structures amidst fluctuating sputter parameters [20]. Additionally, the meticulous management of crystallinity, thickness consistency, and stress levels is required to meet the demands of high-volume manufacturing. The formation of the transparent AlN phase is susceptible to variations in reactant gases like nitrogen (N_2) and the flow rates during sputtering procedures [26]. Achieving superior quality AlN films necessitates adherence to specific processing prerequisites, such as an optimized nitrogen and argon ratio (N_2/Ar) and reduced pressure during growth [30]. Furthermore, depositing AlN onto diverse substrates can induce divergent stress states within the film, influencing its inherent properties [31]. These challenges underscore the importance of precise control and optimization in the magnetron sputtering processes for the successful fabrication of high-quality AlN films. One significant challenge is the incorporation of oxygen in vacuum systems, which can adversely affect the crystallinity and properties of the films. Magnetron sputtering, particularly RF magnetron sputtering, mitigates oxygen incorporation by enabling precise control over sputtering parameters, thus ensuring the growth of high-quality crystalline AlN films, as demonstrated by studies such as [32] using radio frequency (RF) magnetron sputtering. Through meticulous parameter control—including adjusting the N_2/Ar gas flow ratio, sputtering power, substrate temperature, and deposition pressure—researchers achieved high-quality films with optimal stoichiometry and minimal oxygen content. Control was attained by precisely regulating gas flow rates, maintaining substrate temperatures within an optimal range of 200–500 °C, and fine-tuning sputtering power and deposition pressure. This comprehensive approach underscores the effectiveness of magnetron sputtering in addressing oxygen incorporation challenges and ensuring the fabrication of crystalline AlN films with the desired properties.

Apart from the need for optimizing operational parameters in magnetron sputtering, achieving desired orientation growth in the (100), (110), or (001) planes is critically dependent on the specific methods used within the magnetron sputtering domain. Investigating parameters related to the type of magnetron sputtering method is essential, and a combined discussion of these methods for AlN growth has not been explored in previous reviews. Therefore, this review aims to provide a comprehensive understanding of the sequential growth of AlN films in light of recent advances in magnetron sputtering. Additionally, we examine how different magnetron sputtering methods influence AlN film fabrication and application, focusing on sputtering conditions and parameter controls.

Although several parameters are essential for the control of the thin-film growth of AlN, some superior parameter control in magnetron sputtering involves the precise adjustment of several key deposition parameters, such as gas flow ratios (N_2/Ar), sputtering power, substrate temperature, deposition pressure, and ion bombardment control through magnetic field adjustments. These parameters are crucial for optimizing aluminum nitride (AlN) film quality. For instance, the N_2/Ar ratio influences the film's stoichiometry and properties, while sputtering power affects the energy of the sputtered atoms and the deposition rate. Maintaining an optimal substrate temperature, typically between 200–500 °C, enhances atom mobility and promotes desirable crystalline structures. Additionally, controlling the deposition pressure allows for better management of the mean free path of the sputtered particles, essential for achieving uniform film growth. Finally, regulating ion bombardment through magnetic field configuration significantly impacts film density and crystalline quality. Together, these factors facilitate the fabrication of high-quality AlN films with tailored properties for various applications.

The paper is structured as follows: The section on AlN growth describes the growth mechanisms, focusing on the desired crystallographic growth of aluminum nitride (AlN) and offering a comprehensive overview of the process. This foundational understanding is essential as it sets the stage for the subsequent sections. Given that the growth mechanism necessitates various fabrication methods through magnetron sputtering, the Fabrication section delves into these requirements, detailing the system architecture involved in magnetron chambers. It emphasizes the significance of different sputtering techniques, thoroughly analyzing their contributions and ultimately recommending the most suitable methods for specific applications within the magnetron sputtering domain. This section also discusses the doping mechanisms for AlN with other materials, highlighting how these modifications can enhance the material's properties. Furthermore, the final section introduces the applications of AlN-based products. The schematic depiction in Figure 1 encapsulates the sequential progression inherent in developing and deploying AlN-based sensors, commencing from the foundational magnetron sputtering process and extending to their varied applications across diverse fields.

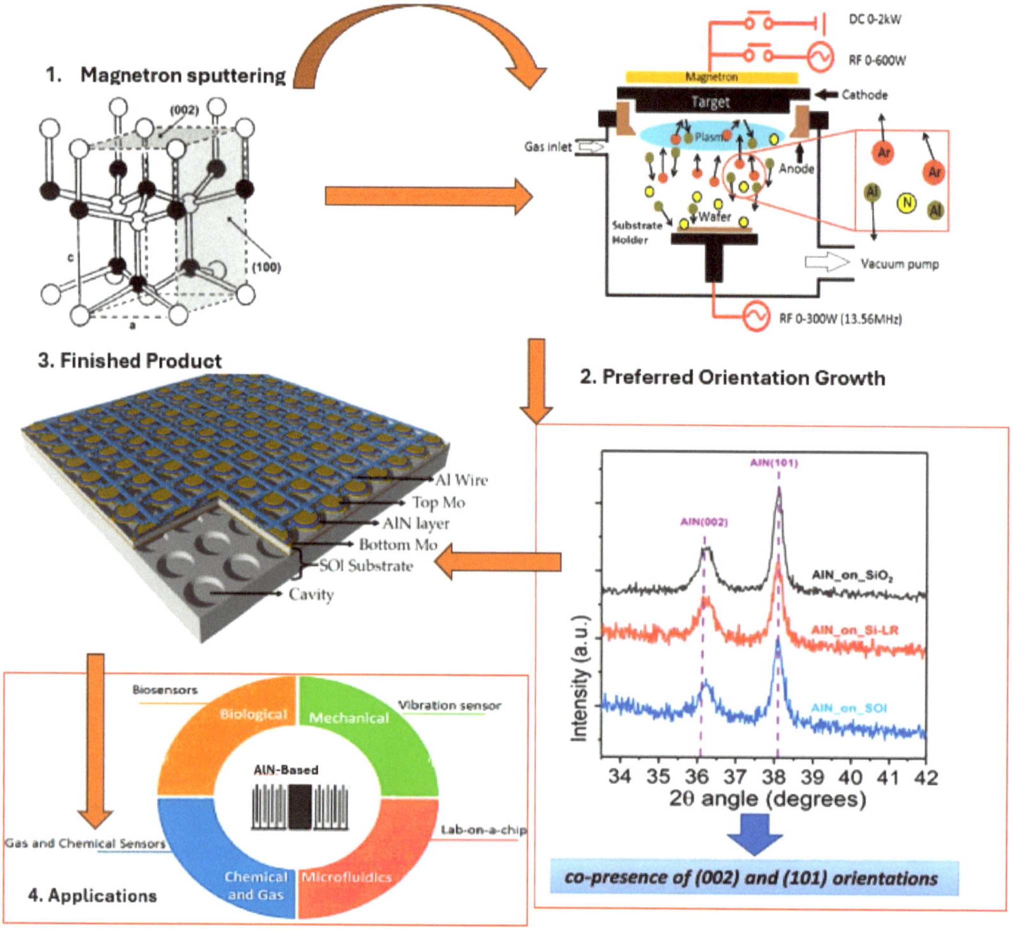

Figure 1. Summary of ALN-based thin-film fabrication process via magnetron sputtering. (1) Magnetron sputtering setup for AlN deposition, depicting plasma generation and target sputtering. The crystallographic planes (002) and (100) represent different orientations within the AlN crystal lattice. (2) X-ray diffraction (XRD) results showing preferred orientation growth, including peaks at AlN(002) and AlN(101), indicating the coexistence of these orientations on various substrates (SOI, Si-LR, SiO_2). (3) Schematic of the fabricated device, comprising an SOI substrate, AlN layer, Mo electrodes (top and bottom), and an underlying cavity. (4) Overview of application areas for AlN-based devices, including biological sensors, mechanical sensors, chemical/gas sensors, microfluidics, and lab-on-a-chip technologies Images reproduced with permission [33–36].

2. AlN Growth and Mechanism

2.1. AlN Crystal Structure and Growth Mechanism

The orientation of aluminium nitride (AlN) growth on a substrate is determined by which plane of the AlN unit cell becomes parallel to the substrate surface. The covalent bonds between aluminium (Al) and nitrogen (N) atoms form a triangular prism base, resulting in different lattice parameter values. An illustration of the lattice structure with these parameter values is depicted in Figure 2b, while the AlN unit cell, forming a hexagonal wurtzite structure, is shown in Figure 2a. Detailed studies have shown that substrate temperature and sputtering power significantly affect this orientation, highlighting the importance of precise control over these parameters [1].

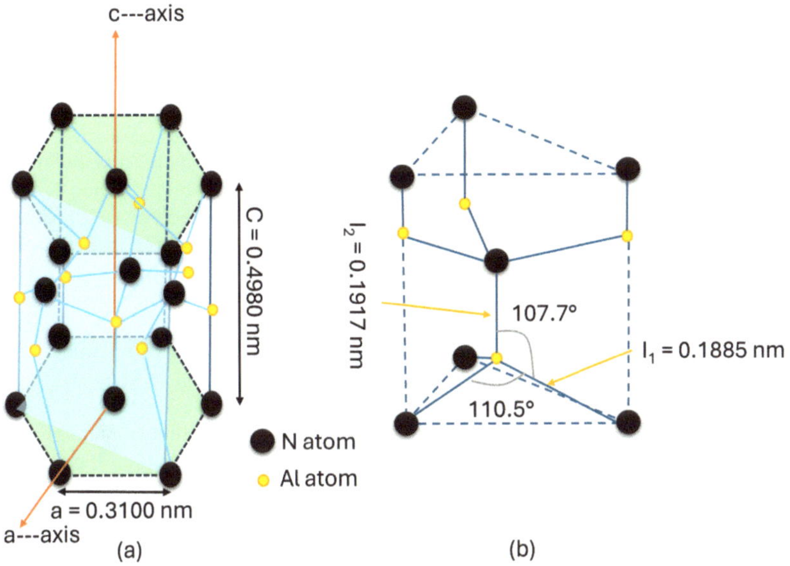

Figure 2. (**a**) illustrates the hexagonal lattice structure and the a, c axes, which are parallel to planes (001) and (100), respectively, and (**b**) illustrates the covalent bond angle and length between Al, and N atoms and the tetrahedron structures [1].

The potential crystal orientations for AlN film growth primarily include the AlN(100) and AlN(001) planes. The preferred orientation is determined by the ratio of energy absorbed by the two vibrational phonon modes. The c-axis lies in the (100) plane, while the a-axis is parallel to the (001) plane. Detailed studies have shown that when the target-to-substrate distance is significantly larger than the mean free path, the AlN(100) orientation is favored due to lower energy atoms. In contrast, high-energy nitrogen and aluminium species, with a considerable mean free path relative to the target–substrate distance, favor the AlN(001) orientation. The mean free path during AlN growth is influenced by the working pressure, affecting the c-axis orientation of the deposited films. At higher working pressures, the mean free path of the sputtered species is reduced, resulting in lower-energy atoms reaching the substrate and favoring the AlN(100) orientation. Conversely, at lower working pressures, the mean free path increases, allowing higher-energy species to reach the substrate, supporting the AlN(001) orientation [37]. The key parameters influencing these orientations include RF power, working pressure, substrate temperature, and substrate-to-target distance, which control the energy of these species. The precise adjustment of these parameters allows for tailored film properties [38].

The deposition of a thin film on a substrate is an iterative process involving several stages. Initially, atoms reach the substrate's surface, where they are adsorbed and diffuse across it. Concurrently, these atoms collide and coalesce with atomic clusters, leading to nucleation. This nucleation progresses to form larger atomic clusters, eventually evolving into substantial islands. Through successive iterations of this process, these islands coalesce to form continuous layers. The accumulation of these layers results in the formation of a thin film [1]. Understanding each stage of this process is critical for optimizing film properties, particularly through the control of deposition parameters like substrate temperature and sputtering power.

2.2. Effect of Operational Parameters and Development Techniques on Grown AlN Film Properties

The operational parameters of film deposition/growth technique, such as working pressure, gas flow rate, energy supplied (RF power, temperature), as well as factors like substrate orientation and seed materials, influence the properties of the film developed and are widely studied. Detailed investigations by [39] have shown that optimizing these parameters can enhance film quality, such as by increasing crystalline orientation and reducing surface roughness.The orientation of the substrate and the choice of seed material determine the nanostructure formation. Chirumamilla et al. conducted a study investigating the normal and glancing angle deposition of AlN using reactive magnetron sputtering on silicon (Si) substrates, employing various seed materials. Their research aimed to characterize and compare the resulting nanostructures, and the study revealed that employing silver (Ag) as a seed material led to the formation of longer nanostructures and a higher deposition rate compared to gold (Au) and aluminium (Al). The use of Ag as a seed material significantly influenced the process temperature by enhancing the local energy transfer during sputtering. This led to an increase in adatom mobility, facilitating higher deposition rates and affecting the overall thermal dynamics of the sputtering process. The investigation concluded that the observed nanostructures primarily consisted of preferential c-axis lamellae growth, highlighting the potential for engineering nanostructure growth via reactive magnetron sputtering [39]. The utilization of glancing angle deposition (GLAD) for AlN holds promise, particularly in shaping the thin-film morphology for diverse applications in optics, gas sensing, and biocompatibility [37]. Bairagi et al. conducted a similar comparative study examining the growth of columnar AlN films using GLAD configuration and regular deposition, while maintaining consistent parameters such as pressure and temperature. Their investigation revealed the dependency of the AlN growth rate on working pressure and flux direction. Notably, a transition in nanocolumn growth direction from 2° to 38° was observed with decreasing working pressure, alongside an abrupt change in c-axis tilt from 0° to 6° with increasing pressure from 3 to 5 mTorr [37,40]. The importance of these findings underscores the role of working pressure and flux direction in determining the film's final structure and properties.

Sandager [41] investigates the impact of processing parameters on AlN thin-film growth, identifying the optimal conditions conducive to high-quality film production by studying the film structure and stress for a DC magnetron sputtering with the set of configuration of magnetron power, working pressure and N_2/Ar ratio, and found that the c-axis orientation and tensile stress prevalent in all configurations with the increasing trend with magnetron power. This study underscore the characteristics AlN film exhibits with a DC magnetron sputtering technique, which helps to further advance the AlN development technique [41].

The effect of working pressure on the optical properties of AlN thin films has also been studied. Alyousef et al. conducted a characterization of the optical properties of AlN thin films developed via DC reactive magnetron sputtering (DCMS). Their investigation covered the variation in working pressure for the films produced with a substrate bias voltage of −50 V. The study revealed that optical transparency of AlN increased with the decrease in working pressure from 6 to 1 mTorr for a thickness of 200 nm. Additionally, a decrease in the refractive index of AlN from 1.92 to 1.65 was observed as working pressure increased [34]. These findings highlight the importance of optimizing working pressure to enhance the optical properties of AlN thin films for potential applications [42].

Riah et al. explored the hetero-epitaxial growth of AlN on silicon substrates and compared it with AlN deposited on Si(100) and Si(111) planes without buffer layers. Given the criticality of AlN film quality in various applications and the imperative to enable efficient epitaxial growth on silicon substrates for mass production, this experiment holds significant relevance. The study found that the use of an AlN buffer layer enhanced the crystallinity, resulting in superior columnar crystal formation and mitigated in-film stress. These findings were observed in hetero-epitaxial AlN deposited atop a buffer layer utilizing DCMS [26].

The concentration percentage of nitrogen emerges as a critical determinant of AlN surface morphology. Surface roughness affects the crystal structure and lattice parameters of deposited AlN films, influencing dislocation density, grain size, and microstrain of the films [43]. Chen et al. found that an increase in nitrogen concentration up to 70% led to improved surface roughness, attributed to reduced argon ion bombardment resulting in lower-energy aluminum atoms facilitating gap atom growth on sapphire substrates. However, beyond this threshold, a decline in surface roughness was observed due to the poor surface mobility of aluminum atoms caused by excessive nitrogen flow, which resulted in less energy for the effective lateral growth of the film grains. Additionally, the study underscored the influence of deposition rate on film roughness. When the deposition rate is slow, grains have more time to grow laterally, leading to smoother surfaces. Higher deposition rates, conversely, provide less time for this lateral growth, resulting in increased roughness [44].

The deposition rate significantly influences the growth dynamics of AlN films. As reported by Sandager et al., higher deposition rates can lead to increased surface roughness due to insufficient time for lateral grain growth, which is essential for achieving a smooth film surface. Conversely, lower deposition rates allow for better lateral growth and reduced roughness, enhancing the overall crystalline quality of the films [41,45]. Moreover, the manipulation of nitrogen flow during deposition plays a crucial role in modulating both the composition and crystallization behavior of AlN films, further impacting their morphological characteristics. This relationship between deposition rate and film morphology is critical for optimizing AlN films for various applications such as optoelectronics and high-temperature devices [46,47].

The structural and electrical properties of AlN play pivotal roles in its application in surface acoustic wave (SAW) devices and energy harvesting technologies. Sanjeeva et al. investigated these properties on Si (111) substrates concerning sputtering power. Their research demonstrated that increased sputtering power correlated with reduced leakage current, attributed to diminished grain boundaries and larger crystal sizes resulting from higher power (>100 W) settings. Moreover, higher power and faster deposition rates were associated with reduced compressive stress in the films, potentially due to the rearrangement of adatoms and momentum transfer from high-energy particles, as well as the dynamics of vacancies and interstitials [48].

These studies collectively deepen our understanding of the intricate relationship between deposition parameters and the resulting properties of AlN thin films, offering valuable insights for optimizing their performance across various applications and authors have summarized the key points in Table 1.

Table 1. Key points and parameters in AlN growth and mechanism.

Key Points/Parameters	Description
Crystal Orientations	AlN film growth primarily along the AlN(100) and AlN(001) planes determined by vibrational phonon modes.
Operational Parameters	RF power, pressure, temperature, and target distance influence crystal orientation.
Deposition Process	Adsorption, diffusion, nucleation, and coalescence form thin-film layers.
Processing Parameters	Key for high-quality films, optimized via magnetron sputtering.
Nanostructure Growth	Reactive magnetron sputtering enables tailored nanostructure growth.
Substrate Influence	Substrate material and morphology impact film properties.

In the next section, different categories are explored in the sputtering domain for the deposition of aluminum nitride (AlN) thin films, which offer precise control over film properties and deposition conditions. Researchers employ various methods such as DC magnetron reactive sputtering, radio frequency (RF) magnetron sputtering, and pulsed DC magnetron sputtering to achieve the desired film characteristics.

3. Fabrication

3.1. Magnetron Sputtering General Mechanism

Magnetron sputtering is a technique used to deposit thin films by utilizing a gaseous plasma. Plasma, commonly known as the fourth state of matter, differs fundamentally from solids, liquids, and gases. It consists of an ionized gas, often referred to as cold plasma in the context of magnetron sputtering, as the ions and electrons are not in thermal equilibrium and the gas temperature is relatively low compared to thermal plasmas [49]. The presence of these charged particles allows plasma to conduct electricity and respond to magnetic fields, making it essential for processes like magnetron sputtering. The main difference between magnetron sputtering and other sputtering techniques is the use of magnets and magnetic fields, introduced by Penning in 1936 [50], to trap electrons in the plasma near the cathode and extend their lifetime. This increases the probability of collisions between these electrons and argon (Ar) gas atoms, ionizing them into Ar^+. The magnetic field thus helps maintain a high-density plasma necessary for efficient sputtering.

A simple schematic is shown in Figure 3, representing the main components of a magnetron sputtering system, and illustrating the dynamics of ions and electrons. The diagram highlights the vacuum chamber, with a target material (cathode) and an anode, it depicts the spiral movement of electrons and the direct paths of the ions towards the cathode.

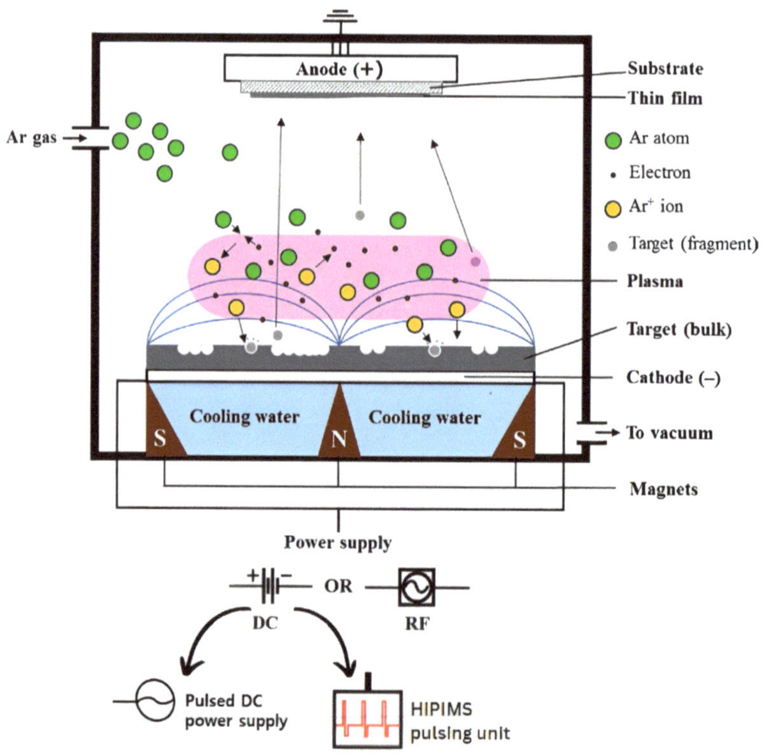

Figure 3. Schematic of the magnetron sputtering process and the main system components for RF, DC, Pulsed DC, and HiPIM DC magnetron sputtering, recreated from [51].

In a standard magnetron sputtering process, the chamber is initially evacuated to achieve a high vacuum, which helps to significantly reduce the partial pressures of any residual gases and contaminants [52]. After achieving the base pressure, argon gas is introduced into the chamber, and the total pressure is regulated. Plasma generation is initiated with the application of a high voltage between the cathode, which is usually positioned directly behind the sputtering target, and the anode, which is often connected to the chamber as the electrical ground. The presence of plasma ensures that there are enough high-energy ions to bombard the target material, as collisions of electrons with nearby argon atoms knock off electrons from these argon atoms, ionizing them into positive argon ions (Ar^+). These Ar^+ ions are then directed towards the negatively charged cathode, which leads to high-energy collisions with the target surface. Each collision has the potential to eject atoms from the target's surface, propelling them into the vacuum environment. Knocked-off atoms traverse the vacuum and settle onto the substrate, creating a thin film [53].

3.1.1. Types of Magnet Configurations and Arrangements

Magnetron sputtering systems utilize various magnet configurations to optimize the deposition process for different applications. The primary configurations include:

1. Planar Magnetron Sputtering: In this configuration, magnets are arranged behind a flat target. The magnetic field lines run parallel to the target surface, creating a closed loop that traps electrons near the target. This is the most common configuration and is widely used for a variety of applications [50,54], depicted in Figure 4.
2. Rotatable Cylindrical Magnetron: In this advanced configuration, the target is a cylindrical tube that rotates around a stationary magnet assembly, as shown in Figure 5. This setup maximizes target utilization and is used for large-area coatings, such as architectural glass and display panels [50,54].
3. Balanced and Unbalanced Magnetrons: Balanced magnetrons have equal magnetic flux from the central and outer magnets, which is the case in Figure 4, in a planner configuration. On the other hand, unbalanced magnetrons have stronger outer magnets, allowing magnetic field lines to extend towards the substrate [50,54].

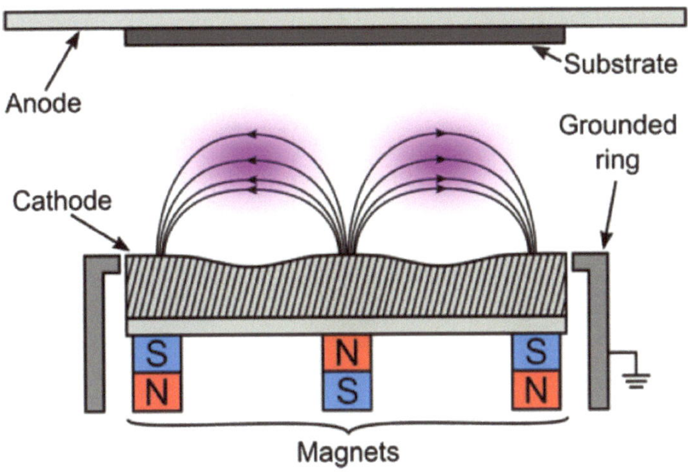

Figure 4. Schematic of a planar magnetron discharge used in the magnetron sputtering processes, where the purple glow represents the plasma, with permission from [50].

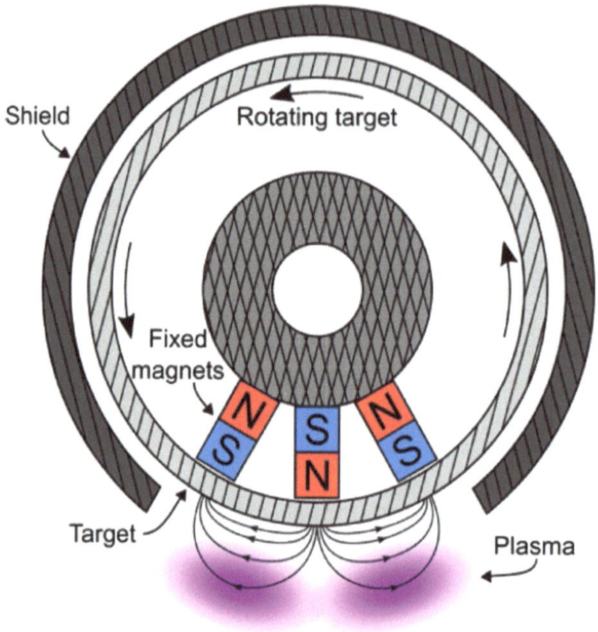

Figure 5. Schematic of a rotating magnetron discharge used in the magnetron sputtering processes, where the purple glow represents the plasma, with permission from [50].

3.1.2. Magnetron Sputtering Categorization

Magnetron sputtering can be categorized based on its power delivery system. In DC magnetron sputtering, a continuous direct current (DC) induces a stream of electrons from the cathode to the anode. This electron flow causes the argon gas to ionize, creating a plasma state, and the target material is sputtered onto the substrate. Pulsed DC magnetron sputtering (PDC MS) employs periodic interruptions in the DC power supply to reduce ion bombardment damage. High-Power Impulse Magnetron Sputtering (HIPIMS) uses short pulses of high power to generate a denser plasma compared to conventional DC sputtering, enhancing film quality and minimizing target damage. For Radio Frequency (RF) magnetron sputtering, an alternating current (AC) at radio frequencies is used to generate a plasma state, allowing for more uniform ionization and deposition of dielectric target materials and the alternate polarity curbs the charge buildup which could cease the discharge of atoms [50].

3.2. Pre-Experimental Steps for AlN Deposition in Magnetron Sputtering Methods

For optimal AlN deposition, certain pre-experimental steps should be carefully carried out, including the preparation and cleaning of the substrate and the aluminium target (either pure or doped), adjusting the sputtering setup, chamber deposition parameters, N_2/Ar ratio, processing pressure, and balancing the magnetron power. The characterization techniques used post deposition are crucial for determining the crystallographic structure, topology, sectional architecture, and piezoelectric characteristics of the AlN films.

3.2.1. Substrate Cleaning

Normally, the cleaning process includes soaking the substrate or wafer in acetone, and subjecting it to an ultrasonic cleaning process for a duration of ten minutes. This is followed by a thorough rinse with isopropanol and a subsequent drying using nitrogen gas. Such surface preparation is aimed at removing contaminants that could compromise the integrity of the deposited AlN films [41]. Depending on the type of substrate in use, slight variations in the cleaning procedure may be employed; for example, C. K. Lee et al. used an ultrasonic bath with isopropyl alcohol for 15 min to remove surface impurities on a glass substrate before depositing the AlN using RF magnetron sputtering [55]. Another cleaning method used by D.L. Maa et al. involved a glow discharge with argon ions to remove surface contaminants on a silicon (Si) wafer before depositing AlN through high-power impulse DC magnetron sputtering [56]. These cleaning methodologies have been shown to improve film adhesion.

3.2.2. Target Preparation and Positioning

The purity of the aluminum (Al) target used in magnetron sputtering is often maintained at a high level, typically around 99.999%, to ensure optimal results [41,55,56]. Prior to the sputtering process, the target is cleaned to remove any surface oxides or contaminants. Sputtering can be used for cleaning the target surface; as for Maa et al., the Al target was cleaned using a 2A DC sputtering for 5 min [56,57]. The distance between the target and the substrate is a critical parameter in DC magnetron sputtering, as it directly influences the kinetic energy and spatial distribution of the sputtered atoms during deposition. Variations in this distance can significantly affect the quality and properties of the deposited films, by affecting their crystal size, lattice constants, and chemical composition [58]. An optimal distance is essential to achieve uniform plasma density and energy distribution, as experienced by Dong-Woo Ko et al. [59]. Typically, in DC magnetron sputtering, the target-to-substrate distance can range from a few centimeters to over a dozen centimeters [41]. In contrast, high-power impulse DC magnetron sputtering (HiPIMS) often employs a shorter target-to-substrate distance, such as the 60 mm specified by Maa et al. [56], which increases the kinetic energy of the sputtered particles due to the reduced mean free path in the chamber. This shorter distance leads to a more focused and directed flux of ions and neutral species, which improves the film's morphology [60].

3.3. Sputtering Conditions and Deposition Parameters for Magnetron Sputtering of AlN Films

Prior to deposition, vacuuming the sputtering environment is crucial for minimizing contamination and ensuring the growth of high-quality films. The specific technology employed in magnetron sputtering significantly influences the process parameters. This section examines the various categories of magnetron sputtering, including RF magnetron sputtering, DC magnetron sputtering, and high-power impulse magnetron sputtering (HiPIMS).

3.3.1. RF Magnetron Sputtering Conditions and Related Work

Lee et al. [55] investigated the RF magnetron sputtering process to deposit AlN films on glass substrates at room temperature. They used a Cryo Vacuum Chamber (CVC) RF magnetron sputtering machine with a high-purity aluminum target, maintaining a target-to-substrate distance of 24 cm. The substrates were ultrasonically cleaned, and the chamber pressure was reduced to approximately 10^{-6} Torr. To explore the effects of different deposition conditions, they varied RF power from 300 W to 800 W at a fixed sputtering pressure of about 5 mTorr in a pure nitrogen (N_2) atmosphere. They also adjusted the sputtering pressure from 3.69 to 9.65 mTorr while keeping RF power constant at 800 W and modified argon (Ar) gas flow rates from 1 to 5 sccm, adjusting N_2 flow to maintain a total flow rate of 10 sccm, with RF power and pressure held constant at 800 W and approximately 5 mTorr, respectively.

X-ray diffraction (XRD) results from [55] indicated that higher RF power promotes the (002) orientation and increases the deposition rate. Varying the sputtering pressure affected the film's structural orientation but not the deposition rate significantly. Introducing argon gas enhanced the deposition rate but caused a shift in orientation from (002) to (101) as the argon ratio increased. The optimal conditions for achieving a highly (002)-oriented AlN film were identified as 800 W RF power and around 6 mTorr sputtering pressure in a pure N_2 atmosphere.

Another investigation by Wei et al. [61] into aluminum nitride (AlN) film fabrication using RF magnetron sputtering focused on understanding distinct orientation preferences on silicon (Si) wafers. The study revealed a shift in preferred orientation from (002) to (100) with increasing distance between target and substrate, as depicted in Figure 6. Cross-sectional SEM images illustrated that the crystalline orientation strongly influenced the film's morphology. The (002)-oriented AlN films exhibited dense columnar grains, while (100)-oriented films lacked a columnar structure. Additionally, mixed orientation growth resulted in a unique rose-like microstructure. The thickness of the AlN films varied with orientation, decreasing from (002) to (100) orientation due to surface energy minimization.

The SEM images in Figure 6 provide crucial insights into the relationship between crystalline orientation and film morphology in aluminum nitride (AlN) thin films deposited via RF magnetron sputtering. Each image visually represents a distinct orientation preference: (002), (100), and mixed orientations. Figure 6a highlights AlN films with a (002) orientation, characterized by dense columnar grains essential for high crystallinity and piezoelectric response. Conversely, Figure 6b showcases films with (100) orientation, lacking columnar grain structure, making them ideal for surface passivation in electronics. Furthermore, Figure 6c presents mixed orientation films, offering tailored properties advantageous for enhancing mechanical stability in thin-film devices. Accompanied by Figure 6d–f, which illustrate the morphology evolution with orientation, these images provide a comprehensive understanding of how different orientations influence film growth. This insight aids in optimizing the deposition processes to achieve the desired film properties, thereby contributing significantly to the advancement of AlN thin-film technology.

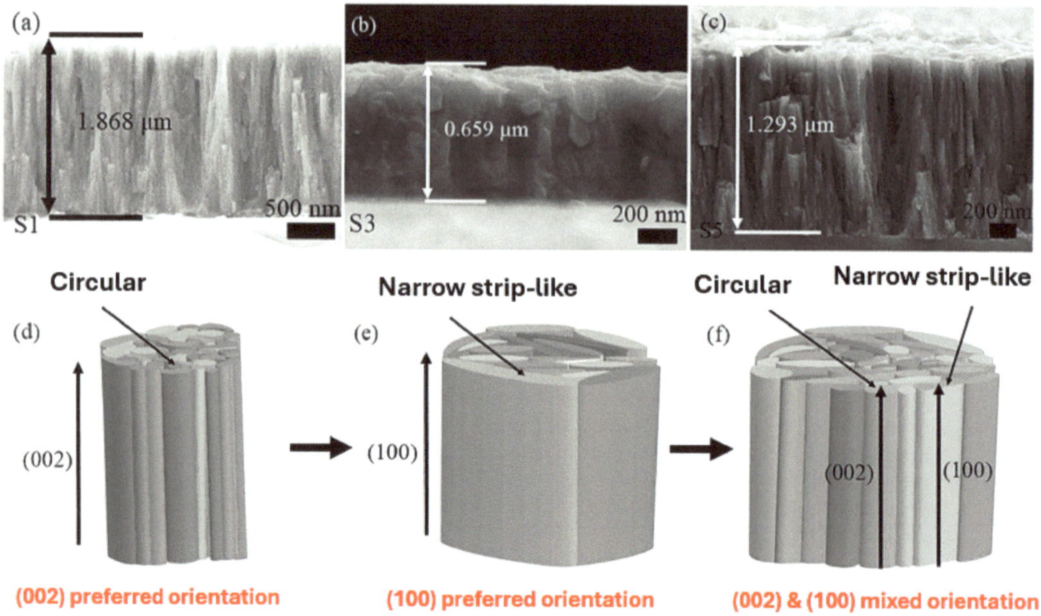

Figure 6. (**a**–**f**) AlN films with varied orientations for three samples alongside schematic morphology diagrams [61].

A novel method was devised for growing highly (100)-oriented aluminium nitride (AlN) thin films on (100) Si substrates under poor vacuum conditions using radio frequency (RF) magnetron sputtering [1]. This method enabled the prediction of optimal deposition conditions, resulting in high-quality films with desirable stoichiometry (Al/N ratio ≈ 1:1) and low oxygen concentration (<10). Throughout the deposition process, the substrate was maintained at floating potential, while parameters such as temperature, nitrogen flow, RF power, and sputtering pressure were held constant. Variations in the target–substrate distance and deposition time were systematically explored, as shown in Figure 7. The results from [1] revealed that films grown outside the cathodic sheath region exhibited good crystallinity and preferred (100) orientation, and under specific working pressure and mean free path conditions, films can grow with the preferred orientation of the c-axis, a direction of [001], parallel to the substrate surface. Within the cathodic sheath region, conditions are less favorable for the formation of high-quality films due to increased ion bombardment and potential thermal effects that can disrupt the growth process. Beyond the cathodic sheath, the mean free path of gas species is significantly smaller than the distance between the target and the substrate, which results in frequent collisions among aluminium and nitrogen molecules. These collisions facilitate the formation of dimers (Al-N), which contribute to a more uniform deposition on the substrate which enhances the crystal structure. The energy of the arriving species is controlled by adjusting the substrate temperature, working pressure, and the target–substrate distance. High energy levels favor the growth of AlN films with the c-axis ([001] direction) parallel to the substrate, while lower energy levels promote the preferred (100) orientation. The choice of our deposition technique allows for precise control over the growth environment, and the minimization of impurities like oxygen, which may adversely affect film quality. Hence, these factors create optimal conditions for achieving AlN films with high crystallinity and desired orientations. These findings underscore the potential of RF magnetron sputtering for fabricating highly (100)-oriented AlN films suitable for surface acoustic wave (SAW) sensing devices.

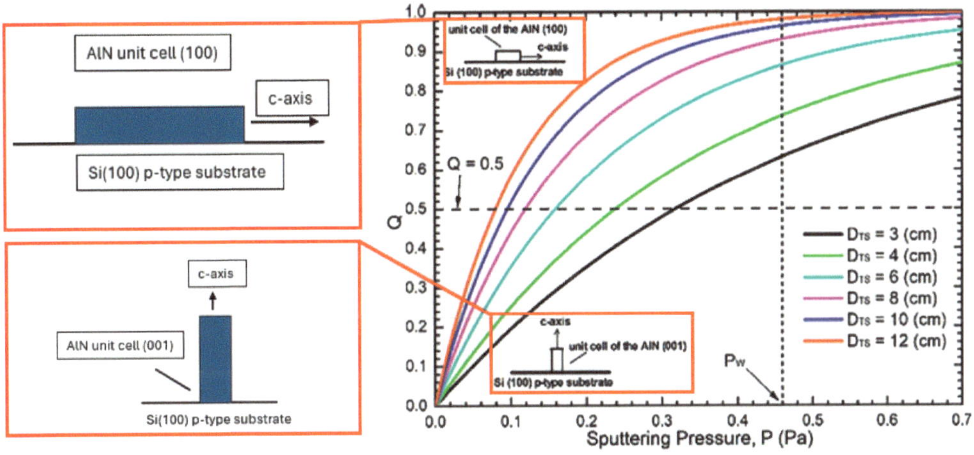

Figure 7. Collision probability vs. sputtering pressure for highly oriented AlN (100) thin film (reproduced from [1]).

Moving forward, adjusting critical parameters such as the mean free path L, temperature T, pressure P, and target–substrate distance allows for control over the crystal orientation in the (100) and (001) directions during AlN growth, as these parameters determine the energy imparted by plasma species onto the substrate. This regulation directly affects both the mean free path L and the collision probability Q of these species within the deposition environment, as illustrated in Figure 7. The mean free path L (measured in meters) represents the average distance a particle travels before colliding with another particle during the deposition process. It is influenced by temperature T, pressure P, and the size of the plasma particles, as described in Equation (1). The theoretical framework linking these parameters is encapsulated in the mean free path equation, where the Boltzmann constant $k_B = 1.38 \times 10^{-23}$ J/K relates the kinetic energy of particles to temperature, expressing energy on a per-particle basis in Joules per Kelvin. Furthermore, the mean free path L

is related to the collision probability Q, which is given in Equation (2) and is calculated after the species pass through a target–substrate distance D_{TS}. The parameter d, defined as the ratio between the target–substrate distance and the target diameter ϕ_T, influences the dependency of the ratio $\frac{D_{TS}}{L}$. For instance, if the substrate is close to the target $(d < 0.5)$, the film tends to exhibit amorphous characteristics, independent of the ratio $\frac{D_{TS}}{L}$. However, when the substrate is further away $(d > 0.5)$, AlN peaks such as (100) and (001) become more prominent as the ratio $\frac{D_{TS}}{L}$ increases. This correlation between film orientation and surface morphology highlights the significance of deposition conditions in shaping film properties [38].

$$L = \frac{k_B T}{\sqrt{2} \cdot (4\pi r_{\text{mean}}^2) P} \quad (1)$$

$$Q = 1 - \exp\left(-\frac{D_{TS}}{L}\right) \quad (2)$$

3.3.2. Reactive Magnetron Sputtering Conditions and Related Work

Reactive magnetron sputtering is a variation of magnetron sputtering where a reactive gas, such as nitrogen or oxygen, is introduced into the vacuum chamber along with the inert gas (typically argon). When the sputtered atoms from the target combine with the reactive gas, they form a compound film on the substrate. In the deposition of AlN thin films via reactive magnetron sputtering, achieving optimal quality involves balancing various parameters such as crystalline quality, surface roughness, residual stress, and deposition rate, alongside factors like target-to-substrate distance, pressure, and temperature, as well as gas ratios. For instance, to enhance crystalline quality and minimize surface roughness, a compromise between high N_2/Ar gas flow ratios and low processing pressures is proposed by Sandager et al. [41], albeit with potential issues like target poisoning and reduced deposition rates, which can pose challenges for industrial applications. Reactive magnetron sputtering exacerbates this with target poisoning, where non-sputterable compounds accumulate on the target surface due to reactions with the reactive gas, diminishing sputtering efficiency and thin-film quality. Adjusting magnetron power can help balance residual stress and deposition rate, while employing lower N_2/Ar ratios can yield acceptable quality films with higher deposition rates, illustrating the trade-offs in optimization efforts. Various studies [24,62,63] underscore these complexities. Figure 8 visually summarizes these compromises, offering guidance for achieving functional AlN films across different operational priorities.

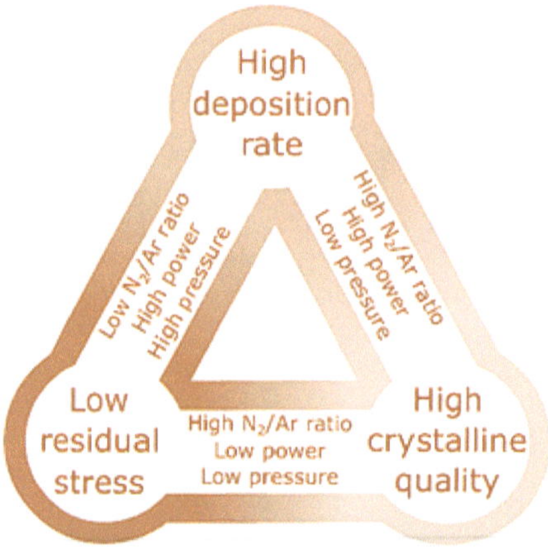

Figure 8. Diagram illustrating the relationship between residual stress, crystalline quality, and deposition rate with respect to sputtering parameters [41].

3.3.3. DC Magnetron Sputtering Conditions and Related Work

The research in the domain of DC magnetron sputtering explores the development and use of aluminum nitride (AlN) thin films to enhance heat dissipation in flexible electronics [64]. The study employs DC magnetron reactive sputtering at ambient temperature to deposit AlN films, analyzing the impact of varying deposition power on the films' characteristics. The results revealed that increasing the deposition power boosts the deposition rate, peaking at 3.3 μm/h at 600 W. X-ray diffraction (XRD) results in [64] showed that thicker films had better crystalline quality, transitioning from disordered to ordered AlN (002) orientation. Thermal conductivity was found to be dependent on film thickness, with measured values of 7.5, 9.8, and 13.3 W/(m·K) for thicknesses of 1, 5, and 10 μm, respectively. The optimal power for achieving high-quality, thermally efficient AlN films was determined to be 600 W. This pivotal finding guided the subsequent investigation into the surface morphology of the deposited films. Illustrated in Figure 9a–d, the atomic force microscopy images offer a comprehensive analysis of the surface characteristics across different power levels. At 300 W (Figure 9a), the films had smaller grains and smoother surfaces, with a mean surface roughness (Ra) of 7.2 nm and a grain diameter (D) of 40.1 nm. When the power was increased to 400 W (Figure 9b), the grain size grew to 138.4 nm and surface roughness increased to Ra 14.2 nm. Further increasing the power to 500 W (Figure 9c) led to a grain diameter of 186.6 nm and a similar surface roughness of around 14.0 nm. At the maximum power of 600 W (Figure 9d), the films exhibited the largest grain size of 250.8 nm and a surface roughness of Ra 14.6 nm. These findings indicate that higher sputtering power results in larger grain sizes and slightly increased surface roughness, suggesting improved crystalline quality and enhanced thermal properties.

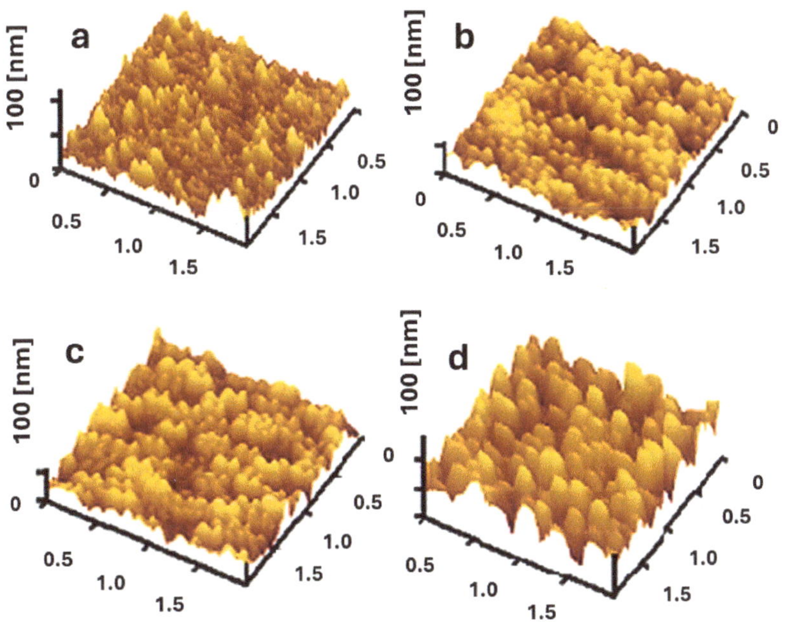

Figure 9. (**a–d**) AlN films grown at room temperature under various power levels: 300 W, 400 W, 500 W, and 600 W for abcd, respectively [64].

In their study, Wu et al. (2024) [65] discuss the growth parameters for aluminum nitride (AlN) thin films. To produce high-quality AlN piezoelectric thin films characterized by a highly c-axis orientation and low stress, 1 μm-thick AlN films were deposited via DC magnetron sputtering on an 8-inch Mo/AlN-interlayer/Si substrate under varied growth conditions. Optimization efforts resulted in achieving a low full width at half the maximum (FWHM) of the AlN (002) rocking curve, reaching as low as 0.905, as shown in Figure 10. The study systematically analyzed the relationship among film quality, morphology, stress, and the underlying growth mechanism of AlN films. Different morphologies and residual stresses were observed in the AlN films, attributed to variations in deposited conditions affecting sputtering particles and film grains. Additionally, the pseudo-Voigt function was employed theoretically to assess grain size and stress in the AlN film, suggesting its feasibility for evaluation when a uniform grain distribution is maintained.

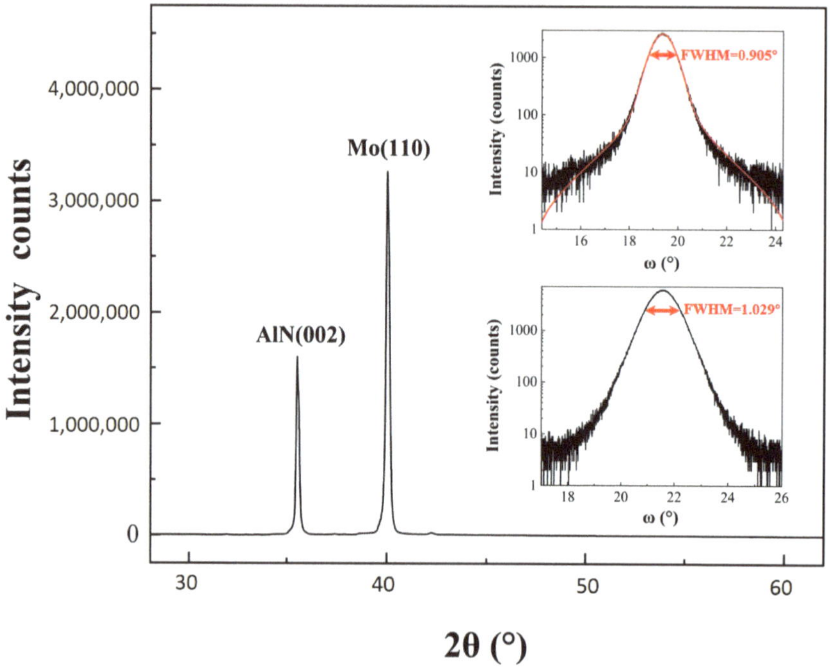

Figure 10. XRD pattern showing the AlN (002) and Mo(110) peaks for growth under a substrate temperature of 450 °C and a sputtering power of 8000 W [65].

3.3.4. High-Impluse DC Magnetron Sputtering Conditions and Related Work

For the high-power impulse magnetron sputtering (HiPIMS) AlN thin-film experiment by Maa D. L et al., the chamber pressure was kept below 8×10^{-4} Pa. The nitrogen (N_2) gas flow rates ranged from 0 to 15 sccm, with argon (Ar) flow kept constant at 40 sccm. The pulsed power supply was set to −800 V, with a frequency of 200 Hz and pulse widths ranging from 50 to 80 μs. Substrates were cleaned using a glow discharge with Ar^+ ions. As N_2 flow increased, the sputtering mode transitioned from metallic (pure aluminum target surface) to transitional (partial nitridation of the target surface) and then to compound mode (complete nitridation of the target surface), resulting in an increased ion-to-neutral ratio of Al species and reduced Al flux. Higher pulse widths (80 μs) led to increased deposition rates compared to shorter ones (50 μs). Films exhibited compressive stress, which increased with N_2 flow, while longer pulse widths resulted in lower residual stresses. Stoichiometric AlN films were achieved at 5 sccm for 50 μs and 8 sccm for 80 μs, confirmed by XPS analysis. They demonstrated that films deposited at 80 μs and higher N_2 flows exhibited a strong (002) texture, as confirmed by XRD analysis. However, excessive ion bombardment shifted the preferred orientation to (100). At lower sputtering pressures, the mean free path of the particles is longer, resulting in higher kinetic energy for the Al and N atoms. This higher energy allows the atoms to migrate more effectively, favoring the growth of the (002) orientation due to its lower surface energy and higher stability. As the sputtering pressure increases, the mean free path decreases, leading to more collisions and reduced kinetic energy of the particles. This reduction in energy makes it more difficult for the atoms to migrate to the (002) sites, causing a shift to the (100) orientation, which requires less energy to form under these conditions.

Additionally, the films showed increased surface roughness in compound mode and a dense, compact morphology. There can be several impacts of this increased surface roughness on the optical, electrical, and mechanical properties of the film. Optically, surface roughness can scatter light, reducing the transparency of the film. This scattering can also affect the refractive index and the optical band gap of the film. For example, a rougher surface can lead to a lower refractive index compared to bulk AlN due to the presence of surface oxides and non-stoichiometric phases. Increased roughness can also introduce defects and grain boundaries, which can act as scattering centres for charge carriers. This can degrade the electrical properties of the film, such as its conductivity and dielectric strength. Rough surfaces can also lead to stress concentration points, which can affect the mechanical integrity of the film. This can be particularly important in applications where the film is subjected to mechanical stress or thermal cycling. Despite the increased surface roughness, AlN films can still exhibit a dense and compact shape. This is because the columnar growth mode, typical in sputtered films, can lead to densely packed columns that are closely aligned. The cross-sectional morphology of the films often shows highly parallel columnar crystals, which contribute to the overall dense structure of the film [56].

Another investigation of AlN thin-film deposition by Jyotish Patidar et al. utilized synchronized HiPIMS, where the substrate bias was timed to coincide with the arrival of Al-rich species during each HiPIMS pulse. The research also included a comparison with the thin films deposited using direct-current magnetron sputtering (DCMS) under similar conditions, and the experimental results were compared to those of HiPIMS. The substrates, p-type silicon wafers (001), were cleaned ultrasonically and maintained at a temperature of 280 °C during deposition. A 2-inch aluminum (Al) target with 99.999% purity was used, positioned 12 cm from the substrate, and angled at 26° relative to the normal substrate. The base pressure was maintained at less than 1×10^{-6} Pa to ensure minimal contamination. The working pressure was set at 0.3 Pa with a gas mixture of argon (Ar) and nitrogen (N_2) flowing at 20 sccm and 10 sccm, respectively. For HiPIMS, the pulse width was 10 µs, and the frequency was 7.5 kHz. A pulsed substrate bias of −30 V was applied, synchronized with the HiPIMS pulse, with a duration of 40 µs and a delay of 15 µs relative to the onset of the HiPIMS pulse. Between pulses, the substrate potential was actively controlled and set to 0. The ion energy distribution functions (IEDFs) for Al, N, and Ar were measured using an EQP-300 Hiden analytical mass spectrometer positioned at the substrate's working distance. Time-resolved measurements were synchronized with the HiPIMS pulse to capture the ion arrival times and their energies. AlN films deposited using HiPIMS were compared to those deposited with DCMS under similar conditions to assess differences in ion flux, kinetic energy, and film growth dynamics.

The HiPIMS process involved applying highly energetic pulses to the sputtering target, resulting in higher ionization rates and increased adatom mobility. This approach significantly improved the texture and crystalline quality of the AlN films, achieving a pronounced c-axis orientation with uniform grain polarization while also reducing defect-related stresses. The process resulted in a denser, more compact columnar structure, crucial for piezoelectric applications such as surface acoustic wave (SAW) devices [66].

Figure 11 compares the ion energy distribution functions (IEDFs) for both DCMS and HiPIMS. The graphs illustrate the intensity of various ion species (such as Al^+, Ar^+, N^+, N_2^+, and Ar^{+2}) as a function of their energy. The key observation is that HiPIMS generates a significantly higher ion flux and higher kinetic energies compared to DCMS. Specifically, the median kinetic energy of Al^+ ions increases from about 6.6 eV in DCMS to 10.7 eV in HiPIMS. This higher ion energy in HiPIMS is attributed to the high-power pulses used, which result in a more energetic plasma and a higher degree of ionization. Consequently,

HiPIMS produced denser and more crystalline films due to the enhanced ion bombardment during deposition [66].

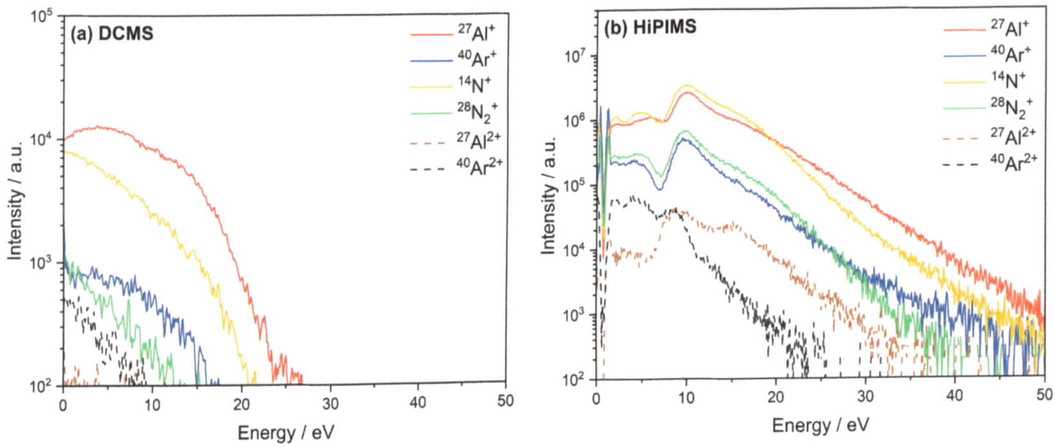

Figure 11. Time-averaged ion energy distribution functions (IEDFs) obtained for (**a**) DC magnetron sputtering (DCMS) and (**b**) high-power impulse magnetron sputtering (HiPIMS) plasmas, as investigated by Jyotish Patidar et al. from [66].

3.4. Comparative Study of Synthesis of AlN with Various Deposition Techniques

Various deposition techniques have been employed to synthesize AlN thin films, each with its advantages and challenges. Here, we provide an overview and comparison of different methods used for forming AlN thin films.

3.4.1. Dual Ion Beam Sputtering

Two ion beams are used in dual ion beam sputtering: one for sputtering the target material and another to help the film grow on the substrate. Improved control over the composition and quality of the film is possible with this technology. In comparison to single ion beam systems, this results in better control over film composition as well as increased film adhesion and density, but it also necessitates more sophisticated and costly equipment [67].

3.4.2. Molecular Beam Epitaxy (MBE)

In the highly regulated technique of molecular or atomic beam epitaxy (MBE), the substrate is exposed to beams of various species in an ultra-high vacuum setting. This process is well known for yielding single-crystal films of excellent purity with exact control over composition and thickness. It produces an epitaxial film of superior quality and allows for exact control over film growth. Low deposition rates, expensive equipment, and ongoing maintenance must all be tolerated as a trade-off [68].

3.4.3. Chemical Vapor Deposition (CVD)

In order to create the desired layer, CVD uses chemical interactions between gaseous precursors and a heated substrate surface. Variants of plasma-enhanced CVD (PECVD) can reduce the necessary substrate temperatures. It has the ability to uniformly and evenly deposit high-quality films across vast regions. Additionally, because of the high temperature and intricate process control, there is a chance of contamination linked to precursors [69].

3.4.4. Pulsed Laser Deposition (PLD)

PLD deposits the film onto the substrate by ablating a target material with high-energy laser pulses, which produce a plasma plume. Complex compounds with stoichiometric transfer from target to substrate can be deposited using this approach. It is able to deposit different elements, such as AlN, and has a high stoichiometric transfer [70].

Each technique offers unique advantages and poses specific challenges. The choice of deposition method depends on the desired film properties, application requirements, and available resources. Below is a summary Table 2 comparing key parameters for these techniques.

Table 2. Comparison of different thin-film deposition techniques.

Deposition Technique	Substrate Temperature (°C)	Growth Rate (Å/min)	Film Quality	Equipment Complexity	Application Suitability
Reactive RF Magnetron Sputtering	500–700	10–20	High	Moderate	Large area, uniform films
DC Magnetron Sputtering	500–700	10–20	Moderate	Low	Conductive targets, large area
Dual Ion Beam Sputtering	300–600	5–15	Very High	High	High-quality films, advanced electronics
Molecular Beam Epitaxy (MBE)	650–850	0.1–1	Very High	Very High	High-purity single-crystal films
Chemical Vapor Deposition (CVD)	600–900	5–10	High	High	Large area, uniform films, electronic packaging
Pulsed Laser Deposition (PLD)	500–750	2-5	High	Moderate	Complex material deposition, research applications

3.5. *Doping and Coating Al Target with Other Materials*

The doping of aluminium target (Al) with other dopant materials during deposition with magnetron sputtering techniques is carried out by combining Al with the other material before sputtering and using the compound as a target instead of the Al target. The introduction of materials like scandium (Sc), erbium (Er) and chromium (Cr) into AlN films has been a focal point of research, due to the resulting films with improved piezoelectric response, surface roughness, high resistance to stress and thermal stability of frequency.

3.5.1. Scandium Doping

Mei Wang et al. investigated the introduction of Sc into the AlN, with 6% Sc and the use of DC Magnetron Sputtering. The target-to-substrate distance maintained during the sputtering process was 8 cm. This doping led to a reduction in surface roughness from 2.1 nm to 1.18 nm and a decrease in film nonuniformity from 1% to 0.4% [71]. Figure 12 illustrates the surface scanning electron microscopy (SEM) image of the undoped aluminium nitride (AlN) film, and the aluminium scandium nitride (AlScN) film, which has been doped with 6% scandium. It provides a clear view of the grain distribution of the film's surface, before and after scandium doping, showing that the Sc doping does not affect the uniformity of grain distribution and the compactness of the surface, while it can reduce the surface roughness, resulting in a smoother film surface [71]. This smoothness enhances piezoelectric performance by minimizing scattering and defects [18]. More importantly, the changes in the crystal structure caused by the incorporation of scandium increase the internal polarization of the material and significantly increase the piezoelectric coefficient (d33), which leads to a more pronounced piezoelectric response, making Sc-doped AlN films more suitable for high-performance applications such as Film Bulk Acoustic Resonator (FBAR) devices [71].

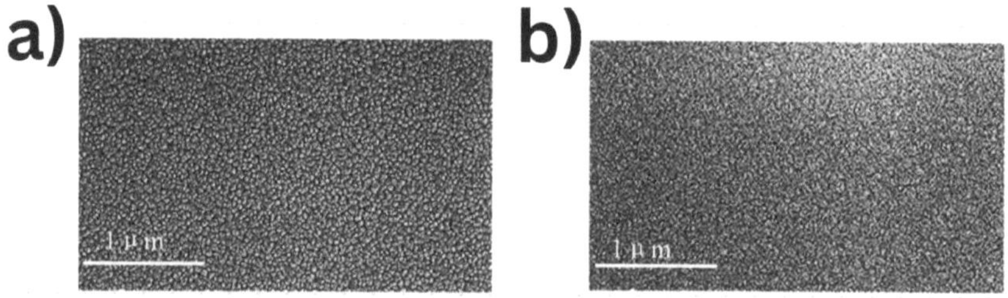

Figure 12. Comparison of SEM images illustrating the grain distribution of AlN films that are: (**a**) undoped and (**b**) scandium-doped. Recreated from [71].

3.5.2. Chromium Doping

Chromium (Cr) doping is another approach used to modify the properties of AlN films, particularly to improve their piezoelectric activity and adjust the temperature coefficient of frequency. This method involves co-sputtering aluminum and chromium targets. V. Felmetsger and M. K. Mikhov carried out the deposition with Cr to produce AlCrN thin films, using RF magnetron sputtering and a dual-target S-gun magnetron system, with setup parameters of 40 kHz AC power applied between the two targets (95.28% aluminum and 4.72% chromium) leading to a plasma discharge. These films of AlCrN were assessed for their piezoelectric activity using solidly mounted resonator bulk acoustic wave (SMR BAW) devices and contour mode resonator (CMR) devices, in which they demonstrated lower values of Q factor and an electromechanical coupling coefficient, with values approximately at 250 and 0.55%, respectively, while the temperature coefficient of frequency (TCF) was higher (at about $-68\,\text{ppm/K}$) compared with pure AlN films. Moreover, the surface morphology of the AlCrN film was notably different across various substrates, showing distinct variations on silicon and platinum substrates [72].

3.5.3. Chromium Coating

Another way of adding chromium to the AlN film is through a coating mechanism, which involves depositing a layer of chromium onto the AlN film surface. In a study carried out by Y.L. Su et al., the films of Cr-Al-N were prepared by unbalanced magnetron sputtering under a base pressure of 1.59×10^{-3} Pa and a working pressure of 0.37 Pa. As a result, the corrosion resistance and the electrical conductivity of the films were improved, as the Cr-Al-N coating scored a corrosion current density of 5.35×10^{-7} A/cm^2 and an electrical conductivity of 5.5 ohm.cm. This makes the coatings suitable as protective coatings for the bipolar plates of proton exchange membrane fuel cells (PEMFCs) [73].

4. Applications of AlN

The multifaceted attributes of aluminum nitride (AlN) render it highly suitable for a diverse array of applications, encompassing resonators, energy harvesting, and thermal management. Within wave resonators, such as Film Bulk Acoustic Wave Resonators (FBARs), AlN's characteristics of low dielectric loss, high bulk acoustic wave velocity, chemical and thermal stability, moderate electromechanical coupling constant, and wide band gap, facilitate the efficient propagation of signals and resonant behavior, both of which are imperative for operational efficacy.

Aluminum nitride (AlN) exhibits several characteristics that distinguish it from other piezoelectric materials such as PZT and ZnO. One significant advantage is its high thermal stability and ability to operate effectively at elevated temperatures, making it suitable for applications in harsh environments where other materials may fail. AlN has a wide band gap of approximately 6.2 eV, which allows it to be used in high-frequency devices without significant energy loss due to dielectric heating. Unlike PZT, which is ferroelectric and can exhibit fatigue over time, AlN's non-ferroelectric nature contributes to its durability and long-term reliability in piezoelectric applications. AlN's lower density compared to PZT enhances its suitability for lightweight applications in micro-electromechanical systems (MEMSs) [45]. As also discussed previously, recent studies have also shown that doping AlN with elements like scandium can significantly improve its piezoelectric properties without compromising its thermal stability [74]. These unique characteristics position AlN as a superior choice for next-generation devices, requiring robust performance across various operating conditions.

In energy harvesting contexts, AlN's favorable combination of a moderate electromechanical coupling coefficient (k) and high-quality factor (Q) position it as an optimal choice for the development of proficient piezoelectric generators, outperforming alternatives like lead zirconate titanate (PZT). There also exist examples where AlN plays a pivotal component in power electronics and high-frequency circuitry, one of the reasons being its efficacy in dissipating heat in various applications. The following section delves into the applications and utilization of AlN, particularly when produced through the magnetron sputtering technique.

4.1. Application in Wave Resonators

In the construction of resonators like the Film Bulk Acoustic Wave Resonator (FBAR), a configuration typically involves a thin piezoelectric film positioned between two electrodes, as depicted in Figure 13. When a voltage is applied across these electrodes, a resonant acoustic wave is generated within the piezoelectric layer. In this context, aluminum nitride (AlN) is the piezoelectric material for FBAR construction. AlN facilitates the conversion of electrical signals into mechanical vibrations and vice versa, thereby fostering the resonant behavior essential for the functioning of FBARs.

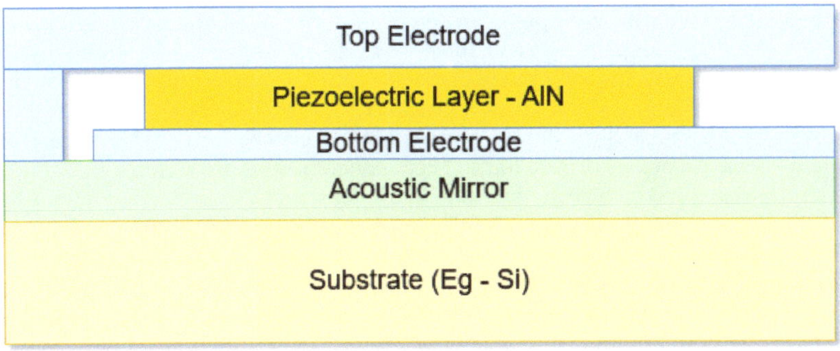

Figure 13. AlN—based cross-section of FBAR structure.

Aluminum nitride (AlN) facilitates efficient signal propagation due to a bulk acoustic wave velocity of approximately 11,000 m/s, the highest among alternatives like PZT and ScAlN. Its moderate electro-mechanical coupling constant indicates efficient conversion between electrical and mechanical energy. Additionally, due to its low dielectric constant, it reduces energy dissipation during signal transmission. AlN boasts excellent chemical and thermal stability, with a melting point of 2100 °C, further contributing to its suitability. Its wide band gap of 6.2 eV renders it suitable for high-frequency operations. The piezoelectric properties of AlN depend on its crystalline structure and orientation along the c-axis. Proper orientation along the c-axis enhances its piezoelectric performance [33], which can be sufficiently obtained using the sputtering process.

High-quality crystalline AlN films can be produced at lower temperatures (in the range of 500 °C) through sputtering. Parameters such as the distance between the target and substrate, pressure, N2 concentration, sputtering power, and substrate temperature influence the quality and orientation of the deposited film. Fine-tuning these parameters optimizes the performance of Film Bulk Acoustic Resonator (FBAR) devices, aiding in the optimization of deposition conditions for FBAR fabrication. The degree of crystallization, preferred orientation, and surface morphology of the deposited AlN layer serve as indicators of FBAR device effectiveness [75].

4.2. AlN-Based Energy Harvesting Application

To develop small sensors and liberate them from the bounds of battery and their limitations, alternative power sources are being developed. Among them is the concept of energy harvesting, where residual or ambient energy is utilized to harness the surrounding energy for the system. Aluminium nitride (AlN) is a material widely used in the application of energy harvesting, leveraging its conversion property from mechanical to electrical energy. While zinc oxide (ZnO), polyvinylidene fluoride (PVDF), and lead zirconate titanate (PZT) are other promising candidates, AlN stands out in certain areas.

AlN possesses a high mechanical quality factor (Q) with a moderate electromechanical coupling coefficient (k), making it a good option for designing efficient piezoelectric generators. Unlike lead zirconate titanate (PZT), AlN lacks ferroelectric properties. AlN has a low crystallization temperature, which makes its fabrication compatible with CMOS processes [76]. The piezoelectric coefficient and permittivity for AlN are approximately $\approx$1.55 C/m^2 and 10, respectively [77]. This results in a good energy-harvesting figure of merits not on par with that of PZT, but one that can be utilized [78]. Additionally, AlN is not harmful to humans health-wise. These unique characteristics make it suitable for applications where a non-ferroelectric piezoelectric material is required.

Xianming et al. [79] conducted a study on the micro-electromechanical system (MEMS) cantilever-based aluminium nitride (AlN) vibration energy harvester, as shown in Figure 14, where the c-axis-oriented AlN piezo thin film was deposited using pulsed direct-current magnetron sputtering. The prototype achieved a maximum output RMS voltage of 4.66 V at 1 g and 210.85 Hz, with an output average power of 56.4 μW and a power density of 854.55 μW/(cm$^3 \cdot$ g^2), aligning well with theoretical predictions. It provides effective guidance for design and optimization, particularly for devices with long tip mass, with potential applications in wireless sensor network nodes, such as structural health monitoring systems. A procedure of scandium doping was used by Stephan Barth et al. [24] to improve the piezoelectric properties of AlN films. These enhancements are ascribed to changes in elastic constants and an increase in piezoelectric constants. For instance, doping AlN films with scandium leads to a notable increase in the piezoelectric coefficient d_{33} compared to pure AlN. It indicates that the piezoelectric coefficient d_{33} increases with a higher scandium concentration in AlScN films, reaching a peak at around 40% scandium content. This demonstrates the efficacy of these films in generating power from defined vibrations, with significant power output. Tiny AlN harvesters were introduced by Z Cao et al. [80] to harvest energy from low-frequency, two-dimensional vibrations. They used a stainless steel substrate with high fracture toughness and yield strength to enhance output power and reduce resonant frequency. AlN films were deposited on the substrate to fabricate the harvesters. This demonstrates that magnetron sputtering is a reliable and repeatable process for depositing good-quality AlN. In this way, novel harvesters using AlN are being developed to enhance the applicability of AlN. The sputtering technique offers an advantage over other techniques due to its low price point, relatively simple process with fewer precursor chemicals involved, moderate uniformity, and high scalability compared to CVD and PVD techniques.

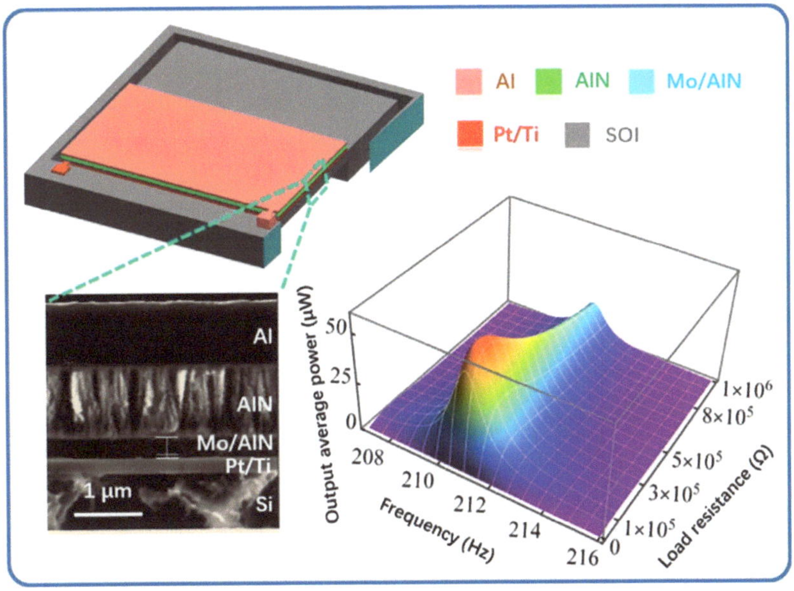

Figure 14. AlN-based energy harvester MEMS device showing layers of materials used, along with the variation in resistance, frequency of operation and output power. Reproduced from [79].

4.3. AlN Applications in Thermal Management

Aluminium nitride (AlN) possesses a high thermal conductivity similar to metals like stainless steel, ranging from 2 to 200 $W.m^{-1}K^{-1}$. This notable thermal conductivity facilitates efficient heat transfer away from heat-generating components, aiding in preventing overheating, as AlN can withstand temperatures well above 1000 °C without significant degradation. The unique combination of thermal properties makes AlN suitable for various applications such as heat sinks, substrates, thermal interface materials, and packaging components, effectively dissipating heat and maintaining device performance [81,82]. For example, Elladan et al. [83] developed AlNB for effective thermal management applications in solid-state devices like LEDs at varying power levels. They achieved a lower thermal resistance of 8.5 (K/W).

A technique for depositing thin films of aluminium nitride (AlN) with different thicknesses, ranging from 100 nm to 1.7 μm at low temperatures (<100 °C), was presented by C. Perez et al. [84]. By controlling deposition conditions, they achieved a wide range of thermal conductivity (36–104 $Wm^{-1}K^{-1}$) in 600 nm films. This highlights that there are methods to optimize AlN for specific thermal management applications in integrated circuits.

Aluminium nitride (AlN) thin films, deposited using sputtering techniques, present diverse applications. They can serve as dielectric layers, barrier coatings, or passivation layers in electronic devices, enhancing their performance and reliability, particularly in high-power and high-frequency applications [85]. Additionally, AlN thin films find utility in optoelectronics and photonics as waveguides, optical filters, or substrates for epitaxial layer growth, owing to their wide band gap and UV and visible spectrum transparency. Thus, AlN thin films offer a wide range of possibilities across thermal, optical, and mechanical domains, making them promising materials for advancing technology and magnetron sputtering a reliable deposition process to accomplish it.

5. Summary and Outlook

In conclusion, our review provides a comprehensive evaluation of magnetron sputtering for the deposition of high-quality AlN films based on the literature, with a focus on controlling orientation and optimizing film properties. By systematically examining the effects of key deposition parameters—such as working pressure, RF power, substrate temperature, and target-to-substrate distance, we have demonstrated the critical role these factors play in determining the quality and orientation of AlN films. Our results underscore that the precise control and optimization of these parameters are crucial for tailoring film properties to meet specific application requirements in electronics, optics, and sensor technologies.

The adaptability of sputtering parameters, combined with conventional temperature-dependent structural growth, enables the formation of advanced structures with improved deposition rates. This positions magnetron sputtering as a valuable technique in semiconductor manufacturing, offering opportunities to produce high-quality AlN films through various methods, including RF, DC, and impulse magnetron sputtering. Each technique provides unique advantages that can be harnessed to enhance AlN growth, making it suitable for a wide range of applications, such as wave resonators, energy harvesting, thermal management, and photonics.

Despite the significant progress made in understanding the relationship between deposition parameters and the resulting film properties, further research is necessary to explore AlN films' behavior under extreme or elevated environmental conditions. Investigating the performance of AlN films in such challenging environments could unveil new applications and further expand their utility.

Future research should aim at further refining the deposition process to push the boundaries of AlN film quality and functionality. This includes investigating alternative seed materials to enhance nanostructure formation, optimizing deposition conditions for a variety of substrate types, and systematically studying the impact of additional parameters such as gas flow rates and substrate bias voltage. Additionally, the development and implementation of in-situ monitoring techniques during deposition can provide real-time insights into the growth mechanisms, enabling the production of more uniform and high-quality films. Such advancements could significantly enhance the reproducibility and scalability of the deposition process.

Author Contributions: Conceptualization, N.A.K.J., V.P., M.A.H.M., A.S. and K.W.; Investigation, N.A.K.J., V.P., M.A.H.M., A.S. and K.W.; Project administration, N.A.K.J. and K.W.; Supervision, K.W.; Visualization, N.A.K.J., V.P., M.A.H.M., A.S. and K.W.; Writing—original draft, N.A.K.J., V.P., M.A.H.M., A.S. and K.W. All authors contributed equally to this work. All authors have read and agreed to the published version of the manuscript.

Funding: This research received no external funding.

Conflicts of Interest: The authors declare no conflicts of interest.

References

1. Cunha, C.L.A.; Pimenta, T.C.; Fraga, M.A. Development and Applications of Aluminum Nitride Thin Film Technology. In *Thin Films*; Yang, D., Ed.; IntechOpen: London, UK, 2022. [CrossRef]
2. Yen, T.T.; Hirasawa, T.; Wright, P.K.; Pisano, A.P.; Lin, L. Corrugated aluminum nitride energy harvesters for high energy conversion effectiveness. *J. Micromechanics Microengineering* **2011**, *21*, 085037. [CrossRef]
3. Elfrink, R.; Kamel, T.M.; Goedbloed, M.; Matova, S.; Hohlfeld, D.; van Andel, Y.; van Schaijk, R. Vibration energy harvesting with aluminum nitride-based piezoelectric devices. *J. Micromechanics Microengineering* **2009**, *19*, 094005. [CrossRef]
4. Elfrink, R.; Renaud, M.; Kamel, T.M.; de Nooijer, C.; Jambunathan, M.; Goedbloed, M.; Hohlfeld, D.; van Andel, Y.; van Schaijk, R. Vacuum-packaged piezoelectric vibration energy harvesters: Damping contributions and autonomy for a wireless sensor system. *J. Micromechanics Microengineering* **2010**, *20*, 104001. [CrossRef]
5. Bu, G.; Ciplys, D.; Shur, M.; Schowalter, L.J.; Schujman, S.; Gaska, R. Surface acoustic wave velocity in single-crystal AlN substrates. *IEEE Trans. Ultrason. Ferroelectr. Freq. Control* **2006**, *53*, 251–254. [CrossRef] [PubMed]
6. Shelton, S.; Chan, M.L.; Park, H.; Horsley, D.; Boser, B.; Izyumin, I.; Przybyla, R.; Frey, T.; Judy, M.; Nunan, K.; et al. CMOS-compatible AlN piezo electric micromachined ultrasonic transducers. In Proceedings of the 2009 IEEE International Ultrasonics Symposium, Rome, Italy, 20–23 September 2009; pp. 402–405.
7. Horsley, D. Aluminum nitride pMUT based on a flexurally-suspended membrane. In Proceedings of the 2011 Solid State Sensors, Actuators and Microsystems Conference, Beijing, China, 5–9 June 2011.

8. Yokoyama, T.; Iwazaki, Y.; Nishihara, T.; Ueda, M. Analysis on electromechanical coupling coefficients in AlN-based bulk acoustic wave resonators based on first-principle calculations. In Proceedings of the 2012 IEEE International Ultrasonics Symposium, Dresden, Germany, 7–10 October 2012; pp. 551–554. [CrossRef]
9. Hartnett, T.M.; Bernstein, S.D.; Maguire, E.A.; Tustison, R.W. Optical properties of ALON (aluminum oxynitride). In Proceedings of the Window and Dome Technologies and Materials V , Orlando, FL, USA, 27 June 1997.
10. Fan, Z.; Qin, Z.; Sun, Z.; Wu, H. Broad Spectrum Detector Based on AlN Crystal. *J. Phys. Conf. Ser.* **2022**, *2350*, 012013. [CrossRef]
11. Bickermann, M.; Epelbaum, B.M.; Filip, O.; Heimann, P.J.; Nagata, S.; Winnacker, A. UV transparent single-crystalline bulk AlN substrates. *Phys. Status Solidi* **2010**, *7*, 21–24. [CrossRef]
12. Xu, X.; Huang, X.; Wang, X.; Kang, J.; Zheng, Y.; Chen, X. Aluminum Nitride Film and Preparation Method and Application Thereof. CN111690907A. 22 September 2020. Available online: https://www.iprdb.com/patent/CN111690907A.html#/basic (accessed on 2 September 2023).
13. Park, J.H.; Steckl, A.J. Laser action in Eu-doped GaN thin-film cavity at room temperature. *Appl. Phys. Lett.* **2004**, *85*, 4588–4590. [CrossRef]
14. Pearton, S.J.; Zolper, J.C.; Shul, R.J.; Ren, F. GaN: Processing, defects, and devices. *J. Appl. Phys.* **1999**, *86*, 1–78. [CrossRef]
15. Furqan, C.M.; Ho, J.Y.L.; Kwok, H.S. GaN thin film: Growth and Characterizations by Magnetron Sputtering. *Surfaces Interfaces* **2021**, *26*, 101364. [CrossRef]
16. Anjum, F.; Ahmad, R.; Afzal, N.; Murtaza, G. Characterization of InN films prepared using magnetron sputtering at variable power. *Mater. Lett.* **2018**, *219*, 23–28. [CrossRef]
17. Ishibe, T.; Tomeda, A.; Watanabe, K.; Kamakura, Y.; Mori, N.; Naruse, N.; Mera, Y.; Yamashita, Y.; Nakamura, Y. Methodology of Thermoelectric Power Factor Enhancement by Controlling Nanowire Interface. *ACS Appl. Mater. Interfaces* **2018**, *10*, 37709–37716. [CrossRef] [PubMed]
18. Polewczyk, V.; Magrin Maffei, R.; Vinai, G.; Lo Cicero, M.; Prato, S.; Capaldo, P.; Dal Zilio, S.; di Bona, A.; Paolicelli, G.; Mescola, A.; et al. ZnO Thin Films Growth Optimization for Piezoelectric Application. *Sensors* **2021**, *21*, 6114. [CrossRef] [PubMed]
19. van Deurzen, L.; Singhal, J.; Encomendero, J.; Pieczulewski, N.; Chang, C.S.; Cho, Y.; Muller, D.A.; Xing, H.G.; Jena, D.; Brandt, O.; et al. Excitonic and deep-level emission from N- and Al-polar homoepitaxial AlN grown by molecular beam epitaxy. *APL Mater.* **2023**, *11*, 081109. [CrossRef]
20. Popov, V.; Chirumamilla, M.; Krekeler, T.; Ritter, M.; Pedersen, K. Magnetron Sputter Grown AlN Nanostructures with Giant Piezoelectric Response toward Energy Generation. *ACS Appl. Nano Mater.* **2023**, *6*, 8849–8856. [CrossRef]
21. Hasegawa, F.; Takahashi, T.; Kubo, K.; Nannichi, Y. Plasma CVD of amorphous AlN from metalorganic Al source and properties of the deposited films. *Jpn. J. Appl. Phys.* **1987**, *26*, 1555. [CrossRef]
22. Hisyam, M.I.; Shuhaimi, A.; Norhaniza, R.; Mansor, M.; Williams, A.; Hussin, M.R.M. Study of AlN Epitaxial Growth on Si (111) Substrate Using Pulsed Metal–Organic Chemical Vapour Deposition. *Crystals* **2024**, *14*, 371. [CrossRef]
23. Sanz-Hervás, A.; Vergara, L.; Olivares, J.; Iborra, E.; Morilla, Y.; García-López, J.; Clement, M.; Sangrador, J.; Respaldiza, M.A. Comparative study of c-axis AlN films sputtered on metallic surfaces. *Diam. Relat. Mater.* **2005**, *14*, 1198–1202. [CrossRef]
24. Barth, S.; Bartzsch, H.; Glöß, D.; Frach, P.; Modes, T.; Zywitzki, O.; Suchaneck, G.; Gerlach, G. Magnetron sputtering of piezoelectric AlN and AlScN thin films and their use in energy harvesting applications. *Microsyst. Technol.* **2016**, *22*, 1613–1617. [CrossRef]
25. Fu, Y.Q.; Luo, J.K.; Nguyen, N.T.; Walton, A.J.; Flewitt, A.J.; Zu, X.T.; McHale, G.; Matthews, A.; Iborra, E.; Rowland, D.J.; et al. Advances in piezoelectric thin films for acoustic biosensors, acoustofluidics and lab-on-chip applications. *Prog. Mater. Sci.* **2017**, *89*, 31–91. [CrossRef]
26. Badis, R.; Camus, J.; Ayad, A.; Rammal, M.; Zernadji, R.; Rouag, N.; Abdou, M.; Hetero, D. Epitaxial Growth of AlN Deposited by DC Magnetron Sputtering on Si(111) Using a AlN Buffer Layer. *Coatings* **2021**, *11*, 1063. [CrossRef]
27. Jiang, J.; Fu, Y.; Zhang, J.; Wei, Y.; Zhang, G.; Cui, J. Growth of c-axis oriented AlN thin films on titanium alloy substrate by middle frequency magnetron sputtering. *J. Vac. Sci. Technol.* **2015**, *33*, 041509. [CrossRef]
28. Uchiyama, S.; Nishi, N.; Makishima, A.; Miyamoto, S.; Ohmori, A.; Araki, Y. Growth of AlN films by magnetron sputtering. *J. Cryst. Growth* **1998**, *189*, 448–451. [CrossRef]
29. Ke, G.; Lu, P.; Jiang, L.; Zhao, W.; Li, Q.; Yu, Z.; Liu, Y. Highly c-axis oriented AlN film grown by unbalanced magnetron reactive sputtering and its electrical properties. *J. Alloy. Compd.* **2015**, *646*, 446–453. [CrossRef]
30. Mishin, S.; Oshmyansky, Y. Magnetron Deposition of AlN and ScAlN for Massproduction for BAW Devices and MEMS. In Proceedings of the 2019 IEEE International Ultrasonics Symposium, Glasgow, UK, 6–9 October 2019. [CrossRef]
31. Behera, L.; Pandey, N.; Gupta, M. Synthesis and characterization of AlN thin films deposited using DC and RF magnetron sputtering. *J. Appl. Phys.* **2020**. [CrossRef]
32. Foroughi-Abari, A.; Cadien, K.C. Growth, structure and properties of sputtered niobium oxide thin films. *Thin Solid Film.* **2011**, *519*, 3068–3073. [CrossRef]
33. Mwema, F.M.; Akinlabi, E.T.; Oladijo, O.P. A systematic review of magnetron sputtering of AlN thin films for extreme condition sensing. *Mater. Today Proc.* **2020**, *26*, 1546–1550. [CrossRef]
34. Iqbal, A.; Mohd-Yasin, F. Reactive Sputtering of Aluminum Nitride (002) Thin Films for Piezoelectric Applications: A Review. *Sensors* **2018**, *18*, 1797. [CrossRef]

35. Liu, W.; He, L.; Wang, X.; Zhou, J.; Xu, W.; Smagin, N.; Toubal, M.; Yu, H.; Gu, Y.; Xu, J.; et al. 3D FEM Analysis of High-Frequency AlN-Based PMUT Arrays on Cavity SOI. *Sensors* **2019**, *19*, 4450. [CrossRef]
36. Signore, M.A.; Velardi, L.; De Pascali, C.; Kuznetsova, I.; Blasi, L.; Biscaglia, F.; Quaranta, F.; Siciliano, P.; Francioso, L. Effect of silicon-based substrates and deposition type on sputtered AlN thin films. *Appl. Surf. Sci.* **2022**, *599*, 154017. [CrossRef]
37. Bairagi, S.; Järrendahl, K.; Eriksson, F.; Hultman, L.; Birch, J.; Hsiao, C.L. Glancing Angle Deposition and Growth Mechanism of Inclined AlN Nanostructures Using Reactive Magnetron Sputtering. *Coatings* **2020**, *10*, 768. [CrossRef]
38. Cunha, C.L.A.; Pimenta, T.C.; Fraga, M.A. Growth and properties of sputtered highly (100)-oriented oxygenated AlN thin films for SAW sensing applications. *Microsyst. Technol.* **2021**, *27*, 3773–3782. [CrossRef]
39. Chirumamilla, M.; Krekeler, T.; Wang, D.; Kristensen, P.K.; Ritter, M.; Popok, V.N.; Pedersen, K. Magnetron Sputter Deposition of Nanostructured AlN Thin Films. *Appl. Nano* **2023**, *4*, 280–292. [CrossRef]
40. Ababneh, A.; Dagamseh, A.M.K.; Albataineh, Z.; Tantawi, M.; Al-Bataineh, Q.M.; Telfah, M.; Zengerle, T.; Seidel, H. Optical and structural properties of aluminium nitride thin-films synthesized by DC-magnetron sputtering technique at different sputtering pressures. *Microsyst. Technol.* **2021**, *27*, 3149–3159. [CrossRef]
41. Sandager, M.K.; Kjelde, C.; Popok, V. Growth of Thin AlN Films on Si Wafers by Reactive Magnetron Sputtering: Role of Processing Pressure, Magnetron Power and Nitrogen/Argon Gas Flow Ratio. *Crystals* **2022**, *12*, 1379. [CrossRef]
42. Alyousef, H.A.; Hassan, A.M.; Zakaly, H.M.H. Reactive magnetron sputtered AlN thin films: Structural, linear and nonlinear optical characteristics. *J. Mater. Sci. Mater. Electron.* **2023**, *34*, 1088. [CrossRef]
43. Yin, J.; Zhou, B.; Li, L.; Liu, Y.; Guo, W.; Talwar, D.N.; He, K.; Ferguson, I.T.; Wan, L.; Feng, Z.C. Optical and structural properties of AlN thin films deposited on different faces of sapphire substrates. *Semicond. Sci. Technol.* **2021**, *36*, 045012. [CrossRef]
44. Chen, G.; Li, H.; Xie, X.; Xie, L.; Wang, E.; Liu, G.; Zhang, H.; Lu, B.; Li, C.; Pei, H. Influence of nitrogen flow ratio on properties of c-axis oriented AlN films grown by RF magnetron sputtering. *Appl. Phys. A* **2021**, *127*, 819. [CrossRef]
45. Desideri, D.; Bernardo, E.; Corso, A.J.; Moro, F.; Pelizzo, M.G. Electrical Properties of Aluminum Nitride Thick Films Magnetron Sputtered on Aluminum Substrates. *Materials* **2022**, *15*, 2090. [CrossRef]
46. Uesugi, K.; Shojiki, K.; Xiao, S.; Kuboya, S.; Miyake, H. Effect of the Sputtering Deposition Conditions on the Crystallinity of High-Temperature Annealed AlN Films. *Coatings* **2021**, *11*, 956. [CrossRef]
47. Liu, S.; Li, Y.; Tao, J.; Tang, R.; Zheng, X. Structural, Surface, and Optical Properties of AlN Thin Films Grown on Different Substrates by PEALD. *Crystals* **2023**, *13*, 910. [CrossRef]
48. Sanjeeva, S.; Rudresh, J.; Vinayakumar, K.B.; Nagaraja, K.K. A strong dependence of sputtering power on c-axis oriented aluminium nitride on Si (111): A structural and electrical study. *IET Nanodielectrics* **2024**, *7*, 7–17. [CrossRef]
49. Wang, F.; Wu, J. Chapter 3—Fundamentals of Plasma Physics. In *Modern Ion Plating Technology*; Elsevier: Amsterdam, The Netherlands, 2023; pp. 29–89. [CrossRef]
50. Gudmundsson, J.T.; Lundin, D. Introduction to magnetron sputtering. In *High Power Impulse Magnetron Sputtering*; Elsevier: Amsterdam, The Netherlands, 2020; pp. 1–48. [CrossRef]
51. Hishimone, P.; Nagai, H.; Sato, M. Methods of Fabricating Thin Films for Energy Materials and Devices. In *Lithium-Ion Batteries—Thin Film for Energy Materials and Devices*; IntechOpen: London, UK, 2020. [CrossRef]
52. Forge Project. High-Throughput Experiments in PVD Coating Technology. 2021. Available online: https://www.forge-project.eu/news/2021/high-throughput-experiments-pvd-coating-technology (accessed on 28 May 2024).
53. Aijaz, A.; Sarakinos, K.; Lundin, D.; Brenning, N.; Helmersson, U. HiPIMS-Based Novel Deposition Processes for Thin Films. Linköping Studies in Science and Technology. Ph.D. Thesis, Linköping University Electronic Press, Linköping, Sweden, 2012.
54. Gudmundsson, J.T. Physics and technology of magnetron sputtering discharges. *Plasma Sources Sci. Technol.* **2020**, *29*, 113001. [CrossRef]
55. Lee, C.; Placido, F.; Kirk, K.F. RF Magnetron Sputtered AlN Films on Glass Substrate without Substrate Heating. 2008. Available online: https://www.researchgate.net/publication/272091562_RF_magnetron_sputtered_AlN_films_on_glass_substrate_without_substrate_heating (accessed on 2 September 2023).
56. Ma, D.; Liu, H.; Deng, Q.; Yang, W.; Silins, K.; Huang, N.; Leng, Y. Optimal target sputtering mode for aluminum nitride thin film deposition by high power pulsed magnetron sputtering. *Vacuum* **2019**, *160*, 410–417. [CrossRef]
57. Bakri, A.S.; Nafaziazi, A.S.; Abu Bakar, M.M.I.; Megat Hasan, N.A.; Raship, W.I.; Wan Omar, Z.; Azman, R.A.; Mohamed Ali, W.H.; Abd Majid, M.K.A. Electrical and Structural Comparison of (100) and (002) Oriented AlN Thin Films Deposited by RF Magnetron Sputtering. *J. Mater. Sci. Mater. Electron.* **2022**, *33*, 12271–12280. [CrossRef]
58. Witit-Anun, N.; Buranawong, A. Effect of Substrate-Target Distance on the Structure of TiCrN Films Deposited from Mosaic Target by Reactive DC Magnetron Sputtering. *Key Eng. Mater.* **2019**, *798*, 163–168. [CrossRef]
59. Ko, D.W.; Jang, S.H.; Kim, J.H.; Lee, K.Y.; Kim, Y.J. Improvement on Plasma Intensity Uniformity in Rectangular DC Magnetron Sputter by Optimizing Structures of Substrate Electrode. *J. Korean Soc. Manuf. Technol. Eng.* **2023**, *32*, 57–62. [CrossRef]
60. Suresh Babu, S.; Fischer, J.; Barynova, K.; Rudolph, M.; Lundin, D.; Gudmundsson, J.T. High power impulse magnetron sputtering of a zirconium target. *J. Vac. Sci. Technol. A* **2024**, *42*, 043007. [CrossRef]
61. Wei, Z.; Shen, L.; Kuang, Y.; Wang, J.; Yang, G.; Lei, W. The evolution of preferred orientation and morphology of AlN films under various sputtering parameters. *J. Cryst. Growth* **2024**, *625*, 127439. [CrossRef]
62. Dubois, M.; Muralt, P. Stress and piezoelectric properties of aluminum nitride thin films deposited onto metal electrodes by pulsed direct current reactive sputtering. *J. Appl. Phys.* **2001**, *89*, 6389–6395. [CrossRef]

63. Ababneh, A.; Roma, P.I.; Georgarakis, A.; Tichelaar, F.D.; Arnoldbik, W.M.; van den Berg, A.; van der Zaag, P.J. The influence of sputter deposition parameters on piezoelectric and mechanical properties of AlN thin films. *Mater. Sci. Eng. B* **2010**, *172*, 253–258. [CrossRef]
64. Bian, Y.; Liu, M.; Ke, G.; Chen, Y.; DiBattista, J.; Chan, E.; Yang, Y. Aluminum nitride thin film growth and applications for heat dissipation. *Surf. Coatings Technol.* **2015**, *267*, 65–69. [CrossRef]
65. Wu, S.; Xu, R.; Guo, B.; Ma, Y.; Yu, D. Influence of Growth Parameters and Systematical Analysis on 8-Inch Piezoelectric AlN Thin Films by Magnetron Sputtering. *Mater. Sci. Semicond. Process.* **2024**, *169*, 107895. [CrossRef]
66. Patidar, J.; Sharma, A.; Zhuk, S.; Lorenzin, G.; Cancellieri, C.; Sarott, M.F.; Trassin, M.; Thorwarth, K.; Michler, J.; Siol, S. Improving the Crystallinity and Texture of Oblique-Angle-Deposited AlN Thin Films Using Reactive Synchronized HiPIMS. *Surf. Coat. Technol.* **2023**, *468*, 129719. [CrossRef]
67. Han, S.; Chen, H.Y.; Cheng, C.H.; Lin, J.H.; Shih, H. Aluminum nitride films synthesized by dual ion beam sputtering. *J. Mater. Res.* **2004**, *19*, 3521–3525. [CrossRef]
68. Davis, C.; Novikov, S.; Cheng, T.; Campion, R.; Foxon, C. Surface reconstruction patterns of AlN grown by molecular beam epitaxy on sapphire. *J. Cryst. Growth* **2001**, *226*, 203–208. [CrossRef]
69. Guerrero, R.; Vargas-Garcia, J.R. Growth of AlN Films by Chemical Vapor Deposition. *Superf. Vacío* **1999**, *9*, 82–84.
70. Vispute, F.; Narayan, J.; Wu, H.; Jagannadham, K. Epitaxial growth of AlN thin films on silicon (111) substrates by pulsed laser deposition. *J. Appl. Phys.* **1995**, *77*, 4724–4727. [CrossRef]
71. Wang, M.; Bo, H.; Wang, A.; Cheng, Z.; Li, S.; Zou, W.; He, J.; Ma, X. High-quality c-axis oriented Al(Sc)N thin films prepared by magnetron sputtering. *Thin Solid Film.* **2023**, *781*, 140000. [CrossRef]
72. Felmetsger, V.V.; Mikhov, M.K. Reactive Magnetron Sputtering of Piezoelectric Cr-doped AlN Thin Films. In Proceedings of the 2011 IEEE International Ultrasonics Symposium, Orlando, FL, USA, 18–21 October 2011; pp. 835–839. [CrossRef]
73. Su, Y.; Kao, W.; Horng, J.; Hsueh, H.; Cian, Y.; Hsieh, T. Corrosion resistance and conductivity behavior of Cr-Al-N MAX phase coatings prepared by magnetron sputtering technology. *Thin Solid Film.* **2024**, *793*, 140274. [CrossRef]
74. Nian, L.; Qu, Y.; Gu, X.; Luo, T.; Xie, Y.; Wei, M.; Cai, Y.; Liu, Y.; Sun, C. Preparation, Characterization, and Application of AlN/ScAlN Composite Thin Films. *Micromachines* **2023**, *14*, 557. [CrossRef]
75. Chiu, K.H.; Chen, J.H.; Chen, H.R.; Huang, R.S. Deposition and characterization of reactive magnetron sputtered aluminum nitride thin films for film bulk acoustic wave resonator. *Thin Solid Film.* **2007**, *515*, 4819–4825. [CrossRef]
76. Zhang, J.; Cao, Z.; Kuwano, H. Fabrication of Low-Residual-Stress AlN Thin Films and Their Application to Microgenerators for Vibration Energy Harvesting. *Jpn. J. Appl. Phys.* **2011**, *50*, 09ND18. [CrossRef]
77. Chung, G.S.; Lee, B.C. Fabrication and characterization of vibration-driven AlN piezoelectric micropower generator compatible with complementary metal-oxide semiconductor process. *J. Intell. Mater. Syst. Struct.* **2015**, *26*, 1971–1979. [CrossRef]
78. Xu, R.; Kim, S.G. Figures of merits of piezoelectric materials in energy harvesters. In Proceedings of the PowerMEMS, Atlanta, GA, USA, 2–5 December 2012; Volume 12.
79. He, X.; Wen, Q.; Lu, Z.; Shang, Z.; Wen, Z. A micro-electromechanical systems based vibration energy harvester with aluminum nitride piezoelectric thin film deposited by pulsed direct-current magnetron sputtering. *Appl. Energy* **2018**, *228*, 881–890. [CrossRef]
80. Pandey, A.; Dutta, S.; Kaushik, J.; Gupta, N.; Gupta, G.; Raman, R.; Kaur, D. Growth, structural and electrical properties of AlN/Si (111) for futuristic MEMS applications. *Mater. Sci. Semicond. Process.* **2021**, *123*, 105567. [CrossRef]
81. Duquenne, C.; Besland, M.P.; Tessier, P.Y.; Gautron, E.; Scudeller, Y.; Averty, D. Thermal conductivity of aluminium nitride thin films prepared by reactive magnetron sputtering. *J. Phys. D Appl. Phys.* **2011**, *45*, 015301. [CrossRef]
82. Duquenne, C.; Popescu, B.; Tessier, P.Y.; Besland, M.P.; Scudeller, Y.; Brylinski, C.; Delage, S.; Djouadi, M.A. Magnetron Sputtering of Aluminium Nitride Thin Films for Thermal Management. *Plasma Process. Polym.* **2007**, *4*, S1–S5. [CrossRef]
83. Hamza El-Ladan, A.; Subramani, S. Development of AlNB alloy in (Al/AlN/B) stacking sequence using RF reactive sputtering towards thermal management application. *J. Mater. Sci. Mater. Electron.* **2021**, *32*, 1–13. [CrossRef]
84. Perez, C.; McLeod, A.J.; Chen, M.E.; Yi, S.i.; Vaziri, S.; Hood, R.; Ueda, S.T.; Bao, X.; Asheghi, M.; Park, W.; et al. High Thermal Conductivity of Submicrometer Aluminum Nitride Thin Films Sputter-Deposited at Low Temperature. *ACS Nano* **2023**, *17*, 21240–21250. [CrossRef] [PubMed]
85. Cassella, C.; Piazza, G. AlN Two-Dimensional-Mode Resonators for Ultra-High Frequency Applications. *IEEE Electron Device Lett.* **2015**, *36*, 1192–1194. [CrossRef]

inorganics

MDPI

Article

Polysilane–Barium Titanate Polymeric Composite Obtained through Ultrasonication

Răzvan Rotaru [1], Maria Emiliana Fortună [1,*], Elena Ungureanu [2,*] and Liviu Sacarescu [1]

[1] "Petru Poni" Institute of Macromolecular Chemistry, Department of Inorganic Polymers, 41A Grigore Ghica Voda Alley, 700487 Iasi, Romania; rotaru.razvan@icmpp.ro (R.R.); livius@icmpp.ro (L.S.)

[2] Department of Exact Sciences, "Ion Ionescu de la Brad" Iasi University of Life Sciences, 3 Mihail Sadoveanu Alley, 700490 Iasi, Romania

* Correspondence: fortuna.maria@icmpp.ro (M.E.F.); elena.ungureanu@iuls.ro (E.U.)

Abstract: This work describes the synthesis of a polysilane (PSH)–barium titanate (BT) ferroelectric polymer composite that keeps stable in the presence of ultraviolet light (UV). To evaluate the stability in the presence of UV radiation and the mechanism of interaction between the PSH matrix and BT, FTIR measurements were carried out. The UV/VIS absorption measurement reveals that PSH absorbs strongly in the ultraviolet range, while the composite behaves similarly to BT. Although PSH is a semiconductor, the dielectric spectrometry analysis determined that BT is a ferroelectric material due to its high dielectric constant and low dielectric losses. In contrast to the polymer matrix, the composite polymer has a greater dielectric constant and a lower loss permittivity. PSH is a semiconductor, as indicated by its electrical conductivity of 10^{-5} S/cm; nevertheless, the UV-irradiated polymer has antistatic properties (10^{-8} S/cm). Irradiated or not, the polymer composite is a semiconductor, with conductivity of 10^{-6} S/cm, significantly lower than that of PSH. The interaction with electromagnetic radiation indicates electromagnetic shielding behavior for both BT (highest absorption magnitude of −57 dB) and the polymer composite (maximum absorption magnitudes range from 8.4 to −15.2 dB). Based on these research results, the novel composite with specific characteristics may be used in electronic applications in UV-irradiated conditions.

Keywords: barium titanate (BT); composite; polysilane (PSH); semiconductor; ultrasonication

1. Introduction

Composites represent a class of heterogeneous materials with at least two components, varying physical and/or chemical properties, and significant structural barriers. The components complement each other, resulting in a new material with superior properties.

Barium titanate, $BaTiO_3$ (BT), is a ferroelectric material with a perovskite structure, a high permittivity (dielectric constant above 1000), and a resistivity higher than 108 Ω·cm. It is probably the most well-known and widely used ferroelectric material in the electrical and electronics industries (ceramic condensers, multi-layer condensers, piezoelectric appliances, pyroelectric sensors, thermistors, electromagnetic shields with electric field absorption, and biomedical applications) [1–3].

Polysilanes (in the scientific literature they are also called organo-polysilanes, polysilylenes, or catena-silicon polymers) were first synthesized in 1920 by Kipping through condensation of dichlorosilanes using molted sodium metal dispersion in toluene. At that time, the highly crystalline material had little practical interest as it was insoluble and non-fusible. After 1975, Fujima and coworkers restarted research and expanded their interest in polysilanes. They discovered uses in optoelectronics such as photoconductor polymers, electric charge transport, and a precursor to generate heat-resistant materials (β-SiC) [4–6]. The interest in optoelectronics resulted from the σ electron delocalization effect along the main chain, which leads to the high mobility of the electric charge carriers, similar to the semiconductive polyenes. S. Nespurek et al. reported values of mobility of about 106 $m^2/V \cdot s$ [6]. A common issue with this kind of macromolecular compound is its structural instability (degradation) caused by the scission of the Si-Si bond under UV radiation. For polysilanes, the UV light absorption is stronger in the near ultraviolet region within 300–400 nm. Moreover, excitation of polysilanes within this range usually results in strong emission at wavelengths that are sometimes located beyond 400 nm depending on the electronic character of the substituents attached to the silicon atoms in the main chain. Polysilanes can be considered a collection of cromophores in trans-planar conformation. Usually, the UV absorption maximum shifts toward a greater wave length if the macromolecular chain has more than 40 atoms [6–8]. The temperature affects the maximum wave length, which shifts to higher values as the temperature drops (a process known as thermochromism). Various types of photostabilizing agents can improve the stability of polysilanes under UV exposure. Common photostabilizing additives include Hostavin PR 25 ($C_{13}H_{14}O_5$), Sanduvor EPU ($C_{22}H_{28}N_2O_3$), and Tinuvin 1577 ($C_{27}H_{27}N_3O_2$) [6].

In this research, we study the possibility of replacing such photostabilizing additives with ferroelectric barium titanate, which is known to act as an electromagnetic shield against UV radiation. The electromagnetic shielding property of a material is related to its capacity to refract, adsorb, or reflect the electromagnetic field. From this point of view, there are three kinds of electromagnetic shields: shields with reflection of radiation, shields with absorption of electromagnetic radiation, and shields with multiple reflective properties. The first class, representing shielding by reflecting radiation, consists of materials that contain charge carriers (electrons or holes) that will interact with the electromagnetic field's electric component. In the second case, the shielding material is usually dielectric and has high electric and/or magnetic dipoles (ferroelectric and/or ferromagnetic materials). Absorption loss occurs when the electromagnetic waves traversing the medium induce electronic cloud shifting, which in turn produces ohmic losses and heating (heat loss). Multiple reflective shields involve the presence of large specific surfaces (porous foam materials) or a large interface (nanofibers, nanotubes, or nanometric pyramidal structures) [9–11]. Common examples include graphene, multi-walled carbon nanotubes, carbon fibers for multiple reflective shields, nickel foam, aluminum foam, carbon-based foam, NiO, Fe_3O_4 nanocrystals, and ferroelectric materials for shields with reflection and absorption of electromagnetic radiation [12,13].

2. Results and Discussion

2.1. Fourier-Transform Infrared (FTIR) Spectroscopy

As observed in Figure 1, the infrared spectrum of the BT sample shows characteristic bands for pure barium titanate at 438 cm^{-1} and 559 cm^{-1} [1]. The band observed within the lower wavelength range (438 cm^{-1}) is characteristic of barium titanate, and was assigned to the Ba-O and Ti-O bond vibrations in metal oxides [1]. The presence of the band within the higher wavelength region (559 cm^{-1}) and assigned to Ti-O bond stretching normal vibration is representative of the formation of the $BaTiO_3$ tetragonal phase with direct consequences for the dielectric properties (high dielectric constant) [1,14]. Figure 2 presents the FTIR spectra for PSH and PSH/BT composites.

The infrared spectra for PSH present characteristic bands at 3604 cm^{-1} (OH vibrations), 3067 and 3046 cm^{-1} (C-H aromatic), 2961 and 2903 cm^{-1} (C-H aliphatic), 2090 (Si-H vibrations), 1649 cm^{-1} (OH vibrations), 1427 and 1096 cm^{-1} (Si-C_6H_5 vibrations), 1269 cm^{-1} (Si-CH_3 vibrations), 1026 cm^{-1} (Si-O-C vibrations), 734 cm^{-1} (C-C aromatic), 696 cm^{-1} (Si-C vibrations), and 472 cm^{-1} (Si-Si vibrations) [7,8]. In PSH/BT_1 and PSH/BT_2 composites, the addition of barium titanate does not change the polysilane's chemical structure. In general, the wave numbers stay the same, but the spectral intensities change.

To observe the influence of barium titanate particles in the degradation of polysilanes under UV radiation, PSH, PSH/BT1, and PSH/BT_2 samples were subjected to ultraviolet irradiation. Figure 3 shows the IR spectra of the UV-irradiated samples (PSH_{uv}, PSH/BT_{1uv}, and PSH/BT_{2uv}).

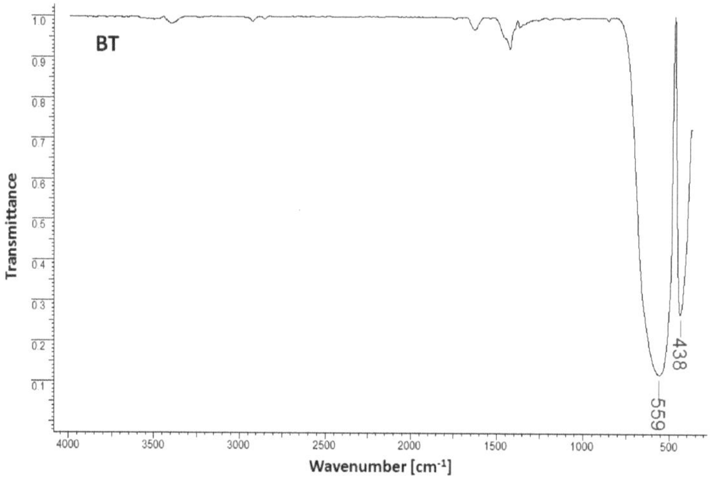

Figure 1. FTIR spectra for BT.

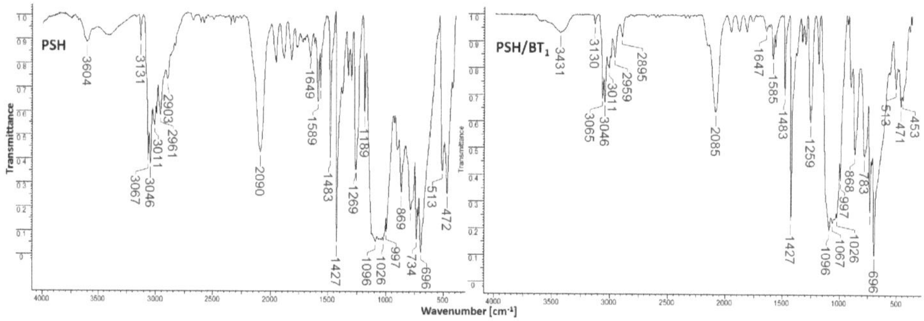

Figure 2. FTIR spectra for PSH and PSH/BT_1 composites.

According to Miller [15], the degradation of polysilanes under UV radiation leads to a reduction in the number of Si-Si bonds while the number of Si-O-C and/or Si-O-Si bonds increases. Quantitative investigations were conducted using the IR absorption band ratios [7]. The representative IR absorption band ratios for the studied system are shown in Table 1.

Table 1. The representative IR absorption band ratios.

Relative Ratio/Sample	PSH	PSH_{UV}	PSH/BT_1	PSH/BT_{1UV}	PSH/BT_2	PSH/BT_{2UV}
Si-Si/Si-O-C	0.78	0.36	0.74	0.69	0.76	0.74
Si-Si/Si-H	1.30	1.11	1.29	1.20	1.28	1.26
Si-Si/Si-CH_3	1.14	1.06	1.13	1.09	1.14	1.13
Si-Si/Si-C_6H_5	0.78	0.53	0.81	0.75	0.82	0.79
Si-H/Si-CH_3	0.60	0.63	0.59	0.61	0.61	0.60
Si-H/Si-C_6H_5	0.59	0.53	0.56	0.50	0.56	0.56
Si-H/Si-O-C	0.60	0.36	0.57	0.48	0.55	0.55
Si-CH_3/Si-C_6H_5	0.68	0.62	0.69	0.63	0.70	0.69

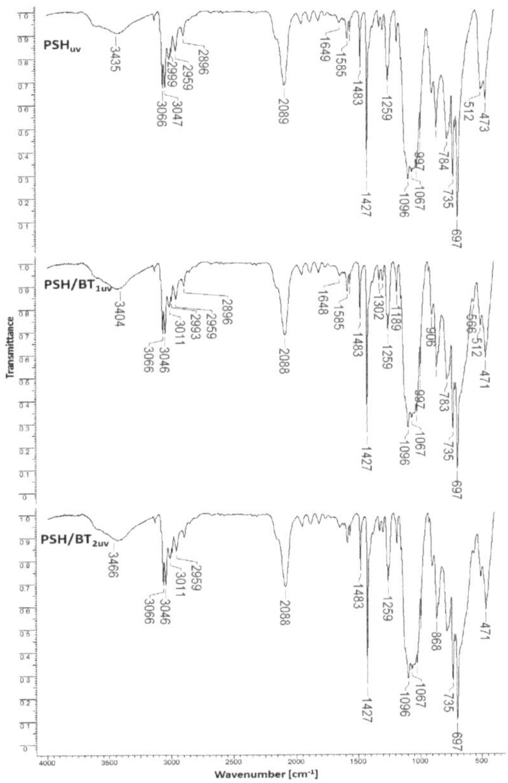

Figure 3. FTIR spectra for UV irradiated samples (PSH_{uv}, PSH/BT_{1uv}, and PSH/BT_{2uv}).

In the case of the irradiated polymer (PSH_{uv}), a drastic reduction in the Si-Si/Si-O-C ratio compared to the non-irradiated polymer (PSH) from 0.78 to 0.36 was observed. This result shows a decrease in the vibration energy of the Si-Si bond compared to the Si-O-C bond. This reduction is strongly attenuated in the case of the PSH/BT_{1uv} composite (from 0.74 to 0.69) and almost insignificant for the composite with a high amount of barium titanate, PSH/BT_{2uv} (from 0.76 to 0.74). The same phenomenon was observed for the Si-Si/Si-H, Si-Si/Si-CH_3, Si-Si/Si-C_6H_5, and Si-H/Si-O-C ratios. This behavior suggests that BT acts as a protective UV electromagnetic shield.

2.2. The UV/VIS Absorption Study in the Solid State

UV/VIS absorption spectroscopy was further used to study the behavior of polysilanes when exposed to UV light. The UV/VIS absorption spectra of PSH, BT, PSH/BT1, and PSH/BT2 are shown in Figure 4. PSH shows a strong absorption maximum in the ultraviolet spectrum. Due to the sample preparation (solid pellets), the intensity of the absorption band was outside the measurement range. The UV spectrum of BT displays a strong UV absorption band around 210 nm. The electron band structure of $BaTiO_3$ has low-lying narrow conduction bands from Ti^{3+}-3d states and valence bands from O^{2-}-2p states. This is a normal behavior for ferroelectric structures [14]. Regarding the composites, it is evident (Figure 4) that their behavior is more like that of BT since the UV absorption band characteristic for PSH is no longer present.

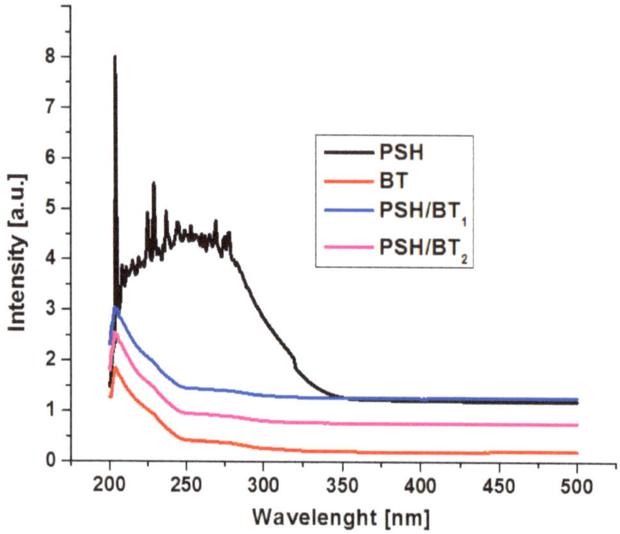

Figure 4. UV/VIS absorption spectra for PSH, BT, and the composites.

The components of the complex dielectric permittivity (ε′ and ε″) were measured in the frequency range of 10^{-2}–10^{6} Hz at room temperature. In general, ε′ and ε″ of the insulating materials are related to five types of polarizations of the electrons and molecules: electronic, vibrational (atomic), directional (dipolar), ionic, and interfacial. Electronic and vibrational polarizations are ideal for materials used as capacitors, while directional, ionic, and interfacial polarizations are perfect for materials used for electromagnetic shielding [1,16,17]. The variation in the dielectric constant and permittivity loss versus frequency for BT, PSH, and their composites is depicted in Figures 5 and 6, respectively.

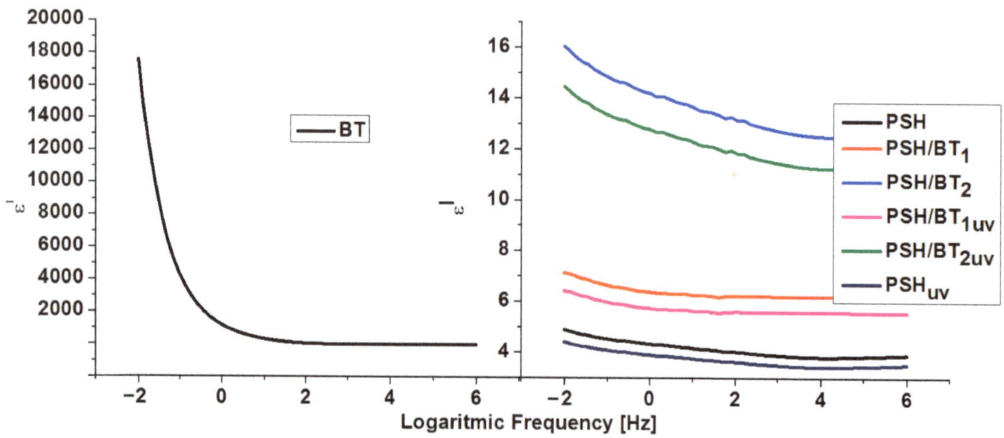

Figure 5. Dielectric constant (ε′) for BT, PSH, and PSH/BT composite.

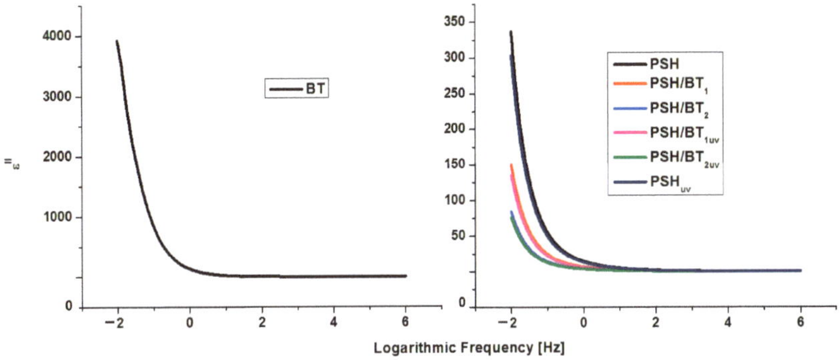

Figure 6. Dielectric loss for BT, PSH, and PSH/BT composite.

It is obvious that in the case of BT, with a frequency increase, ε′ and ε″ decrease and remain constant at higher frequencies, indicating a dielectric dispersion (ionic and directional polarizabilities). This was assigned to dipole formation due to the changes in the valence states of cations and space-charge polarization [1,16,18]. The decrease for ε′ stops at frequencies higher than 10^2 Hz. Electronic polarization is always higher than ionic and directional polarization (ε′ > ε″). This behavior is normal for ferroelectric materials [1,19]. The values for ε′ ranged from 1.4 (for 10^6 Hz) to 17,604 (for 10^{-2} Hz), while ε″ registered values from 0.005 (for 10^6 Hz) to 3921 (for 10^{-2} Hz). These values were similar to values evidenced for barium titanate in the literature [20,21].

Polysilane shows a variation in the dielectric constant that is typical for dielectric polymers. However, the variation in the dielectric losses and the fact that they are higher than the dielectric constant could be related to the intrinsic semiconducting nature of the semiconductor materials [20]. Thus, ε′ varies from 3.8 to 4.8 in the frequency range 10^6–10^{-2}, and ε″ varies from 0.11 to 336.5 in the same interval. In the case of polysilanes/barium titanate composites, the dielectric constant increases and the permittivity loss decreases. Thus, for the PSH/BT_1 sample, ε′ varied from 6.18 to 7.12 and ε″ from 0.05 to 149.57 in the frequency range 10^6–10^{-2} Hz, and for the PSH/BT2, ε′ varied from 12.75 to 16.04 and ε″ from 0.02 to 84.13. For semiconductor materials, reducing the dielectric loss directly influences the electrical conductivity of the composite [22,23].

Figure 7 displays the variation in electrical conductivity (σ) of PSH, PSH_{uv}, and the BT composites within the frequency range of 10^6–10^{-2} Hz (BT is not shown because it is an insulator and has insignificant electrical conductivity [1]).

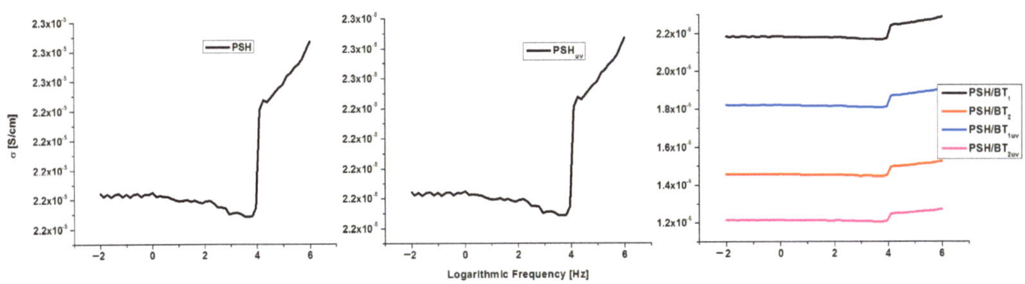

Figure 7. Electrical conductivity (S/cm) for PSH, PSH_{uv}, and BT/PSH samples.

The conductivity of polysilane slightly varies within the studied frequency range (10^{-2}–10^6 Hz), with values between 2.18· × 10^{-5} and 2.28· × 10^{-5} S/cm. This range of conductivity variation indicates that polysilane belongs to the intrinsic semiconductor polymer class. Such materials could be used in the semiconductor industry to fabricate various devices (transistors, diodes, and photovoltaic cells) for electronic applications. The UV irradiation of polysilane strongly modifies the electrical conductivity to 10^{-8} S/cm, which is usual for polymeric electrical insulators.

On the other hand, doping with BT lowers the conductivity of the PSH/BT composite to 10^{-6} S/cm. Furthermore, the concentration of BT within the composite modifies the electrical conductivity since σ PSH/BT_1 > σ PSH/BT_2. The UV irradiation of these composite samples leads to only a tempered decrease in electrical conductivity. From this point of view, both irradiated composites (PSH/BT_{1uv} and PSH/BT_{2uv}) remained in the semiconductor materials range, having conductivities around 10^{-6} S/cm. Therefore, BT acts as an electromagnetic shield that absorbs most of the UV radiation and protects polysilane from altering its chemical structure. In order to evidence the influence of UV radiation on polysilane and its composite with BT, the magnitude of the absorbance losses for these materials was calculated. Negative values indicate that the material behaves like an UV shield, absorbing the incident UV radiation through displacements of the electronic cloud [3,9]. The variation in the magnitude of the incident UV absorbance loss versus the frequency of the incident radiation is shown in Figure 8.

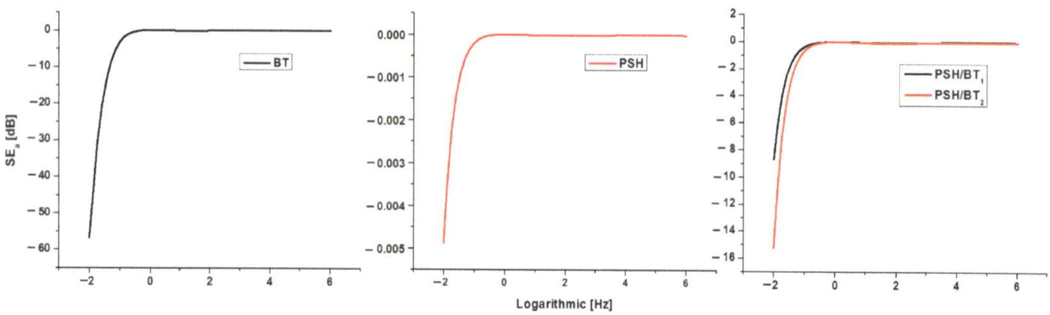

Figure 8. Plot of the absorbance loss for BT, PSH, PSH/BT_1, and PSH/BT_2 versus frequency of the incident radiation.

In general, BT, which has high dielectric constants and a maximum absorption of −57 dB, is currently a good absorber of electromagnetic radiation and is usually used as an electromagnetic shield for various materials [3,9]. On the other hand, polysilane has a very low absorption, the maximum being −0.0048 dB. The polysilane/BT composite proved to be a good absorber of electromagnetic radiation, having absorption maxima of −8.4 dB for PSH/BT_1 and −15.2 dB for PSH/BT_2. Materials with electromagnetic shielding properties caused by absorption cause a temporary displacement of the electronic cloud. The energy of the incident electromagnetic radiation is consumed in this way and, in the case of composites with polysilanes, electronic delocalization σ no longer occurs. This phenomenon is the explanation for the resistance to UV degradation of barium titanate composites compared to polysilane.

3. Materials and Methods

3.1. Materials

Barium carbonate (Sigma-Aldrich, St. Louis, MO, USA, 99% purity), titanium dioxide (Chemical Company, Iasi, Romania, 99% purity), diphenyldichlorosilane, methyl (hydrogen) dichlorosilane (Sigma-Aldrich, 99% purity), toluene (Lachner, Czech Republic, M = 92.14 g/mol), sodium (Sigma-Aldrich, 99% purity), and Milli-Q ultrapure distilled water were employed as dispersion mediums in ultrasonication experiments.

3.2. Methods

Synthesis of the BT and PSH/BT composites was performed using an ultrasound generator from Sonics Vibracell (750W, 20 kHz). A microwave oven (800 W electrical power, 2.5 GHz microwave frequency) and a Vulcan A 130 laboratory oven were used for BT synthesis. Irradiation of the polymer composite samples was conducted using a Spectroline-Spectronics lamp with two wavelengths of emission: short-wave (254 nm) and long-wave (365 nm).

The structures of the precursors and composites were investigated by FTIR spectroscopy on potassium bromide pellets using a Bruker Vertex 70 spectrometer (2 cm^{-1} resolution).

^{1}H-NMR spectra were registered with a Bruker NMR instrument (Model DRX 400 MHz).

The molecular weight distribution of the polymers synthesized in this work was measured in THF on a Spectra Physics 8800 gel permeation chromatograph with two PL-gel packed columns. All molecular weights are reported versus the polystyrene standard.

To observe the influence of UV radiation, three samples (PSH, PSH/BT1, and PSH/BT2) were prepared as 13 mm pellets and then exposed to ultraviolet irradiation (365 nm wavelength) for 30 min (see Table 1). Specord 200 Analytik Jena UV/VIS (Jena, Germany) was used for the UV/VIS absorption study. The samples were studied in an ethyl alcohol solution at a 1% wt. concentration.

Barium titanate with a very high dielectric constant can act like an electromagnetic shield since it has the capacity to convert the incident electromagnetic radiation into thermal energy. For such materials, the absorption magnitude of the incident electromagnetic radiation can be calculated using Equation (1) [3,10]:

$$\text{SEA [dB]} = -8.68(t\sigma\omega\mu r/2)^{1/2} \quad (1)$$

where t = material thickness; σ = total electrical conductivity of the material; ω = angular frequency (Hz), $\omega = 2\pi f$, f = linear frequency (Hz), frequency of electromagnetic field; μr = relative magnetic permeability of the material.

The dielectric measurements (dielectric constant, loss permittivity, and electrical conductivity) were performed on a Concept 40 Novocontrol Dielectric Spectrometer (Montabaur, Germany) at room temperature with silver electrodes in the 10^6–10^{-1} Hz frequency range. The contact surfaces with spectrometer electrodes were covered with silver conductive paste to avoid interfacial polarization and possible edge effects. BT, PSH, and the two analyzed composites (PSH/BT_1 and PSH/BT_2) were prepared in the form of pellets with a thickness of 3 mm by pressing with a Specac laboratory press.

For the calculation of SE_A, the following working assumptions were taken into account: t = 3 mm, $\sigma = \omega \cdot \varepsilon_0 \cdot \varepsilon''$ (ε_0: free space permittivity: $8.854 \cdot 10^{-12}$ F/m, and ε'' is the absorption or permittivity loss), and μ_r = 1, material without magnetic properties [3].

3.2.1. Barium Titanate (BT) Particle Preparation

The BT preparation followed a method described in previous work [1]. Briefly, a 1/2 $TiO_2/BaCO_3$ mixture dispersion in Milli-Q water was ultrasonicated for 60 min. The resulting powder was then separated, dried in a microwave oven for 10 min, and thermally treated at 500 °C for 3 h to obtain perovskite-like BT submicronic particles. The final product, BT, was obtained as a white, solid powder.

3.2.2. Poly(diphenyl-co-methyl(hydrogen))silane (PSH) Preparation

Poly(diphenyl-co-methyl(hydrogen))silane (PSH) was obtained by the conventional Wurtz coupling method [24] from diphenyldichlorosilane and methyl(H) dichlorosilane in the presence of molted sodium metal dispersion in toluene. The condensation reaction was carried out at reflux temperature. The reaction product, PSH, results in a white-yellowish solid powder.

^{1}H-NMR ($CDCl_3$, δ, ppm): 0.799-0.979 (Si-CH_3); 3.22-4.00 (Si-H); 6.93-7.26 (Si-C_6H_5).

FT-IR (KBr, cm^{-1}): 3065 and 3045 (C-Harom.); 2955 (C-Haliph.); 2085 (Si-H); 1482; 1426 and 1093 (Si-C_6H_5); 1257 and 869 (Si-CH_3); 695 (Si-C); 468 (Si-Si).

GPC: Mw = 4495 g/mol; Mw/Mn= 1.5.

UV absorption (toluene): 310 nm.

3.2.3. PSH/BT Composite Preparation

Two samples of the PSH/BT composite were prepared with 10% (PSH/BT1) and 20% PSH/BT2 barium titanate.

For this purpose, 50 mL of Milli-Q water was mixed with 1 g of PSH, 0.1 g of BT (PSH/BT1), and 1 g of PSH, 0.2 g of BT (PSH/BT2). The resulting suspensions were ultrasonicated for 10 min at 20 kHz ultrasound frequency (Table 2) into a glass vial of 100 cc. These technical conditions were established in accordance with Santos [25] and other previous works [26–28] with the ultrasound probe immersed in liquid for at least 40% of its length and the distance to the vial walls as minimal as possible. The aqueous dispersions were then filtered, and the solid samples were vacuum dried overnight at 50 °C. The two composite samples resulted in solid yellowish-white powders.

Table 2. Samples and the terms of the synthesis.

Sample	Ratio PSH/BT w/w	Homogenization Time by Ultrasonication [min]	Total Energy Dissipated in the Form of Ultrasonic Waves [kJ]	Temperature at the End of Synthesis [°C]	UV Irradiation (30 min at 365 nm)
PSH/BT_1	1/0.1	10	24.7	78	no
PSH/BT_2	1/0.2	10	26.4	81	no
PSH_{uv}	-	-	-	-	yes
PSH/BT_{1uv}	1/0.1	10	24.7	78	yes
PSH/BT_{2uv}	1/0.2	10	26.4	81	yes

4. Conclusions

A composite polysilane–barium titanate was synthesized in two concentrations through an ultrasound homogenization procedure. Following the dielectric measurements, it was found that the composites have a dielectric constant that is much lower than that of the ferroelectric precursor (barium titanate) but higher than that of the polymer precursor (polysilane). Additionally, the new composite shows a degradation in the presence of ultraviolet radiation that is much more attenuated in comparison with the pristine polysilane. Following UV/VIS absorption measurements, it was found that PSH absorbs strongly in the ultraviolet range, while the composite behaves similarly to BT. Although PSH behaves almost like a semiconductor, the dielectric spectrometry analysis determined that BT is a ferroelectric material due to its high dielectric constant and low dielectric losses. In contrast to the polymer matrix, the composite polymer has a higher dielectric constant and a lower loss permittivity. Following these results, the new composite could be used in electronic applications used in UV-irradiated environments.

Author Contributions: Conceptualization, R.R. and L.S.; methodology, E.U. and M.E.F.; software, R.R. and L.S.; validation, R.R., M.E.F. and L.S.; formal analysis, E.U. and R.R.; investigation, R.R., E.U. and M.E.F.; resources, E.U.; data curation, M.E.F. and E.U.; writing—original draft preparation, R.R. and M.E.F.; writing—review and editing, R.R., E.U. and L.S.; visualization R.R. and M.E.F.; supervision, L.S. and M.E.F. All authors have read and agreed to the published version of the manuscript.

Funding: This research received no external funding.

Data Availability Statement: Data are contained within the article.

Conflicts of Interest: The authors declare no conflicts of interest.

References

1. Nowacki, B.; Mistewicz, K.; Hajra, S.; Kim, H.J. 3D printed triboelectric nanogenerator for underwater ultrasonic sensing. *Ultrasonics* **2023**, *133*, 107045. [CrossRef] [PubMed]
2. Markovic, S.; Mitric, M.; Starcevic, G.; Uskokovic, D. Ultrasonic de-Agglomeration of Barium Titanate Powder. *Ultrason. Sonochem.* **2008**, *15*, 16–20. [CrossRef] [PubMed]

3. Patsidis, A.C. Barium titanate/polydimethylsiloxane nano/microcomposites fabrication, morphology, dielectric response and functionality. *IET Nanodielectr.* **2020**, *1*, 14–19. [CrossRef]
4. Yajima, S.; Hayashi, J.; Okamura, K. Pyrolysis of a polyborodiphenylsiloxane. *Nature* **1977**, *266*, 521. [CrossRef]
5. Miller, R.D.; Michl, J. Poly(di-n-hexyl-silane) in solid solutions: Experimental and theoretical studies of electronic excitations of a disordered linear chain. *Chem. Rev.* **1989**, *89*, 1359. [CrossRef]
6. Nespurek, S.; Pospisil, J.; Kratochvilova, I.; Sworakowski, J. Polymeric Composites Based on Polysilanes for Plastic Electronics. *Mol. Cryst. Liq. Cryst.* **2008**, *484*, 265–631. [CrossRef]
7. Feigl, A.; Bockholt, A.; Weis, J.; Rieger, B. Modern Synthetic and Application Aspects of Polysilanes: An Underestimated Class of Materials? In *Silicon Polymers: Advances in Polymer Science*; Muzafarov, A.M., Ed.; Springer: Berlin/Heidelberg, Germany, 2009; Volume 235, pp. 1–31.
8. Kumar, V.B.; Leitao, E.M. Properties and applications of polysilanes. *Appl. Organomet. Chem.* **2020**, *34*, e5402. [CrossRef]
9. Qin, M.; Zhang, L.; Wu, H. Dielectric loss mechanism in electromagnetic wave absorbing materials. *Adv. Sci.* **2022**, *9*, 105553. [CrossRef] [PubMed]
10. Sundar, U.; Lao, Z.; Cook-Chennault, K. Enhanced Dielectric Permittivity of Optimized Surface Modified of Barium Titanate Nanocomposites. *Polymers* **2020**, *12*, 827. [CrossRef]
11. Fortună, M.E.; Ignat, M.; Tudorachi, N.; Ungureanu, E.; Rotaru, R.; Harabagiu, V. Hybrid Siloxane Materials Based on a Mutually Reactive Epoxy–Amine System: Synthesis, Structure, and Thermal Stability Investigations. *Inorganics* **2024**, *4*, 118. [CrossRef]
12. Um, M.H.; Lee, C.T.; Kumasawa, H.J. Preparation and dielectric properties of ferroelectric barium titanate fine particles by hydrothermal method. *J. Ind. Eng. Chem.* **1997**, *3*, 251–256.
13. Liu, P.; Hong Ng, V.M.; Yao, Z.; Zhou, J.; Lei, Y.; Yang, Z.; Lv, H.; Kong, L.B. Facile synthesis and hierarchical assembly of flowerlike NiO structures with enhanced dielectric and microwave absorption properties. *ACS Appl. Mater. Interfaces* **2017**, *9*, 16404–16416. [CrossRef]
14. Liu, G.H.; Wei, C.Y.; Huang, T.; Wang, F.; Chang, J.F.; Sun, Q.; Zhang, X.H. Metal-catalyzed carbon foams synthesized from glucose as highly efficient electromagnetic absorbers. *Materials* **2024**, *17*, 3488. [CrossRef] [PubMed]
15. Miller, R.D.; Wallraff, G.; Clecak, N.; Sooriyakumaran, R.; Michl, J.; Karatsu, T.; McKinle, A.J.; Klingensmith, K.A.; Downing, J. Polysilanes: Solution Photochemistry and Deep-UV Lithography. *Polym. Eng. Sci.* **1989**, *29*, 882–886. [CrossRef]
16. Rotaru, R.; Popescu, C.M.; Dascalu, A.; Timpu, D.; Asanduleasa, M.; Fortuna, M.E.; Harabagiu, V. Influence of Ultrasonic Treatment and Heating/Cooling Under Electric Field on High-K Cellulose-Barium Titanate Composites. *Rev. Roum. Chim.* **2023**, *68*, 175–185. [CrossRef]
17. Astapenko, V. *Polarization Bremsstrahlung on Atoms, Plasmas, Nanostructures and Solids, Chapter Polarization Bremsstrahlung on Nanostructures*; Springer: Berlin/Heidelberg, Germany, 2013; pp. 207–243, ISBN 978-3-642-34082-6.
18. Devan, R.S.; Chougule, B.K. Effect of composition on coupled electric, magnetic, and dielectric properties of two phase particulate magnetoelectric composite. *J. Appl. Phys.* **2007**, *101*, 014109. [CrossRef]
19. Lines, M.E.; Glass, A.M. The Ferroelectric Perovskites—Barium Titanate. In *Principles and Applications of Ferroelectric and Related Materials*; Clarendon Press: Oxford, UK, 2001; pp. 241–245.
20. Paunovic, V.; Mitic, V.V.; Kocic, L. Dielectric characteristic of donor-acceptor modified $BaTiO_3$ ceramics. *Ceram. Int.* **2016**, *42*, 11692–11699. [CrossRef]
21. Mitic, V.V.; Paunovic, V.; Kocic, L. Fractal approach to BaTiO3- ceramics micro-impedances. *Ceram. Int.* **2015**, *41*, 6566–6574. [CrossRef]
22. Wang, Q.; Che, J.; Wu, W.; Hu, Z.; Liu, X.; Ren, T.; Chen, Y.; Zhang, J. Contributing factors of dielectric properties for polymer matrix composites. *Polymers* **2023**, *15*, 590. [CrossRef]
23. Taherian, R. *Electrical Conductivity in Polymer-Based Composites: Experiments, Modelling, and Applications*; Elsevier: Amsterdam, The Netherlands, 2018.
24. Sacarescu, L.; Fortuna, M.; Soroceanu, M.; Cojocaru, C.; Sacarescu, G.; Simionescu, M.; Harabagiu, V. Computational Study of the Electronic Absorption Spectra of Polyhydrosilanes. *Silicon* **2015**, *7*, 343–349. [CrossRef]
25. Santos, H.M.; Lodeiro, C.; Capelo-Martinez, J.L. The Power of Ultrasound. In *Ultrasound in Chemistry: Analytical Applications*; Capelo-Martinez, J.L., Ed.; Wiley-VCH Verlag GmbH & Co., KGaA: Weinheim, Germany, 2008; pp. 1–15.
26. Shokrollahi, H. A review of the magnetic properties, synthesis methods and applications of maghemite. *J. Magn. Magn. Mater.* **2017**, *426*, 74–81. [CrossRef]
27. Fatimah, I.; Fadillah, G.; Yudha, S.P. Synthesis of iron-based magnetic nanocomposites: A review. *Arab. J. Chem.* **2024**, *17*, 105405. [CrossRef]
28. Ahmed, M.A.; Seddik, U.; Okasha, N.; Imam, N.G. One-dimensional nanoferroic rods; synthesis and characterization. *J. Mol. Struct.* **2015**, *1099*, 330–339. [CrossRef]

inorganics

MDPI

Review

Bridgman Method for Growing Metal Halide Single Crystals: A Review

Hui Zhu [1,†], Suqin Wang [1,†], Ming Sheng [1], Bo Shao [1], Yu He [1], Zhuang Liu [2,*] and Guangtao Zhou [1,*]

1 College of Engineering, Shandong Xiehe University, Jinan 250109, China; zhuhui@sdxiehe.edu.cn (H.Z.); wangsuqin@sdxiehe.edu.cn (S.W.); shengming@sdxiehe.edu.cn (M.S.); shaobo@sdxiehe.edu.cn (B.S.); heyu@sdxiehe.edu.cn (Y.H.)

2 School of Materials Science and Engineering, Shandong University, Jinan 250061, China

* Correspondence: liu.zhuang@sdu.edu.cn (Z.L.); charlie@sdxiehe.edu.cn (G.Z.)

† These authors contributed equally to this work.

Abstract: The Bridgman method for single-crystal growth enables the formation of crystals at the lower end of the molten material by cooling it under a precisely controlled temperature gradient. This makes it particularly suitable for producing high-quality single-crystal materials. Over the years, the Bridgman technique has become widely adopted for growing single crystals of semiconductors, oxides, sulfides, fluorides, as well as various optoelectronic, magnetic, and piezoelectric materials. Recently, there has been growing interest in metal halide materials, with the growth of high-quality metal halide single crystals emerging as a major focus for both the scientific community and industry. However, traditional solution-based single-crystal growth methods have several limitations, such as slow growth rates, inconsistent crystal quality, challenges in solvent selection, and difficulties in controlling saturation levels. These issues present significant obstacles, particularly when large, defect-free, high-quality single crystals are needed for certain high-performance materials. As a result, the Bridgman method has emerged as an effective solution to overcome these challenges. This review provides an overview of various categories of metal halide single-crystal systems grown using the Bridgman method in recent years. The systems are classified based on their dimensionality into three-dimensional, two-dimensional, and zero-dimensional metal halide structures. Furthermore, we highlight novel metal halide single crystals developed through the Bridgman technique. Additionally, we offer a brief introduction to the structures, properties, and applications of these single crystals, underscoring the crucial role of the Bridgman method in advancing research in this field.

Keywords: metal halide; single crystal; Bridgman method; photoelectric properties; applications

1. Introduction

The Bridgman method for single-crystal growth is a pivotal technique, which has been extensively utilized in the preparation of single crystals for semiconductor materials, optoelectronic devices, and other high-performance materials [1–3]. First proposed by the American scientist Theodore Bridgman in 1915, it has since evolved into one of the most established technologies for producing single crystals with high quality [4]. The Bridgman method is renowned for its process stability, precise controllability, and stringent material requirements, making it a fundamental technique in materials science, particularly in research focused on high-purity and high-quality single crystals [5–7].

1.1. Basic Principle

As illustrated in Figure 1a, the Bridgman method operates on the principle of the dissolution-crystallization process [8–11]. Through controlling the temperature gradient precisely, the material within the solution gradually cools, facilitating the transition from the liquid solution to solid state, thereby yielding a single crystal. In this process, the material is first heated above its melting point to form a liquid phase, and then it is progressively cooled under a controlled temperature gradient, promoting the crystallization of the material from liquid state to solid status [12,13]. The Bridgman method is particularly effective for materials with high melting points that can maintain stability at elevated temperatures, such as certain semiconductor materials, ceramics, and metal alloys [14].

It should be noted that there are necessary ways to determine the appropriate method in advance for growing different halide single crystals. In other words, not all halide single crystals can be grown using the Bridgman method. Although a large body of previous work suggests that potential halides can be predicted using simulation methods, growing these potential halide single crystals faces significant uncertainty [15,16]. A phase diagram is a concise graphical representation of the relationship between the components, temperature, and pressure of a system at thermodynamic equilibrium. Initially, phase diagrams were obtained through extensive experiments [17]. With the development of computational materials science, computational phase diagrams based on molecular dynamics have become an efficient method for obtaining phase diagrams [18–20]. For halides, by referencing phase diagrams of related components, it is possible to determine whether the single crystal of the compound is likely to grow easily [21]. Taking the binary phase diagram of chlorides as an example, a large number of previous works have summarized the synthesis conditions of approximately one thousand halides, including their melting points, whether they fully crystallize, lattice structures, and space groups [22]. In the relevant literature, researchers have even been able to compile statistics on how many reported works have successfully grown halide single crystals and the preparation methods they used. In the future, as more and more new halides are successfully synthesized, these phase diagrams will continue to expand, becoming a reference handbook for researchers. It should be noted that, compared to solution-based crystal growth, the progress of phase diagrams in this field will be slower because methods for growing large single crystals, such as the Bridgman method, require more expensive and larger-scale growth equipment, which is not always available for researchers.

1.2. Experimental Setup for the Bridgman Method

As exhibited in Figure 1b, the experimental setup for the Bridgman method comprises several essential components: a high-temperature furnace, a crucible, a cooling system, and a temperature gradient control system [8,23]. The high-temperature furnace is used to heat the crucible and create a high-temperature environment for the process. The furnace's temperature control system allows for precise regulation of the heating temperature and distribution, ensuring uniform heating and melting of the material [24]. The crucible serves as the container for the material to be crystallized and is typically constructed from high-temperature-resistant materials such as quartz or platinum [25]. The selection of the crucible material must account for both chemical stability and resistance to high temperatures. The Bridgman method relies on a cooling process to guide the material from liquid phase to a solid state [26]. Therefore, a cooling system is essential to facilitate this phase of the process. Water or air cooling systems are commonly employed to regulate the cooling rate. During this process, precise control over the temperature gradient is critical. The temperature gradient is a key factor influencing the success of the Bridgman method [27]. By controlling the temperature range of the furnace and the cooling system, a suitable temperature gradient can be maintained from the top to the bottom of the crucible, thereby promoting the preparation of high-quality single crystals.

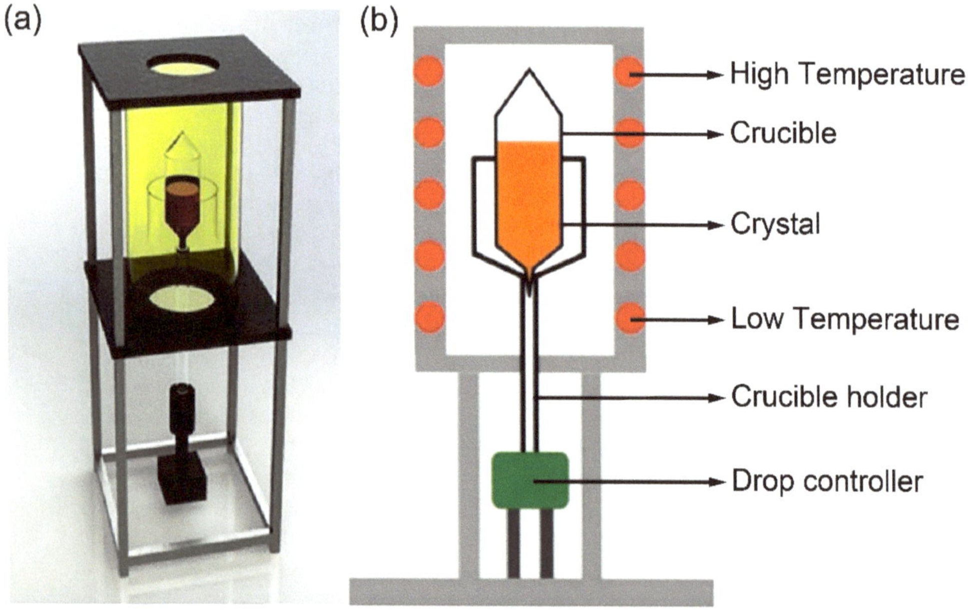

Figure 1. (**a**) Three-dimensional diagram of a typical Bridgman furnace. Reprinted with permission from Ref. [8]. Copyright 2021, Royal Society of Chemistry. (**b**) Planar graph of vertical Bridgman furnace and their key assemblies. Reprinted with permission from Ref. [23]. Copyright 2018, American Chemical Society.

1.3. Growth Process

The growth process using the Bridgman method for single crystals is typically divided into several stages. First, the raw materials intended for crystal growth are placed in the crucible and heated above their melting point to ensure complete transition into the liquid phase. During the melting process, it is crucial to maintain a stable temperature to prevent excessive bubble formation or the introduction of undesirable impurities [28]. Once the material has fully melted, the next step involves establishing a reasonable temperature gradient by adjusting the high/low-temperature zone. This gradient is achieved by gradually moving the crucible from the high-temperature zone to the lower-temperature zone [29]. Generally, the temperature at the bottom zone is lower, while the top remains hotter. This temperature differential prompts the material to begin crystallizing at the bottom of the crucible, initiating the growth of single crystals. As the temperature decreases, the material transitions from a liquid state to a solid phase. Under the influence of the temperature gradient, crystal nuclei begin to form in the lower-temperature region of the crucible. As the solution becomes supersaturated, crystals start to deposit on the liquid's surface. The process must proceed slowly to ensure smooth crystal growth and the formation of a single, high-quality crystal [30]. Over time, the crystal grows steadily out of the liquid and increases in size. When the crystal reaches the desired size, mechanical methods are employed to extract it from the crucible. To minimize the risk of damaging the crystal during removal, the temperature is typically lowered gradually to room temperature, and protective measures are implemented [31]. After extraction, further processing, such as cutting and polishing, is often conducted to achieve the certain shape and surface quality of the crystal [32].

1.4. Advantages of the Bridgman Method

The Bridgman method, as a traditional technique for single-crystal growth, offers several distinct advantages. First and foremost, it provides precise control over the temperature gradient, a critical factor in achieving high-quality single crystals [33]. The accuracy of temperature regulation directly impacts the quality of as-prepared crystals, making this feature particularly significant. Additionally, the Bridgman method is versatile, suitable for single-crystal growth from wide ranges of materials, including metals, semiconductors, ceramics, and more [34,35]. It is especially advantageous for growing materials with high melting points. Moreover, as the Bridgman method involves a transition from liquid state to solid phase, impurities in the liquid solution are effectively excluded, resulting in single crystals with high purity [36].

The Czochralski method, as another popular technology for single crystal growth, has often been compared to Bridgman method [37]. The Czochralski method is widely used in semiconductor materials and the single crystal growth of high-tech industries such as silicon and germanium [38]. Its advantages include higher growth rate, that the Czochralski method stretches the crystal while it is in a molten state, which results in a relatively fast growth rate, making it suitable for large-scale production [39]. Moreover, since the crystal is gradually stretched in a molten state, more uniform crystals with higher purity can be obtained. In addition, this method is capable of growing large single crystals, suitable for industrial production [40]. However, the Czochralski method has several disadvantages. For example, the equipment for this method is more complex and requires precise control of temperature, stretching rate, and other parameters, thus making it more expensive. In addition, since the crystal is grown by stretching, the process may introduce more defects, especially when not operating under suitable conditions [41].

In summary, the Bridgman method is suitable for high-melting-point materials and precision-demanding applications, but it has a slower growth rate, limited crystal size, and lower cost. The Czochralski method is better suited for large-scale production and single crystals with high purity and uniformity, but the equipment is more complex, costly, and has a limited range of applicability.

1.5. Current Mainstream Applications

The Bridgman method is extensively utilized for synthesis of single crystals for semiconductor materials, ceramics, and corresponding optoelectronic devices [34,35,42]. In the semiconductor industry, for instance, the Bridgman method is employed to grow high-quality single crystals such as gallium arsenide (GaAs), indium phosphide (InP), and other compounds [43,44]. The aforementioned materials are widely applied in fields such as integrated circuits, laser diodes, and solar cells. Additionally, the Bridgman method plays a crucial role in the research and development of high-temperature superconducting materials, ferroelectric materials, and other advanced materials [45–49].

Metal halide materials have garnered considerable attention due to their outstanding photoelectronic properties, particularly their promising potentials in photovoltaics, photocatalysis, and light-emitting devices [15,16,50,51]. For example, all-inorganic perovskite nanocrystal $CsPbBr_3$ has been used as scintillators to provide convenient visualization tool for X-ray radiography [52]. In addition, recent studies indicate that $CsPbBr_3$ nanocrystals can also be applied in quantum computing and quantum communications owing to their excellent single-photon feature when assembled to be superlattice [51]. Metal halides with lower dimensions possess unique properties. For example, zero-dimensional Cs_2MoCl_6

and Cs_2WCl_6 single crystals emit near-infrared light, which can be used in safety inspection and night vision [53]. Low-dimensional organic–inorganic hybrid metal halides as subgroup halide compounds have been widely used in optoelectronic domains. For example, one-dimensional $(C_5H_{11}N_3)MnCl_2Br_2 \cdot H_2O$ exhibited water-molecule-induced reversible photoluminescence transformation, which has been used in anti-counterfeiting field [54].

Metal halide materials are characterized by high melting points and a strong sensitivity to impurities, which means that the single-crystal growth process demands precise control over factors including temperature, liquid concentration, and cooling speed [55–58]. The Bridgman method operates by gradually cooling the material under a controlled temperature gradient, facilitating the transition from liquid state to solid phase. This method allows for effective regulation of both the crystal growth rate and its quality. Moreover, the crystal quality and growth rate of metal halide single crystals are closely dependent on the temperature gradient [59]. The Bridgman method excels in providing a uniform and stable temperature gradient throughout the growth process, making its temperature control capabilities essential for producing high-quality single crystals. Additionally, the cooling process inherent in the Bridgman method effectively excludes impurities, ensuring the formation of high-purity crystals. The removal of impurities is crucial for enhancing the optoelectronic performance of metal halide materials [60]. Furthermore, given the high melting points of many metal halide materials, the Bridgman method's capacity to operate at elevated temperatures makes it particularly well-suited for growing perovskite materials, which require high-temperature conditions [61].

2. Bridgman Method for Growing Three-Dimensional Metal Halide Single Crystals

Three-dimensional (3D) metal halides represent the largest subgroup of metal halides, distinguished by their unique crystal structures and exceptional optoelectronic properties [62–66]. In recent years, they have attracted widespread attention in fields such as optoelectronic devices, photovoltaics, photocatalysis, and light-emitting diodes (LEDs) [67,68]. Three-dimensional metal halides are known for their excellent light absorption properties, including broad absorption spectra and efficient absorption cross-sections, which make them highly effective in energy conversion applications, particularly in photovoltaic devices [69,70]. Notably, in solar cells, 3D metal halides can achieve high energy conversion efficiencies even under low light intensities, positioning them as a key area of research in solar energy. In addition to their optical properties, 3D metal halides exhibit large carrier mobility and excellent carrier diffusion, which makes them highly efficient for use in electronic devices [71–73].

However, 3D metal halides are easily to be degraded when exposed to water and oxygen, which can significantly impair their optoelectronic performance [74,75]. As a result, enhancing the environmental stability and longevity of these materials has become a major focus of ongoing research. Fortunately, large-sized 3D metal halide single crystals prepared by the Bridgman method have demonstrated significantly improved stability [76–78]. Research in this area has experienced rapid growth in recent years, offering promising prospects for addressing these challenges.

Three-dimensional metal halides often undergo nondestructive phase transitions at relatively low temperatures [79]. Taking $CsPbBr_3$ as an example, as shown in Figure 2a, the first phase transition occurs around 403 K, where the material shifts from the cubic phase to tetragonal configuration, with the feature of first-order phase transition. After that, a second-order transition occurs at 361 K to the orthorhombic phase, which finally becomes stable at room temperature [80]. However, residual stress still exists after these transitions, which can induce mechanical deformation and crack formation. Such imperfections can significantly degrade the properties of detectors, highlighting the importance of carefully managing these transitions during the crystal growth process.

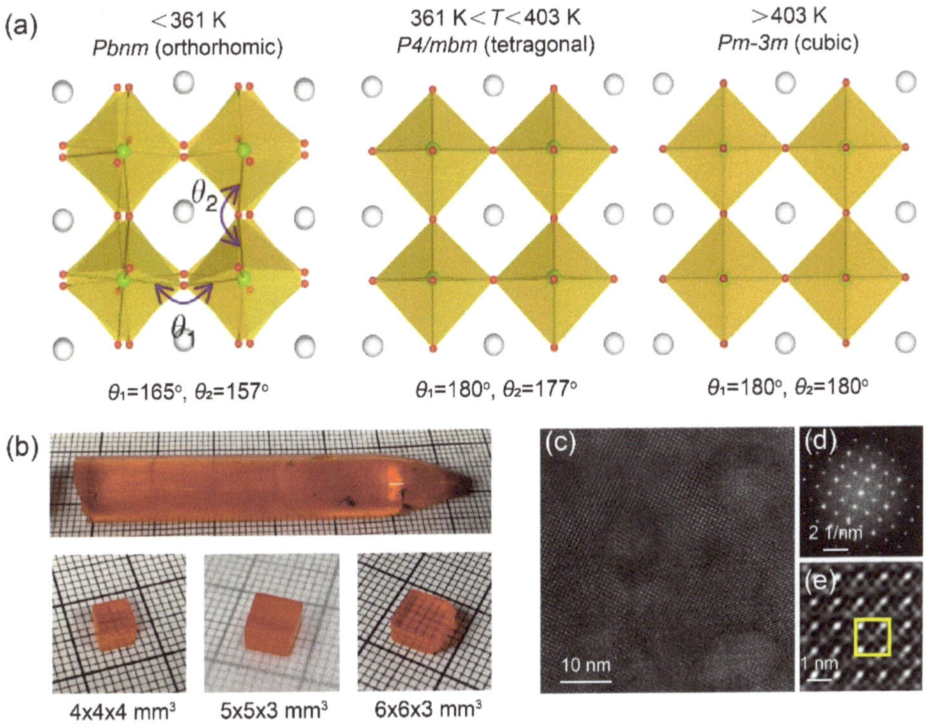

Figure 2. (**a**) Phase transitions in $CsPbBr_3$, during which the $PbBr_6$ octahedra exhibit distortion. (**b**) Photos of as-prepared large-sized $CsPbBr_3$ single crystal. Bottom pictures are the single crystal wafers of different sizes. (**c**) HRTEM picture of 3D $CsPbBr_3$ single crystal. (**d**) The corresponding SAED pattern and (**e**) magnified lattice image. The yellow box is drawn to compare HRTEM image. Reprinted with permission from Ref. [80]. Copyright 2018, Nature Publishing Group.

In a recent study, researchers presented an enhanced Bridgman method for growing $CsPbBr_3$ single crystals, significantly improving their optoelectronic performance [80]. The as-grown $CsPbBr_3$ single crystals, with dimensions reaching several centimeters, exhibit exceptionally low impurity concentrations (below 10 ppm across a total of 69 elements). The critical process in this work is the modification of growth parameters, which leads to large-sized, crack-free single crystals, as shown in Figure 2b. It is important to highlight that the precise control of kinetic parameters during the crystal growth process is critical. Parameters including temperature gradient and cooling rate play pivotal roles, and have been meticulously optimized through growing many times. Notably, the impurity concentration in final products is below 10 ppm, with impurities exceeding 0.5 ppm likely being introduced from the precursor materials. Additionally, the optical transmittance is higher than 65% for wavelengths greater than 600 nm and exceeds 80% for wavelengths above 1000 nm. These findings confirm that the $CsPbBr_3$ single crystals exhibit exceptional chemical purity, which is one of the distinct advantages of the Bridgman method [81]. The single-crystallinity and outstanding uniformity of the $CsPbBr_3$ crystals were verified through selected area electron diffraction (SAED) and high-resolution transmission electron microscopy (HRTEM) with fast Fourier transform (FFT) analysis, as shown in Figure 2c–e. No precipitates were detected in these single crystals, further confirming their high chemical purity. Consequently, the scalable growth of 3D $CsPbBr_3$ single crystals underscores the potential of metal halides as exceptional next-generation materials for photoelectric applications [82,83].

In another study, large-sized 3D $CsPbBr_3$ single crystals were grown using an optimized Bridgman method, as shown in Figure 3a [23]. After determining the crystallographic directions using the X-ray orientation technique, the anisotropic optoelectronic properties of the material were investigated for the first time. To examine the optoelectronic anisotropy, photoresponse measurements were performed on the (100), (010), and (001) crystallographic planes, as shown in Figure 3b. The three $CsPbBr_3$ single crystal devices possessed distinct anisotropic optoelectronic properties. The responsivity and external quantum efficiency (EQE) of the best plane demonstrated a three-fold improvement compared to other planes. This unique behavior can be attributed to the structural anisotropy of $CsPbBr_3$, which determines the charge carrier's transport. The anisotropic electron-transport properties in $CsPbBr_3$ single crystals were further characterized, as shown in Figure 3c [84]. In the current–voltage curves, Ohmic, trap-filling, and Child regions were detected as the bias voltage increased. These results reveal a pronounced anisotropy in the electron transport properties, with superior electron transport observed along a certain direction. This is due to the variations in the carrier transport within the crystal structure [85–87]. Consequently, the exceptional optoelectronic properties of $CsPbBr_3$ single crystals prepared via the Bridgman method are expected to inspire further exploration of their potential for high-performance optoelectronic applications.

The detection of γ-rays with high energy resolution remains the most challenging applications today [88–90]. Even the smallest presence of defects can significantly degrade the signal generated by γ-rays, making semiconductor detectors relatively rare. However, lead halide perovskite semiconductors exhibit exceptional defect tolerance, resulting in outstanding and unique optoelectronic properties [91–93]. As a result, these materials are poised to have a transformative impact on applications in photoelectric conversion and radiation detection. In a recent study, researchers demonstrated that $CsPbCl_3$ single crystals grown by the Bridgman method can act as high-performance detectors for γ-ray radiation [94]. Similarly to $CsPbBr_3$, $CsPbCl_3$ undergoes two phase transitions: from the cubic to the tetragonal phase at 325 K, followed by a transition to the orthorhombic phase at 316 K, which finally becomes stable at room temperature (Figure 4a) [95]. Large-scale $CsPbCl_3$ single crystals were synthesized using the Bridgman method, as shown in Figure 4b. The resulting $CsPbCl_3$ crystal exhibits pale-yellow with high optical transparency. The optical transmission exceeds 80% for wide light wavelengths, as shown in Figure 4c. This high transparency indicates that $CsPbCl_3$ single crystals contain few impurities. The optical absorption edge was around 425 nm, corresponding to a wide bandgap of 2.90 eV. Therefore, the wide-bandgap $CsPbCl_3$ single crystals obtained through the Bridgman method hold significant potential for γ-ray detection [96–98].

The Bridgman method can also be used to grow 3D metal halide single crystals with mixed compositions [99–102]. In a recent study, series of large-sized perovskite single crystals $CsPbBr_{3-3n}X_{3n}$ were fabricated by the modified Bridgman method through precisely controlling the synthesis parameters, as exhibited in Figure 5a [8]. The sharp absorption edges in the absorption spectra indicate the direct-band-gap feature of these perovskite single crystals, as depicted in Figure 5b. Furthermore, the absorption edges of them are continuously tunable, corresponding to wide bandgaps, by only adjusting the halide composition [103]. Consequently, perovskite single crystals with mixed halogens exhibit considerable potential as filter-free photodetectors.

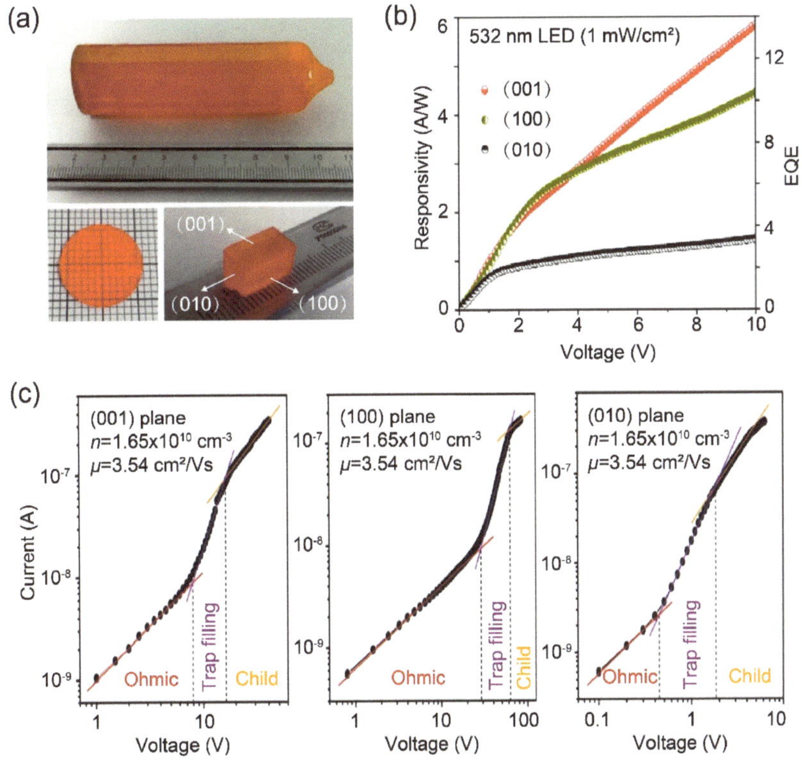

Figure 3. (**a**) Pictures of large-sized $CsPbBr_3$ single crystals (top), cut wafer (bottom left), and cuboid $CsPbBr_3$ single crystals (bottom right). (**b**) Responsivity and EQE for the three crystallographic planes. (**c**) Dark current–voltage features of $CsPbBr_3$ single crystals. Reprinted with permission from Ref. [23]. Copyright 2018, American Chemical Society.

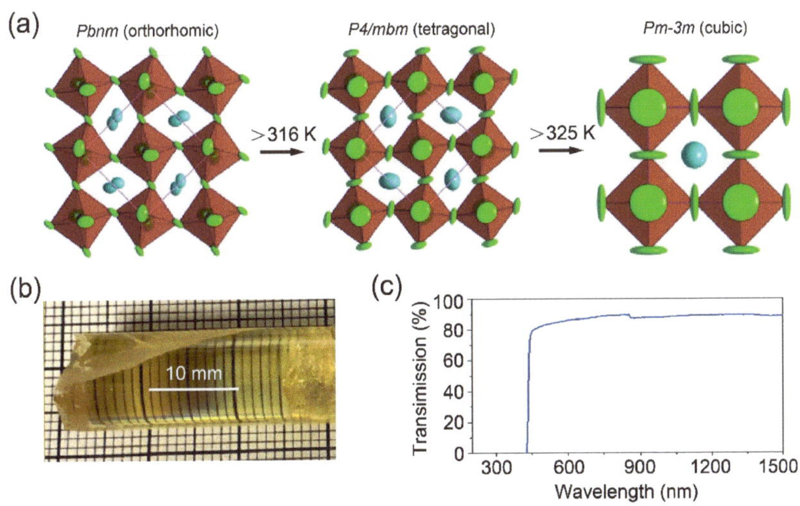

Figure 4. (**a**) Structural evolution of $CsPbCl_3$ near ambient temperature. (**b**) Photograph of $CsPbCl_3$ single crystal grown by the Bridgman method and (**c**) its optical transmission. Reprinted with permission from Ref. [94]. Copyright 2021, American Chemical Society.

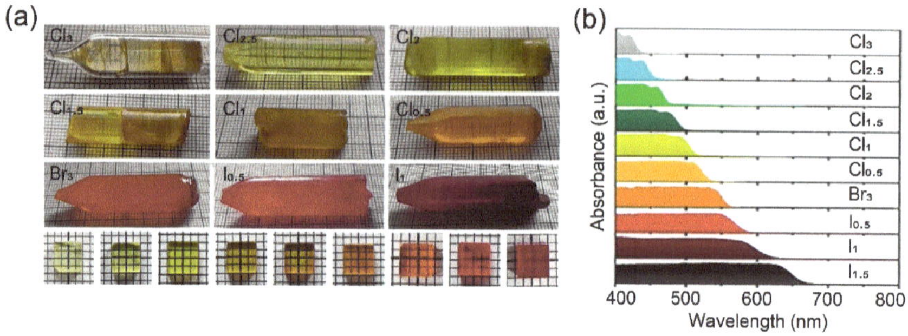

Figure 5. (**a**) Photographs of $CsPbBr_{3-3n}X_{3n}$ single crystals with mixed halogens and (**b**) their corresponding UV–Vis absorption spectra. Reprinted with permission from Ref. [8]. Copyright 2021, Royal Society of Chemistry.

3. Bridgman Method for Growing Two-Dimensional Metal Halide Single Crystals

The two-dimensional (2D) Ruddlesden–Popper (RP) phases constitute a significant class of metal halides, characterized by their diverse and remarkable optoelectronic properties [104–106]. To date, the majority of hybrid RP phases incorporate long organic spacers [107,108]. However, conventional synthesis methods are predominantly limited to the growth of 2D RP metal halides with organic components. In another word, all-inorganic 2D RP metal halides cannot be prepared. In other words, synthesizing all-inorganic 2D RP metal halides presents significant challenges.

Recent studies have successfully synthesized all-inorganic 2D RP $Cs_2PbI_2Cl_2$ single crystals using the Bridgman method, as shown in Figure 6a [79,109]. These crystals are nearly centimeter-sized, significantly larger than those obtained through solution-based methods. $Cs_2PbI_2Cl_2$ adopts the K_2NiF_4-type configuration and crystallizes in tetragonal phase, as illustrated in Figure 6b–d [110]. The 2D $[PbI_2Cl_2]^{2-}$ plane is composed of corner-sharing Pb-centered $[PbI_2Cl_4]^{4-}$ units. In addition, Cl^- ions occupy in-plane sites, behaving as shared corners of octahedra, while I^- ions occupy out-of-plane sites as terminal ligands. Cs^+ spacer ions balance the charge, resulting in a standard single-layer (n = 1) RP structure. Density functional theory calculations, along with experimental findings, support the thermodynamic stability of these single crystals. Further computational analysis indicates that $Cs_2PbI_2Cl_2$ behaves as a direct bandgap semiconductor with low effective carrier mass towards in-plane direction, which aligns with the experimentally obtained in-plane photoresponse [111,112]. The successful growth of $Cs_2PbI_2Cl_2$ by the Bridgman method not only expands the family of 2D metal halides but also opens new possibilities for applications beyond photovoltaics.

Subsequent efforts were made by researchers to grow larger single crystals using the Bridgman method. Through meticulous adjustment of synthesis parameters, including the cooling rate, temperature gradient, and post-heat treatment, large-sized single crystals were successfully achieved, as exhibited in Figure 7a [113]. Structural analysis reveals that the single-layer RP structure is composed of corner-sharing $[PbI_2Cl_4]^{4-}$ octahedral units, as depicted in Figure 7b. Further investigations confirm that the remarkable phonon transport properties of $Cs_2PbI_2Cl_2$, driven by lone-pair-driven octahedral rotations and anharmonic lattice dynamics, which demonstrates the potential for novel thermal transport behavior across a broad range of low-dimensional metal halide perovskites [114].

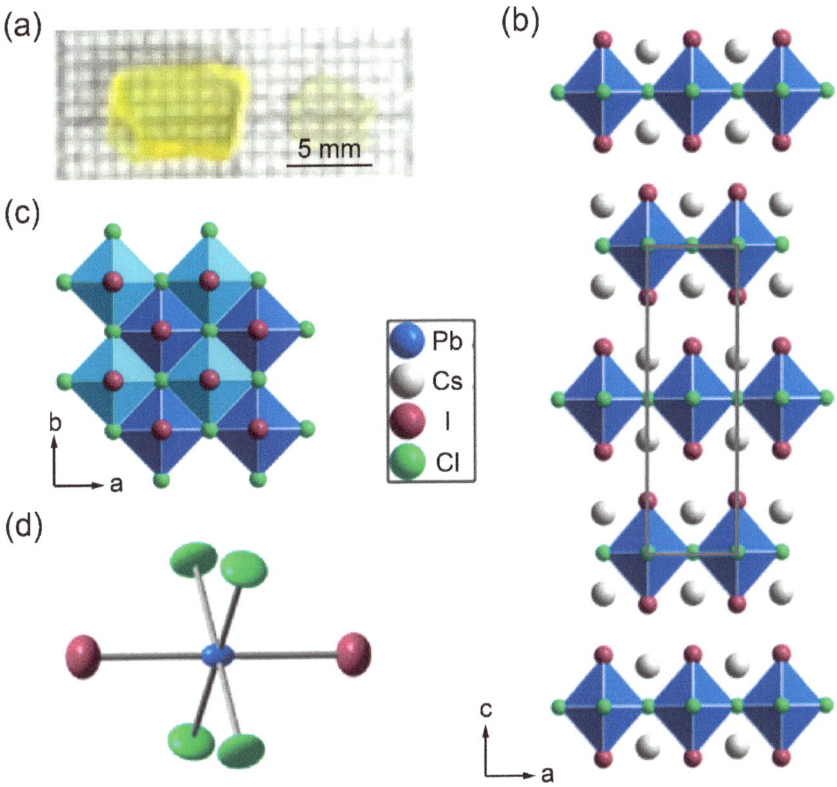

Figure 6. (**a**) Picture of $Cs_2PbI_2Cl_2$ single crystal grown using the Bridgman method. Crystal structure of $Cs_2PbI_2Cl_2$, shown from (**b**) the side view and (**c**) the top-down view. Additionally, (**d**) depicts the elongated $[PbI_2Cl_4]^{4-}$ octahedral unit. Reprinted with permission from Ref. [109]. Copyright 2018, American Chemical Society.

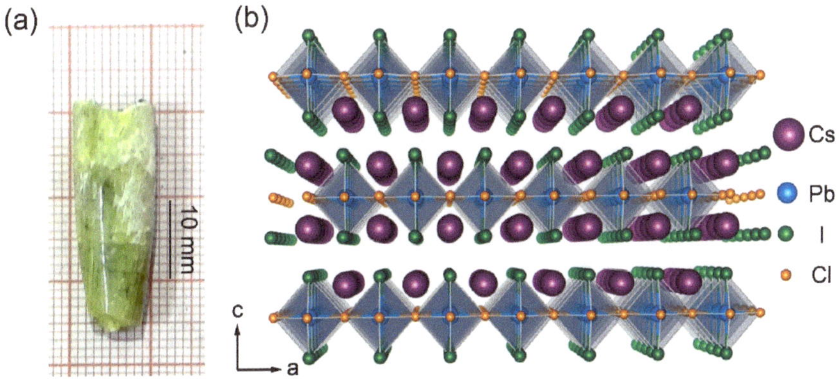

Figure 7. (**a**) Picture of $Cs_2PbI_2Cl_2$ single crystals prepared by the Bridgman method. (**b**) Crystal structure of the 2D RP metal halide $Cs_2PbI_2Cl_2$. Reprinted with permission from Ref. [113]. Copyright 2020, American Chemical Society.

4. Bridgman Method for Growing Zero-Dimensional Metal Halide Single Crystals

In recent years, less toxic metals such as Sn, Sb, and Bi have attracted considerable attention as environmentally friendly alternatives to Pb [115–119]. Among these, low-dimensional Bi-based halide materials have emerged as promising candidates for optoelectronic applications [120–122]. Notably, all-inorganic zero-dimensional (0D) bismuth-halide single crystals of $Cs_3Bi_2X_9$ (X = Br and I) have been successfully synthesized using the Bridgman method.

As shown in Figure 8a, a custom-designed semitransparent Bridgman furnace has been developed to enable real-time monitoring and precise adjustment of crystal growth parameters during the synthesis process [123]. This innovative furnace facilitated the optimization of the thermal profile for growing high-quality $Cs_3Bi_2Br_9$ single crystals, as illustrated in Figure 8b. Using this modified Bridgman method, large-sized single crystals $Cs_3Bi_2Br_9$ were successfully synthesized. To minimize structural defects and prevent crystal cracking, small temperature gradient and slow growth rates have been proved essentially. Specifically, slower rates were employed to grow transparent, crack-free $Cs_3Bi_2Br_9$-2 crystals, while faster rates resulted in cracked $Cs_3Bi_2Br_9$-1 crystals, as shown in Figure 8c. Figure 8d displays the optical transmittance spectra of these two single crystals. Both exhibit a broad transparency range with no significant absorption features. Notably, $Cs_3Bi_2Br_9$-2 demonstrates an average transmittance of approximately 80% in the range of 2 μm to over 18 μm, surpassing the ~70% transmittance of $Cs_3Bi_2Br_9$-1. At room temperature, $Cs_3Bi_2Br_9$ crystallizes into trigonal system, with its crystal structure depicted in Figure 8e [124]. The layered perovskite structure of $Cs_3Bi_2Br_9$ can be regarded as a tripling of typical perovskite unit cell, where only partial octahedra are fully occupied. The remaining octahedra form corrugated, vacant layers that separate the occupied layers. This arrangement enables a corner-sharing configuration of $[BiBr_6]^-$ octahedra [125].

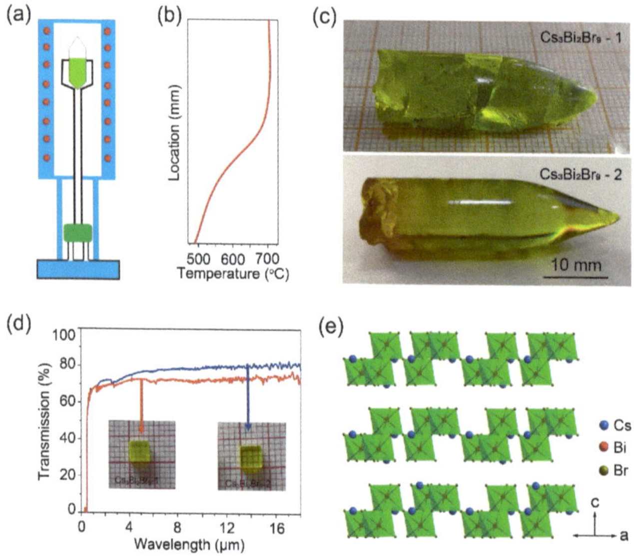

Figure 8. (**a**) Schematic illustration of the Bridgman method used for the synthesis of $Cs_3Bi_2Br_9$ single crystals. (**b**) The optimized thermal profile. (**c**) Pictures of $Cs_3Bi_2Br_9$-1 (top) and $Cs_3Bi_2Br_9$-2 (bottom). (**d**) Transmission spectra of $Cs_3Bi_2Br_9$-1 and $Cs_3Bi_2Br_9$-2. Insets show the oriented (−1 2 0) planes of these two single crystals. (**e**) Crystal structure of $Cs_3Bi_2Br_9$. Reprinted with permission from Ref. [123]. Copyright 2021, Science China Press.

Subsequently, researchers proposed a mixed-halogen strategy to fine-tune the structural dimensions and optoelectronic properties of $Cs_3Bi_2I_{9-n}Br_n$ [126]. High-quality single crystals, with dimensions reaching up to ten centimeters, were successfully grown using the Bridgman technique, as depicted in Figure 9a. By incorporating iodine into the 0D $Cs_3Bi_2Br_9$ framework, the optoelectronic properties were systematically modified. Among these, $Cs_3Bi_2I_8Br$ emerged as the most promising candidate, demonstrating negligible ion migration, moderate resistivity, and superior carrier transport capabilities. The halogen-mixing strategy offers a pathway to balance carrier transport properties with resistivity in targeted compositions, optimizing performance for applications such as X-ray detection. Notably, $Cs_3Bi_2I_8Br$ was identified as the optimal composition, achieving a record-breaking sensitivity of 1.74×10^4 μC Gy^{-1} cm^{-2}. This sensitivity is second only to the 6.3×10^4 μC Gy^{-1} cm^{-2} achieved by Pb-based $CsPbBr_{2.9}I_{0.1}$ single crystals under 120 keV hard X-ray detection. Furthermore, X-ray detectors fabricated from $Cs_3Bi_2I_8Br$ single crystals demonstrated exceptional long-term stability in ambient air, ultralow dark current drift, and robust high-resolution imaging performance even at elevated temperatures. This combination of properties underscores the material's potential for next-generation X-ray detection technologies [127,128].

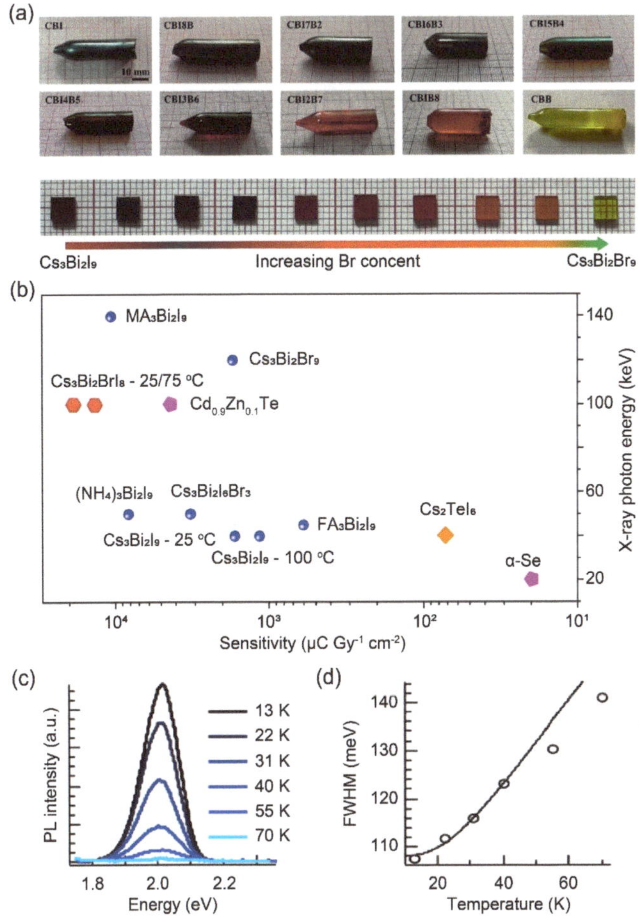

Figure 9. (**a**) Photographs of as-grown $Cs_3Bi_2I_{9-n}Br_n$ single crystals and their corresponding (001)-oriented plates. (**b**) Comparison of sensitivities between Pb-free perovskite single crystals, commercial

CdZnTe single crystals, and commercial α-Se. Reprinted with permission from Ref. [126]. Copyright 2023, Wiley-VCH GmBH. (**c**) Temperature-dependent photoluminescence spectra of $Cs_3Sb_2I_9$ at 5 mW excitation power. (**d**) Temperature-dependent FWHM fitted using *H-S* equation. Reprinted with permission from Ref. [129]. Copyright 2017, American Chemical Society.

The optical and electronic properties of Bridgman-grown single crystals of wide-bandgap semiconducting halide perovskites $A_3M_2I_9$ (A = Cs, Rb; M = Bi, Sb) have been extensively studied [130]. Intense Raman scattering was observed at room temperature for all compounds, indicating high polarizability and pronounced electron–phonon coupling [129]. Strong evidence of electron–phonon interactions, comparable to those found in alkali halides, was revealed through phonon broadening in the photoluminescence emission spectra, as shown in Figure 9c,d. Effective phonon energies, derived from temperature-dependent photoluminescence measurements, aligned well with the observed Raman peak energies. Based on these findings, a model has been proposed wherein electron–phonon interactions in $Cs_3Sb_2I_9$, $Rb_3Sb_2I_9$, and $Cs_3Bi_2I_9$ generate small polarons, resulting in exciton trapping by the lattice. The recombination of these self-trapped excitons accounts for the broad photoluminescence emission. Additionally, $Rb_3Bi_2I_9$, $Rb_3Sb_2I_9$, and $Cs_3Bi_2I_9$ exhibit high resistivity and strong photoconductive responses under laser photoexcitation. These characteristics suggest that these compounds are promising candidates for semiconductor applications, particularly in the detection of hard radiation.

5. Bridgman Method for Growing Novel Metal Halide Single Crystals

5.1. $Cs_3Cu_2I_5$ Single Crystals

Low-dimensional Cu(I) metal halides, known for their efficient exciton emissions, have recently gained attention as promising scintillation materials for X-ray and gamma-ray detection [131–133]. A recent study demonstrated the potential of 0D $Cs_3Cu_2I_5$ as a sensitive thermal neutron detector with effective neutron–gamma discrimination achieved through lithium doping [134,135]. Single crystals of $Cs_3Cu_2I_5$ doped with 95% enriched ^6Li were successfully grown using the Bridgman method, as illustrated in Figure 10a. As shown in Figure 10b, $Cs_3Cu_2I_5$:^6Li exhibits impressive performance metrics, including an energy resolution of 4.8% for 662 keV ^{137}Cs gamma rays, a high light yield of 30,000 photons/MeV for gamma rays, and 96,000 photons/neutron for thermal neutrons. Additionally, it achieves a robust neutron–gamma pulse shape discrimination figure of merit of 2.27 using Power Spectral Density (PSD) waveforms. This breakthrough discovery of 0D metal halide single crystals, specifically $Cs_3Cu_2I_5$:^6Li, represents a significant advancement in the development of stable, high-performance scintillators for neutron–gamma detection applications [136,137].

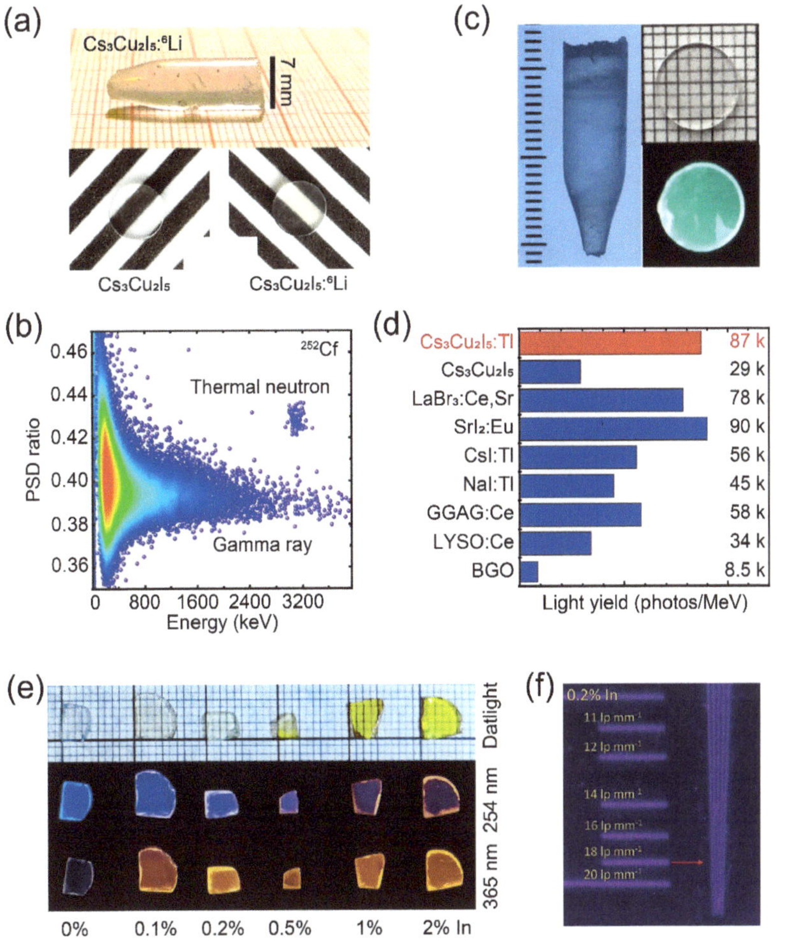

Figure 10. (**a**) As-grown $Cs_3Cu_2I_5$:^{6}Li crystal ingot with a 7 mm diameter (top) and polished $Cs_3Cu_2I_5$ (bottom left) and $Cs_3Cu_2I_5$:^{6}Li (bottom right) slabs. (**b**) PSD scatter plot and one-dimensional histogram of a $Cs_3Cu_2I_5$:^{6}Li single crystal upon excitation with a ^{252}Cf source and a ^{137}Cs source. Reprinted with permission from Ref. [134]. Copyright 2022, American Chemical Society. (**c**) As-grown 7 mm diameter Tl-doped $Cs_3Cu_2I_5$ crystal ingot, and a 1 mm thick sample under daylight and ultraviolet light. (**d**) Comparison of absolute light yield between Tl-doped $Cs_3Cu_2I_5$:Tl, undoped $Cs_3Cu_2I_5$, and other conventional scintillators. Reprinted with permission from Ref. [138]. Copyright 2021, Wiley-VCH GmbH. (**e**) $Cs_3Cu_2I_5$:In crystal slabs under daylight, 254 nm, and 365 nm UV light illumination. (**f**) X-ray 2D spatial images of the test-pattern plate using 0.2% In-doped $Cs_3Cu_2I_5$ single crystals. Reprinted with permission from Ref. [139]. Copyright 2022, Wiley-VCH GmbH.

In another study, Tl cations were successfully incorporated into the $Cs_3Cu_2I_5$ host using the Bridgman method, leading to the development of an ultrabright and highly efficient scintillator for X-ray and γ-ray detection, as shown in Figure 10c [138]. The Tl-doped $Cs_3Cu_2I_5$ single crystals achieved an impressive photoluminescence quantum efficiency of 79.2%. Under X-ray excitation, the radioluminescence emission of $Cs_3Cu_2I_5$:Tl crystals featured a self-trapped exciton emission at 440 nm and a Tl-related emission at 510 nm at room temperature. Optimized Tl doping resulted in a nearly fivefold enhancement of the steady-state scintillation yield, reaching up to 150,000 photons/MeV, alongside a significant improvement in X-ray detection sensitivity, which decreased from 103.6 to

66.3 nGy s^{-1}. Additionally, the material exhibited an exceptionally low afterglow of just 0.17% at 10 ms after X-ray cutoff. The $Cs_3Cu_2I_5$:Tl scintillator demonstrated outstanding performance, including an energy resolution of 3.4% at 662 keV and an ultrahigh light yield of 87,000 photons/MeV under ^{137}Cs γ-ray radiation, rivaling the capabilities of commercial scintillators, as illustrated in Figure 10d [140,141].

Based on the aforementioned studies, it can be concluded that doping-modified $Cs_3Cu_2I_5$ single crystals exhibit significant potential as X-ray detection materials, primarily due to their efficient exciton emissions arising from strongly localized charge carriers [142–147]. However, many of these materials still display scintillation yields well below their theoretical maxima [148]. A recent study demonstrated that the charge carrier harvesting efficiency could be significantly enhanced by introducing a small amount of indium (In) doping into these highly localized structures [139]. Bright 0D $Cs_3Cu_2I_5$:In single crystals were synthesized via the Bridgman method, exhibiting efficient and tunable dual emissions. All samples were transparent and free of inclusions, as shown in Figure 10e. Under X-ray excitation, the radioluminescence emission of $Cs_3Cu_2I_5$:In crystals featured a self-trapped exciton emission at 460 nm and an In^+-related emission at 620 nm at room temperature. In doping significantly enhanced the photoluminescence quantum efficiency of $Cs_3Cu_2I_5$from 68.1% to 88.4%. To assess the impact of In doping on X-ray imaging performance, undoped and 0.2% In-doped $Cs_3Cu_2I_5$ single crystals were utilized as scintillation screens in an X-ray imaging system. As depicted in Figure 10f, the In-doped $Cs_3Cu_2I_5$ crystals achieved an exceptional spatial resolution of 18 lp mm^{-1}, surpassing commercial CsI:Tl-based flat-panel X-ray detectors, all-inorganic perovskite nanocrystal scintillators, and thermally activated delayed fluorescence (TADF) organic scintillator-based X-ray imaging detectors ($\approx$6 lp mm^{-1}) [52,149]. This remarkable spatial resolution is attributed to the superior transparency, reduced light scattering, and enhanced scintillation yield of $Cs_3Cu_2I_5$ single crystals compared to commercial polycrystalline thick films or fluorescent powders. These advancements position $Cs_3Cu_2I_5$:In as a highly promising candidate for next-generation X-ray imaging applications.

5.2. Tl-Containing Single Crystals

Materials containing thallium (Tl) often exhibit strong fluorescence, making Tl-doped crystals highly suitable as scintillators for high-energy radiation detection [150–153]. In fact, such scintillators are widely utilized today. For instance, Tl-doped sodium antimonite (NaSb:Tl) crystals are known for their high scintillation efficiency and excellent radiation response, making them a common choice for particle detectors [150]. Additionally, Tl-doped magnesium aluminate ($MgAl_2O_4$:Tl) is recognized for its exceptional thermal stability and high light output, while Tl-doped cesium chloride (CsCl:Tl) is frequently employed for neutron detection, as Tl significantly enhances the material's sensitivity to neutrons [154–157]. Similarly, Tl-doped cesium fluoride (CsF:Tl) scintillators exhibit a high light yield, making them ideal for radiation detection and particle physics experiments. As a result, Tl-containing scintillators are extensively applied across diverse fields, including nuclear physics, radiation monitoring, medical imaging, and particle detection, underscoring their importance in advancing high-energy detection technologies [158].

In a recent study, centimeter-sized $TlPbI_3$ single crystals were successfully grown using the Bridgman method, as illustrated in Figure 11a [159]. $TlPbI_3$ exhibits a unique crystal structure within the halides, distinguishing itself from conventional perovskites by belonging to the broader class of perovskite-like compounds. It crystallizes in the orthorhombic space group Cmcm and adopts the $CaIrO_3$ structure type, as shown in Figure 11b [160]. The structure consists of alternating layers of $[PbI_3]^-$ units interspersed with Tl^+ ions. Within the 2D layers, the axially compressed $[PbI_6/2]^-$ octahedra expand anisotropically in the perovskite plane. These octahedra are connected via edge-sharing ($PbI_4/_2$ connectivity) along the shorter a-axis and corner-sharing ($PbI_2/_2$ connectivity) along the longer c-axis. The Tl atoms adopt a trigonal prismatic coordination geometry [161]. The bandgap structure of $TlPbI_3$ is mainly determined by Pb 6s and Tl 6s states deriving from the lone pair electrons in Pb^{2+} and Tl^{+} [159]. As we all know, metal elements often result in quite different electronic structures and properties in halide compounds [162,163]. The 5s electron pair of Pb^{2+} results in high tolerance against defects, therefore better stability of $TlPbI_3$ is often obtained than their counterparts such as $CsPbI_3$ [164]. Another study indicates that stronger bonding between Tl and I is the main cause of better stability of $TlPbI_3$ than $CsPbI_3$, owing to the lower total energy [165]. Furthermore, due to their different configurations of perovskite structure in $CsPbI_3$ and perovskite-like structure in $TlPbI_3$, they possess significantly disparate properties. For example, stronger bonding interactions in TlPbI3 lead to more dispersive band edges, which increases the bandwidth and reduces the bandgap [166]. Moreover, the enhanced dispersion of band edges makes less effective carrier masses and anisotropic transport features.

$TlPbI_3$ demonstrates high electrical resistivity of approximately 10^{12} Ω·cm. Detectors fabricated from these single crystals show robust photoresponsiveness to Ag Kα X-rays (22.4 keV) and are capable of detecting 122 keV γ-rays from a ^{57}Co radiation source. The electron mobility-lifetime product was measured at 1.8×10^{-5} $cm^2 \cdot V^{-1}$. Furthermore, $TlPbI_3$ exhibits a high relative static dielectric constant of 35.0, indicating its strong capability to screen carrier scattering and suppress the effects of charged defects. These characteristics position $TlPbI_3$ as a promising material for high-energy radiation detection applications [164,166,167].

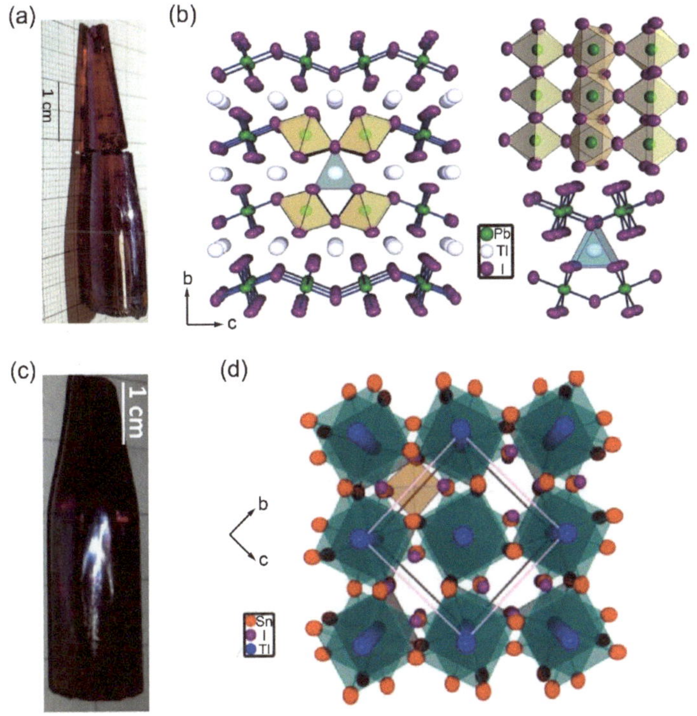

Figure 11. (**a**) Image of a pristine $TlPbI_3$ single crystal. (**b**) The crystal structure of $TlPbI_3$ with the left panel providing an overview of the structure. The right panels focus on the connectivity along the

$[PbI_3]^-$ layers and the coordination environment around the Tl^+ ions, respectively. Reprinted with permission from Ref. [159]. Copyright 2021, Wiley-VCH GmBH. (**c**) Image of the $TlSn_2I_5$ single crystal. (**d**) The antiperovskite structure of $TlSn_2I_5$. Reprinted with permission from Ref. [168]. Copyright 2017, American Chemical Society.

In addition to $TlPbI_3$, the semiconductor $TlSn_2I_5$, which features a two-dimensional crystal structure and an antiperovskite topology, has emerged as a promising novel material for radiation detection applications [169]. $TlSn_2I_5$ incorporates elements with high atomic numbers (Tl: 81, Sn: 50, I: 53) and a relatively high density (6.05 $g{\cdot}cm^{-3}$), ensuring a superior absorption coefficient compared to both halide perovskites and CdZnTe in response to hard radiation. In a recent study, centimeter-sized $TlSn_2I_5$ single crystals were successfully grown using the Bridgman method, as shown in Figure 11c [168]. As illustrated in Figure 11d, when viewed without considering chemical bonding constraints, the lattice formed by I, Tl, and Sn adopts an antiperovskite structure. In this arrangement, iodine is positioned at the center of an axially elongated octahedron, with its equatorial plane occupied by four Sn ions and its polar positions filled by two Tl ions. This results in the formation of a cationic $[ISn_2Tl]^{4+}$ framework. This framework exhibits a distorted perovskite structure, with the octahedra tilting out-of-phase in a manner similar to that observed in the hybrid perovskite $CH_3NH_3PbI_3$, which crystallizes in the non-centrosymmetric I4/mcm space group due to the asymmetric $CH_3NH_3{}^+$cation. However, compared to $CH_3NH_3PbI_3$, $TlSn_2I_5$ demonstrates superior long-term stability, enhanced photon stopping power, higher resistivity (~10^{10} $\Omega{\cdot}cm$), and robust mechanical properties. Additionally, $TlSn_2I_5$ can be utilized to fabricate detector devices capable of detecting Ag Kα X-rays (22 keV), ^{57}Co γ-rays (122 keV), and ^{241}Am α-particles (5.5 MeV). The mobility-lifetime product and electron mobility were estimated to be 1.1×10^{-3} $cm^2{\cdot}V^{-1}$ and 94 ± 16 $cm^2{\cdot}V^{-1}{\cdot}s^{-1}$, respectively. Notably, unlike other halide perovskites, $TlSn_2I_5$ exhibits no evidence of ionic polarization under prolonged high-voltage bias, further highlighting its suitability for long-term, stable operation in radiation detection applications.

5.3. Zero-Dimensional A_2BX_6 Perovskite Single Crystals

Zero-dimensional halide perovskites with the A_2BX_6 (A = Cs, Rb; B = Hf, Zr, Te, Ti; X = Cl, Br, I) structure represent a distinct class of materials, exhibiting unique physicochemical properties [170–178]. These A_2BX_6 halide perovskites are notable for their lack of extension into higher-dimensional crystal networks. Instead, they exist as isolated molecules or clusters. This structural characteristic leads to highly localized electronic states, which in turn result in exceptional optoelectronic properties, including efficient light absorption and emission. Additionally, in contrast to higher-dimensional metal halide perovskites, A_2BX_6 perovskites demonstrate superior chemical stability under certain conditions, owing to their more localized structure.

In a recent study, large-scale Cs_2ZrCl_6:Ce,Li single crystals with exceptional energy resolution were successfully synthesized by optimizing the temperature gradient and growth rate using the Bridgman method, as shown in Figure 12a [179]. Single-crystal diffraction analyses confirm the formation of 0D perovskite structures. Specifically, Zr^{4+} cations are coordinated with Cl^- ions to form discrete octahedra, which are isolated by Cs^+ ions, as depicted in Figure 12b [53]. Further investigations reveal that this effective dual-ion co-substitution strategy makes the Cs_2ZrCl_6 single crystal an environmentally friendly material, with promising applications in solar-blind ultraviolet optoelectronics and X-ray detection [180–182].

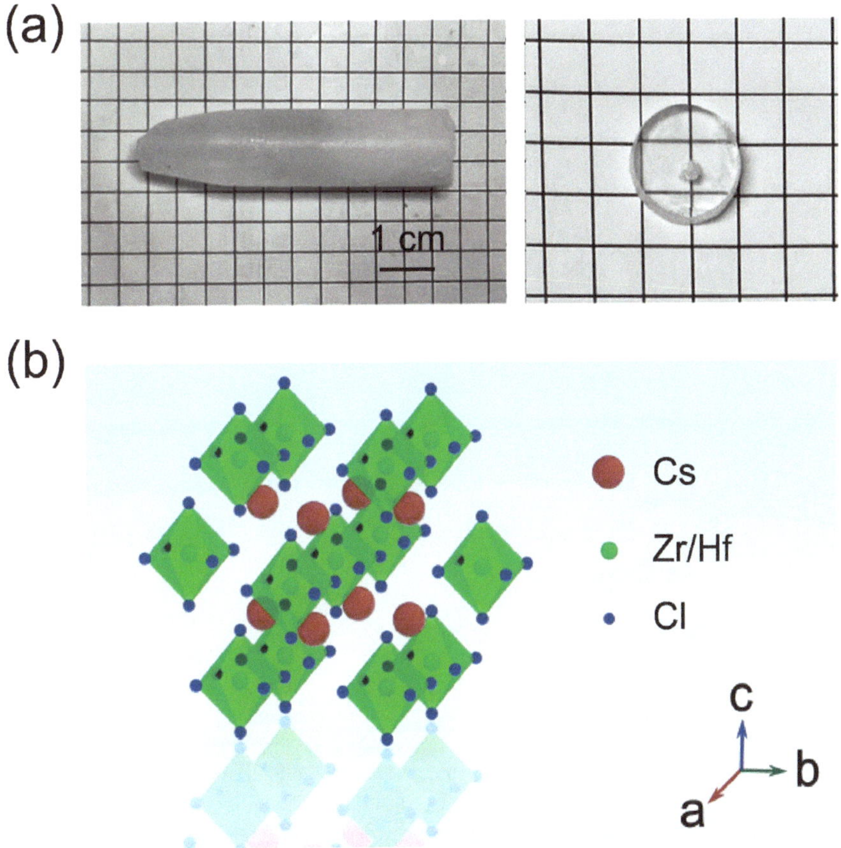

Figure 12. (**a**) As-grown Cs_2ZrCl_6:Ce,Li single crystal (left) and its wafer (right). Reprinted with permission from Ref. [179]. Copyright 2023, American Chemical Society. (**b**) Atomic structure of 0D perovskite crystal. Reprinted with permission from Ref. [53]. Copyright 2022, Wiley-VCH GmbH.

6. Conclusions and Perspectives

As a traditional single-crystal growth technique, the Bridgman method has been extensively employed in the preparation of single crystals for a wide range of materials. In the field of metal halide crystal research, the Bridgman method has played a crucial role, significantly advancing both the fundamental understanding and practical applications of halide materials. Based on the metal halide compositions discussed in this review, we summarize the growth parameters of different metal halide single crystals using the Bridgman method, as shown in Table 1.

Table 1. Growth parameters of metal halide single crystals using the Bridgman method.

Materials	Substance	Temperature Gradient (°C/cm)	Growth Rate	References
$CsPbX_3$	CsX, PbX_2	10–30	0.2–1 mm/hour	[8]
$CsPbBr_3$	CsBr, $PbBr_2$	30–40	1 mm/hour	[23]
$CsPbBr_3$	CsBr, $PbBr_2$	50–70	0.2–0.6 mm/hour	[76,183]
$CsPbBr_3$	CsBr, $PbBr_2$	5–20	0.5–2 mm/hour	[80]
$CsPbCl_3$	CsCl, $PbCl_2$	5–20	0.1–1 mm/hour	[94]

Table 1. *Cont.*

Materials	Substance	Temperature Gradient (°C/cm)	Growth Rate	References
γ-CuCl	CuCl	20	8 mm/day	[30]
γ-CuBr	CuBr	5	6 mm/day	[35]
$CsPbBr_3$	CsBr, $PbBr_2$	10	3–30 mm/hour	[79]
PbI_2	PbI2	22	0.5–1.0 mm/hour	[41]
$TlSn_2I_5$	Sn, I_2, TlI	23		[168]
Tl_2HfX_6	TlX, HfX_4	/	0.2–1 mm/hour	[150]
Tl_2LaX_5	TlX, LaX_3	/	0.2–1 mm/hour	[150]
$A_3M_2I_9$ (A = Cs, Rb; M = Bi, Sb)	$Cs_3Sb_2I_9$: Sb_2O_3, HI, Cs_2CO_3; $Cs_3Bi_2I_9$: Bi_2O_3, HI, CsI; $Rb_3Bi_2I_9$: BiI_3, RbI	/	2 mm/hour	[129,130]
$TlPbI_3$	TlI, PbI_2	12	20 mm/day	[167]
$TlPbI_3$	Tl, Pb, I_2	8	0.5 mm/hour	[159]
Eu-doped $TlSr_2I_5$	TlI, SrI_2, EuI_2	/	10 mm/day	[152]
Bi-doped $CsCdX_3$	CsX, CdX_2, BiX_3	/	2 mm/hour	[162]
Sn-doped $CsPbX_3$	CsX, PbX_2, SnX_2	/	1.4 mm/hour	[61]
Bi-doped $TlCdCl_3$	TlCl, $CdCl_2$, $BiCl_3$	/	1 mm/hour	[163]
$Cs_2PbI_2Cl_2$	CsI, $PbCl_2$	/	0.7 mm/hour	[109]
$Cs_2PbI_2Cl_2$	CsI, $PbCl_2$	/	0.77 mm/hour	[113]
$Cs_3Bi_2Br_9$	CsBr, $BiBr_3$	10–20	0.5–3.0 mm/hour	[123]
$Cs_3Bi_2I_8Br$	CsI, $BiBr_3$	15	1 mm/hour	[126]
^{6}Li-doped $Cs_3Cu_2I_5$	CsI, CuI, ^{6}LiI	/	0.4 mm/hour	[134]
Tl-doped $Cs_3Cu_2I_5$	CsI, CuI, TlI	20–30	0.5–1.0 mm/hour	[138]
Ce,Li-doped Cs_2ZrCl_6	CsCl, $ZrCl_4$, $CeCl_3$, LiCl	15	0.7 cm/day	[179]

Up to now, metal halide single crystals grown by the Bridgman method have been widely used in various fields. For example, high-quality $CsPbBr_3$, $CsPbCl_3$, and $TlPbI_3$ single crystals demonstrate high resolution X-ray detection feature [80,94,159]. ^{6}Li-doped Cs_3CuI_5 single crystals show great potential as scintillator for neutron discrimination [134]. In addition, Er-doped $CsCdBr_3$ single crystals possess unique up conversion capability that can convert low-energy photon to high-energy light, which shows great potential in solar cells [77].

While the Bridgman method offers numerous advantages, it also presents several drawbacks and limitations. One primary disadvantage is the slow cooling process typically employed, which leads to a relatively slow crystal growth rate. As a result, the growth of large single crystals can be time-consuming, potentially increasing production costs. Additionally, although the method allows for precise temperature gradient control, maintaining a stable gradient remains challenging due to factors such as heat conduction, convection, and other operational variables. Furthermore, the Bridgman method may not be suitable for materials with a high thermal expansion coefficient or those prone to volatility. Notably, for materials that require extremely fine structural control, the Bridgman method may lack the necessary precision.

In summary, the Bridgman method for single crystal growth is poised to facilitate the advancement of larger, higher-quality, and higher-performance halide single crystals in future research. This progress will offer substantial support for the application of novel optoelectronic materials, semiconductor materials, and other functional materials. As the process continues to be refined, the Bridgman method will play an increasingly pivotal role in fields such as materials science, energy, and electronic devices.

Funding: This research was funded by Shandong Provincial Natural Science Foundation, grant number ZR2024QE319.

Data Availability Statement: No new data were created or analyzed in this study. Data sharing is not applicable to this article.

Acknowledgments: We would like to express our sincere gratitude to Zhiyan Zhou from the School of Stomatology, Shandong University, for her invaluable assistance with language polishing.

Conflicts of Interest: The authors declare no conflicts of interest.

References

1. Bisson, P.; Groenzin, H.; Barnett, I.L.; Shultz, M.J. High Yield, Single Crystal Ice via the Bridgman Method. *Rev. Sci. Instrum.* **2016**, *87*, 034103. [CrossRef] [PubMed]
2. Derby, J.J.; Yeckel, A. Heat Transfer Analysis and Design for Bulk Crystal Growth: Perspectives on the Bridgman Method. In *Handbook of Crystal Growth*; Elsevier: Amsterdam, The Netherlands, 2015; pp. 793–843, ISBN 978-0-444-63303-3.
3. Rudolph, P.; Kiessling, F. The Horizontal Bridgman Method. *Cryst. Res. Technol.* **1988**, *23*, 1207–1224. [CrossRef]
4. Gaxandet, J.P.; Alboussire, T. Bridgman Growth: Modelling and Experiments. *Prog. Cryst. Growth Charact. Mater.* **1999**, *38*, 133–159. [CrossRef]
5. Murugesan, M.; Sidden, C.; Paulraj, R. Growth, Optical and Thermal Properties of Triphenylamine Single Crystal Grown by Modified Vertical Bridgman Method for Optical Application. *Solid. State Sci.* **2022**, *132*, 106991. [CrossRef]
6. Fedyushkin, A.I.; Burago, N.G.; Puntus, A.A. Convective Heat and Mass Transfer Modeling under Crystal Growth by Vertical Bridgman Method. *J. Phys. Conf. Ser.* **2020**, *1479*, 012029. [CrossRef]
7. Yu, P.; Jiang, B.; Chen, Y.; Zheng, J.; Luan, L. Study on In-Doped CdMgTe Crystals Grown by a Modified Vertical Bridgman Method Using the ACRT Technique. *Materials* **2019**, *12*, 4236. [CrossRef] [PubMed]
8. Zhang, P.; Hua, Y.; Li, X.; Zhang, L.; Liu, L.; Li, R.; Zhang, G.; Tao, X. Filter-Free Color Image Sensor Based on $CsPbBr_{3-3n}X_{3n}$ (X = Cl, I) Single Crystals. *J. Mater. Chem. C* **2021**, *9*, 2840–2847. [CrossRef]
9. Gille, P.; Mühlberg, M.; Parthier, L.; Rudolph, P. Crystal Growth of PbTe and (Pb, Sn)Te by the Bridgman Method and by THM. *Cryst. Res. Technol.* **1984**, *19*, 881–891. [CrossRef]
10. Zhang, G.; Tao, X.; Wang, S.; Shi, Q.; Ruan, H.; Chen, L. Growth Improvement and Quality Evaluation of $ZnGeP_2$ Single Crystals Using Vertical Bridgman Method. *J. Cryst. Growth* **2012**, *352*, 67–71. [CrossRef]
11. Wang, C.; Xia, H.; Feng, Z.; Zhang, Z.; Jiang, D.; Gu, X.; Zhang, Y.; Chen, B.; Jiang, H. Infrared Luminescent Properties of $Na_5Lu_9F_{32}$ Single Crystals Co-Doped Er^{3+}/Yb^{3+} Grown by Bridgman Method. *J. Alloys Compd.* **2016**, *686*, 816–822. [CrossRef]
12. Mooney, R.P.; McFadden, S.; Gabalcová, Z.; Lapin, J. An Experimental–Numerical Method for Estimating Heat Transfer in a Bridgman Furnace. *Appl. Therm. Eng.* **2014**, *67*, 61–71. [CrossRef]
13. He, X.; Li, L.; He, X.; Xie, C. Multi-Physical Field Simulation of Cracking during Crystal Growth by Bridgman Method. *Materials* **2023**, *16*, 3260. [CrossRef] [PubMed]
14. Du, Y.; Jie, W.; Wang, T.; Zheng, X.; Xu, Y.; Luan, L. Solution Growth of In-Doped CdMnTe Crystals by the Vertical Bridgman Method with the ACRT Technique. *J. Cryst. Growth* **2012**, *355*, 33–37. [CrossRef]
15. Sheng, M.; Wang, S.; Zhu, H.; Liu, Z.; Zhou, G. Computational Applications for the Discovery of Novel Antiperovskites and Chalcogenide Perovskites: A Review. *Front. Chem.* **2024**, *12*, 1468434. [CrossRef] [PubMed]
16. Sheng, M.; Zhu, H.; Wang, S.; Liu, Z.; Zhou, G. Accelerated Discovery of Halide Perovskite Materials via Computational Methods: A Review. *Nanomaterials* **2024**, *14*, 1167. [CrossRef] [PubMed]
17. Wojakowska, A.; Krzy, E. Phase Diagram for the RbBr–CuBr System. *J. Therm. Anal. Calorim.* **2011**, *65*, 491–495. [CrossRef]
18. Kanno, R.; Takeda, Y.; Masuyama, Y.; Yamamoto, O. Phase Diagram and High Copper Ion Conductivity of the Copper(I) Chloride-Rubidium Chloride System. *Solid State Ion.* **1983**, *11*, 221–226. [CrossRef]
19. Takahashi, T. Solid-State Ionics—The CuCI-CuI-RbCl System. *Solid State Ion.* **1981**, *3*, 283–287. [CrossRef]
20. Malakhovskaya-Rosokha, T.A.; Barchii, I.E.; Pogodin, A.I.; Kokhan, A.P.; Stercho, I.P.; Peresh, E.Y. Interaction of Components in the RbI-CsI-CuI Quasi-Ternary System. *Russ. J. Inorg. Chem.* **2013**, *58*, 577–580. [CrossRef]
21. Matsui, T.; Wagner, J.B. Investigations on a High Conductivity Solid Electrolyte System, RbCl + CuCl. *J. Electrochem. Soc.* **1977**, *124*, 941–944. [CrossRef]
22. Gut, R.; Iberson, E.; Gruen, D.M. *Compounds in the Binary Phase Diagrams of the Monovalent Chlorides with the Multivalent Chlorides*; Report number ANL-6803; OSTI ID: 4117130; Argonne National Laboratory: Lemont, IL, USA, 1963.
23. Zhang, P.; Zhang, G.; Liu, L.; Ju, D.; Zhang, L.; Cheng, K.; Tao, X. Anisotropic Optoelectronic Properties of Melt-Grown Bulk $CsPbBr_3$ Single Crystal. *J. Phys. Chem. Lett.* **2018**, *9*, 5040–5046. [CrossRef] [PubMed]

24. Yu, X.; Huang, C.; Ni, Y.; Wang, Z.; Wu, H.; Hu, Q.; Liu, G.; Zhou, Q.; Wei, L. Optical Property and Judd-Ofelt Analysis of Dy^{3+}, Na+:$PbGa_2S_4$ Crystal Grown by Bridgman Method. *J. Lumin.* **2023**, *262*, 119951. [CrossRef]
25. Wang, L.-H.; Chang, L.-S. Thermoelectric Properties of P-Type $Ba_8Ga_{16}Ge_{30}$ Type-I Clathrate Compounds Prepared by the Vertical Bridgman Method. *J. Alloys Compd.* **2017**, *722*, 644–650. [CrossRef]
26. Zayachuk, D.M.; Ilyina, O.S.; Pashuk, A.V.; Mikityuk, V.I.; Shlemkevych, V.V.; Csik, A.; Kaczorowski, D. Segregation of the Eu Impurity as Function of Its Concentration in the Melt for Growing of the Lead Telluride Doped Crystals by the Bridgman Method. *J. Cryst. Growth* **2013**, *376*, 28–34. [CrossRef]
27. Lai, J.; Zhang, J.; Wang, L.; Min, J.; Wu, W.; Shen, M.; Liang, W. Comparison of Cadmium Manganese Telluride Crystals Grown by Traveling Heater Method and Vertical Bridgman Method. *Cryst. Res. Technol.* **2015**, *50*, 817–822. [CrossRef]
28. Xiong, W.; Chen, L.; Huang, L.; Guo, F.; Zhou, Y.; Yuan, H. Bridgman Growth and Characterization of Birefringent Crystal $NaNO_3$. *Cryst. Res. Technol.* **2015**, *50*, 250–254. [CrossRef]
29. Jin, M.; Shen, H.; Fan, S.; He, Q.; Xu, J. Industrial Growth and Characterization of Si-doped GaAs Crystal by a Novel Multi-crucible Bridgman Method. *Cryst. Res. Technol.* **2017**, *52*, 1700052. [CrossRef]
30. Yang, C.; Wen, Y.; Xiong, J.; Zhou, H.; Pan, S.; Chen, H.; Pan, J. Crystal Growth and Photoelectric Performance of γ-CuCl by the Vertical Bridgman Method. *J. Cryst. Growth* **2022**, *579*, 126463. [CrossRef]
31. Gul, R.; Roy, U.N.; Camarda, G.S.; Hossain, A.; Yang, G.; Vanier, P.; Lordi, V.; Varley, J.; James, R.B. A Comparison of Point Defects in $Cd_{1-x}Zn_xTe_{1-y}Se_y$ Crystals Grown by Bridgman and Traveling Heater Methods. *J. Appl. Phys.* **2017**, *121*, 125705. [CrossRef]
32. Roy, U.N.; Bolotnikov, A.E.; Camarda, G.S.; Cui, Y.; Hossain, A.; Lee, K.; Lee, W.; Tappero, R.; Yang, G.; Gul, R.; et al. High Compositional Homogeneity of $CdTe_xSe_{1-x}$ Crystals Grown by the Bridgman Method. *APL Mater.* **2015**, *3*, 026102. [CrossRef]
33. Colombier, E.; Braithwaite, D. Simple Adaptation of the Bridgman High Pressure Technique for Use with Liquid Media. *Rev. Sci. Instrum.* **2007**, *78*, 093903. [CrossRef] [PubMed]
34. Li, L.; Zhang, P.; Li, Z.; Chen, Z. Thermal Field Simulation and Optimization of PbF_2 Single Crystal Growth by the Bridgman Method. *Crystals* **2024**, *14*, 473. [CrossRef]
35. Zhou, H.; Chen, X.; Zhao, T.; Chen, C.; Han, J.; Pan, S.; Pan, J. Growth and Luminescence Properties of γ-CuBr Single Crystals by the Bridgman Method. *CrystEngComm* **2023**, *25*, 1669–1674. [CrossRef]
36. Zayachuk, D.M.; Mikityuk, V.I.; Shlemkevych, V.V.; Kaczorowski, D. Complex Formation and Background Impurity of Oxygen in the PbTe: Eu Doped Crystals Grown from Melt by the Bridgman Method. *J. Cryst. Growth* **2012**, *338*, 35–41. [CrossRef]
37. Toufanian, R.; Swain, S.; Becla, P.; Motakef, S.; Datta, A. Cesium Lead Bromide Semiconductor Radiation Detectors: Crystal Growth, Detector Performance and Polarization. *J. Mater. Chem. C* **2022**, *10*, 12708–12714. [CrossRef]
38. Dahlen, A.; Fattah, A.; Hanke, G.; Karthaus, E. Bridgman and Czochralski Growth of Ge Si Alloy Crystals. *Cryst. Res. Technol.* **1994**, *29*, 187–198. [CrossRef]
39. Kliemt, K.; Krellner, C. Crystal Growth by Bridgman and Czochralski Method of the Ferromagnetic Quantum Critical Material $YbNi_4P_2$. *J. Cryst. Growth* **2016**, *449*, 129–133. [CrossRef]
40. Hurle, D.T.J. A Mechanism for Twin Formation during Czochralski and Encapsulated Vertical Bridgman Growth of III–V Compound Semiconductors. *J. Cryst. Growth* **1995**, *147*, 239–250. [CrossRef]
41. Tonn, J.; Danilewsky, A.N.; Cröll, A.; Matuchova, M.; Maixner, J. Czochralski Growth of Lead Iodide Single Crystals: Investigations and Comparison with the Bridgman Method. *J. Cryst. Growth* **2011**, *318*, 558–562. [CrossRef]
42. Akasaka, M.; Iida, T.; Matsumoto, A.; Yamanaka, K.; Takanashi, Y.; Imai, T.; Hamada, N. The Thermoelectric Properties of Bulk Crystalline N- and p-Type Mg_2Si Prepared by the Vertical Bridgman Method. *J. Appl. Phys.* **2008**, *104*, 013703. [CrossRef]
43. Lyubimova, T.P.; Croell, A.; Dold, P.; Khlybov, O.A.; Fayzrakhmanova, I.S. Time-Dependent Magnetic Field Influence on GaAs Crystal Growth by Vertical Bridgman Method. *J. Cryst. Growth* **2004**, *266*, 404–410. [CrossRef]
44. Shimizu, A.; Nishizawa, J.; Oyama, Y.; Suto, K. InP Single Crystal Growth by the Horizontal Bridgman Method under Controlled Phosphorus Vapor Pressure. *J. Cryst. Growth* **2001**, *229*, 119–123. [CrossRef]
45. Li, F.; Cabral, M.J.; Xu, B.; Cheng, Z.; Dickey, E.C.; LeBeau, J.M.; Wang, J.; Luo, J.; Taylor, S.; Hackenberger, W.; et al. Giant Piezoelectricity of Sm-Doped $Pb(Mg_{1/3}Nb_{2/3})O_3$-$PbTiO_3$ Single Crystals. *Science* **2019**, *364*, 264–268. [CrossRef]
46. Li, F.; Lin, D.; Chen, Z.; Cheng, Z.; Wang, J.; Li, C.; Xu, Z.; Huang, Q.; Liao, X.; Chen, L.-Q.; et al. Ultrahigh Piezoelectricity in Ferroelectric Ceramics by Design. *Nat. Mater.* **2018**, *17*, 349–354. [CrossRef]
47. Qiu, C.; Wang, B.; Zhang, N.; Zhang, S.; Liu, J.; Walker, D.; Wang, Y.; Tian, H.; Shrout, T.R.; Xu, Z.; et al. Transparent Ferroelectric Crystals with Ultrahigh Piezoelectricity. *Nature* **2020**, *577*, 350–354. [CrossRef] [PubMed]
48. Chen, H.; Liang, Z.; Luo, L.; Ke, Y.; Shen, Q.; Xia, Z.; Jiang, C.; Pan, J. Bridgman Growth, Crystallographic Characterization and Electrical Properties of Relaxor-Based Ferroelectric Single Crystal PIMNT. *J. Alloys Compd.* **2012**, *518*, 63–67. [CrossRef]
49. Liao, F.; Zhao, Y.; Chen, Z.; Zheng, Y.; Chen, H. Bridgman Growth and Photoelectronic Property of Relaxor-Based Ferroelectric Single Crystal $Pb(Sm_{1/2}Nb_{1/2})O_3$-$Pb(Mg_{1/3}Nb_{2/3})O_3$-$PbTiO_3$. *Crystals* **2021**, *11*, 402. [CrossRef]

50. Liu, Z.; Qin, X.; Liu, X. Luminescence Enrichment in Perovskite-Lanthanide Composites: Complexity and Complementarity. In *Handbook on the Physics and Chemistry of Rare Earths*; Elsevier: Amsterdam, The Netherlands, 2022; Volume 61, pp. 1–29, ISBN 978-0-323-98594-9.
51. Liu, Z.; Qin, X.; Chen, Q.; Jiang, T.; Chen, Q.; Liu, X. Metal–Halide Perovskite Nanocrystal Superlattice: Self-Assembly and Optical Fingerprints. *Adv. Mater.* **2023**, *35*, 2209279. [CrossRef] [PubMed]
52. Chen, Q.; Wu, J.; Ou, X.; Huang, B.; Almutlaq, J.; Zhumekenov, A.A.; Guan, X.; Han, S.; Liang, L.; Yi, Z.; et al. All-Inorganic Perovskite Nanocrystal Scintillators. *Nature* **2018**, *561*, 88–93. [CrossRef] [PubMed]
53. Liu, Z.; Qin, X.; Chen, Q.; Chen, Q.; Jing, Y.; Zhou, Z.; Zhao, Y.S.; Chen, J.; Liu, X. Highly Stable Lead-Free Perovskite Single Crystals with NIR Emission Beyond 1100 Nm. *Adv. Opt. Mater.* **2022**, *10*, 2201254. [CrossRef]
54. Ma, W.; Qian, Q.; Qaid, S.M.H.; Zhao, S.; Liang, D.; Cai, W.; Zang, Z. Water-Molecule-Induced Reversible Fluorescence in a One-Dimensional Mn-Based Hybrid Halide for Anticounterfeiting and Digital Encryption–Decryption. *Nano Lett.* **2023**, *23*, 8932–8939. [CrossRef]
55. Han, T.; Tan, S.; Xue, J.; Meng, L.; Lee, J.; Yang, Y. Interface and Defect Engineering for Metal Halide Perovskite Optoelectronic Devices. *Adv. Mater.* **2019**, *31*, 1803515. [CrossRef]
56. Motti, S.G.; Meggiolaro, D.; Martani, S.; Sorrentino, R.; Barker, A.J.; De Angelis, F.; Petrozza, A. Defect Activity in Lead Halide Perovskites. *Adv. Mater.* **2019**, *31*, 1901183. [CrossRef]
57. Ono, L.K.; Liu, S.; Qi, Y. Reducing Detrimental Defects for High-Performance Metal Halide Perovskite Solar Cells. *Angew. Chem. Int. Ed.* **2020**, *59*, 6676–6698. [CrossRef] [PubMed]
58. Huang, H.; Bodnarchuk, M.I.; Kershaw, S.V.; Kovalenko, M.V.; Rogach, A.L. Lead Halide Perovskite Nanocrystals in the Research Spotlight: Stability and Defect Tolerance. *ACS Energy Lett.* **2017**, *2*, 2071–2083. [CrossRef]
59. Yang, F.; Jie, W.; Wang, M.; Sun, X.; Jia, N.; Yin, L.; Zhou, B.; Wang, T. Growth of Single-Crystal Cd0.9Zn0.1Te Ingots Using Pressure Controlled Bridgman Method. *Crystals* **2020**, *10*, 261. [CrossRef]
60. Macpherson, S.; Doherty, T.A.S.; Winchester, A.J.; Kosar, S.; Johnstone, D.N.; Chiang, Y.-H.; Galkowski, K.; Anaya, M.; Frohna, K.; Iqbal, A.N.; et al. Local Nanoscale Phase Impurities Are Degradation Sites in Halide Perovskites. *Nature* **2022**, *607*, 294–300. [CrossRef] [PubMed]
61. Valueva, A.D.; Novikov, S.A.; Bledsoe, J.; Cai, Y.; Maksimova, A.A.; Locklin, J.; Zhao, Y.; Klepov, V.V. Cubic Halide Perovskites in the $Cs(Pb_{1-x}Snx)(Br_{3-y}Cl_y)$ Solid Solutions for Crack-Free Bridgman Grown Single Crystals. *MRS Commun.* **2024**, *14*, 942–948. [CrossRef]
62. Shi, Y.; Zhou, Y.; Ma, Z.; Xiao, G.; Wang, K.; Zou, B. Structural Regulation and Optical Behavior of Three-Dimensional Metal Halide Perovskites under Pressure. *J. Mater. Chem. C* **2020**, *8*, 12755–12767. [CrossRef]
63. Wong, W.P.D.; Yin, J.; Chaudhary, B.; Chin, X.Y.; Cortecchia, D.; Lo, S.-Z.A.; Grimsdale, A.C.; Mohammed, O.F.; Lanzani, G.; Soci, C. Large Polaron Self-Trapped States in Three-Dimensional Metal-Halide Perovskites. *ACS Mater. Lett.* **2020**, *2*, 20–27. [CrossRef]
64. Huang, X.; Guo, Q.; Kang, S.; Ouyang, T.; Chen, Q.; Liu, X.; Xia, Z.; Yang, Z.; Zhang, Q.; Qiu, J.; et al. Three-Dimensional Laser-Assisted Patterning of Blue-Emissive Metal Halide Perovskite Nanocrystals inside a Glass with Switchable Photoluminescence. *ACS Nano* **2020**, *14*, 3150–3158. [CrossRef] [PubMed]
65. Giappa, R.M.; Selivanov, N.I.; Samsonova, A.Y.; Pantousas, A.; Remediakis, I.N.; Kapitonov, Y.V.; Emeline, A.V.; Kopidakis, G.; Stoumpos, C.C. $(3\text{-}CF_3pyH)_2(3\text{-}CF_3P_y)Pb_3I_8$: A Three-Dimensional Metal Halide Inorganic Framework with Distinctive Kagomé Bands. *Chem. Mater.* **2024**, *36*, 11804–11813. [CrossRef]
66. Krawczyk, R.P.; Hammerl, A.; Schwerdtfeger, P. Coinage Metal Halide Clusters: From Two-Dimensional Ring to Three-Dimensional Solid-State-Like Structures. *ChemPhysChem* **2006**, *7*, 2286–2289. [CrossRef]
67. Zhang, W.; Eperon, G.E.; Snaith, H.J. Metal Halide Perovskites for Energy Applications. *Nat Energy* **2016**, *1*, 16048. [CrossRef]
68. Lei, L.; Dong, Q.; Gundogdu, K.; So, F. Metal Halide Perovskites for Laser Applications. *Adv Funct Mater.* **2021**, *31*, 2010144. [CrossRef]
69. Fu, Y.; Zhu, H.; Chen, J.; Hautzinger, M.P.; Zhu, X.-Y.; Jin, S. Metal Halide Perovskite Nanostructures for Optoelectronic Applications and the Study of Physical Properties. *Nat. Rev. Mater.* **2019**, *4*, 169–188. [CrossRef]
70. Stranks, S.D.; Snaith, H.J. Metal-Halide Perovskites for Photovoltaic and Light-Emitting Devices. *Nat. Nanotechnol.* **2015**, *10*, 391–402. [CrossRef] [PubMed]
71. Lin, Y.; Pattanasattayavong, P.; Anthopoulos, T.D. Metal-Halide Perovskite Transistors for Printed Electronics: Challenges and Opportunities. *Adv. Mater.* **2017**, *29*, 1702838. [CrossRef]
72. Sun, J.; Wu, J.; Tong, X.; Lin, F.; Wang, Y.; Wang, Z.M. Organic/Inorganic Metal Halide Perovskite Optoelectronic Devices beyond Solar Cells. *Adv. Sci.* **2018**, *5*, 1700780. [CrossRef] [PubMed]
73. Liu, X.-K.; Xu, W.; Bai, S.; Jin, Y.; Wang, J.; Friend, R.H.; Gao, F. Metal Halide Perovskites for Light-Emitting Diodes. *Nat. Mater.* **2021**, *20*, 10–21. [CrossRef] [PubMed]
74. Xiang, W.; Liu, S.; Tress, W. A Review on the Stability of Inorganic Metal Halide Perovskites: Challenges and Opportunities for Stable Solar Cells. *Energy Environ. Sci.* **2021**, *14*, 2090–2113. [CrossRef]

75. Cho, H.; Kim, Y.; Wolf, C.; Lee, H.; Lee, T. Improving the Stability of Metal Halide Perovskite Materials and Light-Emitting Diodes. *Adv. Mater.* **2018**, *30*, 1704587. [CrossRef]
76. Chen, X.; Wang, Y.; Song, J.; Li, X.; Xu, J.; Zeng, H.; Sun, H. Temperature Dependent Reflectance and Ellipsometry Studies on a $CsPbBr_3$ Single Crystal. *J. Phys. Chem. C* **2019**, *123*, 10564–10570. [CrossRef]
77. Cockroft, N.J.; Jones, G.D.; Nguyen, D.C. Dynamics and Spectroscopy of Infrared-to-Visible Upconversion in Erbium-Doped Cesium Cadmium Bromide ($CsCdBr_3$:Er^{3+}). *Phys. Rev. B* **1992**, *45*, 5187–5198. [CrossRef] [PubMed]
78. Pilla, O.; Cazzanelli, E.; Blanzat, B.; Andraud, C.; Pelle, F. Comparative Raman Study of Phonon Linewidths in Pure and Lead-Doped $CsCdBr_3$. *Phys. Status Solidi (B)* **1987**, *144*, 845–851. [CrossRef]
79. Stoumpos, C.C.; Malliakas, C.D.; Peters, J.A.; Liu, Z.; Sebastian, M.; Im, J.; Chasapis, T.C.; Wibowo, A.C.; Chung, D.Y.; Freeman, A.J.; et al. Crystal Growth of the Perovskite Semiconductor $CsPbBr_3$: A New Material for High-Energy Radiation Detection. *Cryst. Growth Des.* **2013**, *13*, 2722–2727. [CrossRef]
80. He, Y.; Matei, L.; Jung, H.J.; McCall, K.M.; Chen, M.; Stoumpos, C.C.; Liu, Z.; Peters, J.A.; Chung, D.Y.; Wessels, B.W.; et al. High Spectral Resolution of Gamma-Rays at Room Temperature by Perovskite $CsPbBr_3$ Single Crystals. *Nat. Commun.* **2018**, *9*, 1609. [CrossRef] [PubMed]
81. Dirin, D.N.; Cherniukh, I.; Yakunin, S.; Shynkarenko, Y.; Kovalenko, M.V. Solution-Grown $CsPbBr_3$ Perovskite Single Crystals for Photon Detection. *Chem. Mater.* **2016**, *28*, 8470–8474. [CrossRef] [PubMed]
82. Liu, D.; Hu, Z.; Hu, W.; Wangyang, P.; Yu, K.; Wen, M.; Zu, Z.; Liu, J.; Wang, M.; Chen, W.; et al. Two-Step Method for Preparing All-Inorganic $CsPbBr_3$ Perovskite Film and Its Photoelectric Detection Application. *Mater. Lett.* **2017**, *186*, 243–246. [CrossRef]
83. Tang, X.; Zu, Z.; Zang, Z.; Hu, Z.; Hu, W.; Yao, Z.; Chen, W.; Li, S.; Han, S.; Zhou, M. $CsPbBr_3$/Reduced Graphene Oxide Nanocomposites and Their Enhanced Photoelectric Detection Application. *Sens. Actuators B Chem.* **2017**, *245*, 435–440. [CrossRef]
84. Adinolfi, V.; Yuan, M.; Comin, R.; Thibau, E.S.; Shi, D.; Saidaminov, M.I.; Kanjanaboos, P.; Kopilovic, D.; Hoogland, S.; Lu, Z.; et al. The In-Gap Electronic State Spectrum of Methylammonium Lead Iodide Single-Crystal Perovskites. *Adv. Mater.* **2016**, *28*, 3406–3410. [CrossRef] [PubMed]
85. Kang, J.; Wang, L.-W. High Defect Tolerance in Lead Halide Perovskite $CsPbBr_3$. *J. Phys. Chem. Lett.* **2017**, *8*, 489–493. [CrossRef] [PubMed]
86. Chen, C.; Fu, Q.; Guo, P.; Chen, H.; Wang, M.; Luo, W.; Zheng, Z. Ionic Transport Characteristics of Large-Size $CsPbBr_3$ Single Crystals. *Mater. Res. Express* **2019**, *6*, 115808. [CrossRef]
87. Li, J.; Du, X.; Niu, G.; Xie, H.; Chen, Y.; Yuan, Y.; Gao, Y.; Xiao, H.; Tang, J.; Pan, A.; et al. Rubidium Doping to Enhance Carrier Transport in $CsPbBr_3$ Single Crystals for High-Performance X-Ray Detection. *ACS Appl. Mater. Interfaces* **2020**, *12*, 989–996. [CrossRef]
88. Lee, I.Y.; Deleplanque, M.A.; Vetter, K. Developments in Large Gamma-Ray Detector Arrays. *Rep. Prog. Phys.* **2003**, *66*, 1095–1144. [CrossRef]
89. Liu, H.; Omura, T.; Watanabe, M.; Yamashita, T. Development of a Depth of Interaction Detector for γ-Rays. *Nucl. Instrum. Methods Phys. Res. Sect. A Accel. Spectrometers Detect. Assoc. Equip.* **2001**, *459*, 182–190. [CrossRef]
90. Funk, S. Indirect Detection of Dark Matter with γ Rays. *Proc. Natl. Acad. Sci. USA* **2015**, *112*, 12264–12271. [CrossRef] [PubMed]
91. Xu, J.; Maxwell, A.; Wei, M.; Wang, Z.; Chen, B.; Zhu, T.; Sargent, E.H. Defect Tolerance of Mixed B-Site Organic–Inorganic Halide Perovskites. *ACS Energy Lett.* **2021**, *6*, 4220–4227. [CrossRef]
92. Smart, T.J.; Takenaka, H.; Pham, T.A.; Tan, L.Z.; Zhang, J.Z.; Ogitsu, T.; Ping, Y. Enhancing Defect Tolerance with Ligands at the Surface of Lead Halide Perovskites. *J. Phys. Chem. Lett.* **2021**, *12*, 6299–6304. [CrossRef] [PubMed]
93. Kim, G.; Petrozza, A. Defect Tolerance and Intolerance in Metal-Halide Perovskites. *Adv. Energy Mater.* **2020**, *10*, 2001959. [CrossRef]
94. He, Y.; Stoumpos, C.C.; Hadar, I.; Luo, Z.; McCall, K.M.; Liu, Z.; Chung, D.Y.; Wessels, B.W.; Kanatzidis, M.G. Demonstration of Energy-Resolved γ-Ray Detection at Room Temperature by the $CsPbCl_3$ Perovskite Semiconductor. *J. Am. Chem. Soc.* **2021**, *143*, 2068–2077. [CrossRef] [PubMed]
95. Cohen, M.I.; Young, K.F.; Chang, T.-T.; Brower, W.S. Phase Transitions in $CsPbCl_3$. *J. Appl. Phys.* **1971**, *42*, 5267–5272. [CrossRef]
96. Kobayashi, M.; Omata, K.; Sugimoto, S.; Tamagawa, Y.; Kuroiwa, T.; Asada, H.; Takeuchi, H.; Kondo, S. Scintillation Characteristics of $CsPbCl_3$ Single Crystals. *Nucl. Instrum. Methods Phys. Res. Sect. A Accel. Spectrometers Detect. Assoc. Equip.* **2008**, *592*, 369–373. [CrossRef]
97. Balvanz, A.; Bayikadi, K.S.; Liu, Z.; Ie, T.S.; Peters, J.A.; Kanatzidis, M.G. Unveiling the Monoclinic Phase in $CsPbBr_{3-x}Cl_x$ Perovskite Crystals, Phase Transition Suppression and High Energy Resolution γ-Ray Detection. *J. Am. Chem. Soc.* **2024**, *146*, 31836–31848. [CrossRef] [PubMed]
98. Liu, F.; Wu, R.; Wei, J.; Nie, W.; Mohite, A.D.; Brovelli, S.; Manna, L.; Li, H. Recent Progress in Halide Perovskite Radiation Detectors for Gamma-Ray Spectroscopy. *ACS Energy Lett.* **2022**, *7*, 1066–1085. [CrossRef]
99. Yoon, S.J.; Stamplecoskie, K.G.; Kamat, P.V. How Lead Halide Complex Chemistry Dictates the Composition of Mixed Halide Perovskites. *J. Phys. Chem. Lett.* **2016**, *7*, 1368–1373. [CrossRef] [PubMed]

100. Xiao, Z.; Zhao, L.; Tran, N.L.; Lin, Y.L.; Silver, S.H.; Kerner, R.A.; Yao, N.; Kahn, A.; Scholes, G.D.; Rand, B.P. Mixed-Halide Perovskites with Stabilized Bandgaps. *Nano Lett.* **2017**, *17*, 6863–6869. [CrossRef]
101. Knight, A.J.; Borchert, J.; Oliver, R.D.J.; Patel, J.B.; Radaelli, P.G.; Snaith, H.J.; Johnston, M.B.; Herz, L.M. Halide Segregation in Mixed-Halide Perovskites: Influence of A-Site Cations. *ACS Energy Lett.* **2021**, *6*, 799–808. [CrossRef]
102. Zarick, H.F.; Soetan, N.; Erwin, W.R.; Bardhan, R. Mixed Halide Hybrid Perovskites: A Paradigm Shift in Photovoltaics. *J. Mater. Chem. A* **2018**, *6*, 5507–5537. [CrossRef]
103. Nedelcu, G.; Protesescu, L.; Yakunin, S.; Bodnarchuk, M.I.; Grotevent, M.J.; Kovalenko, M.V. Fast Anion-Exchange in Highly Luminescent Nanocrystals of Cesium Lead Halide Perovskites ($CsPbX_3$, X = Cl, Br, I). *Nano Lett.* **2015**, *15*, 5635–5640. [CrossRef] [PubMed]
104. Yu, Y.; Zhang, D.; Yang, P. Ruddlesden–Popper Phase in Two-Dimensional Inorganic Halide Perovskites: A Plausible Model and the Supporting Observations. *Nano Lett.* **2017**, *17*, 5489–5494. [CrossRef]
105. Pan, D.; Fu, Y.; Spitha, N.; Zhao, Y.; Roy, C.R.; Morrow, D.J.; Kohler, D.D.; Wright, J.C.; Jin, S. Deterministic Fabrication of Arbitrary Vertical Heterostructures of Two-Dimensional Ruddlesden–Popper Halide Perovskites. *Nat. Nanotechnol.* **2021**, *16*, 159–165. [CrossRef] [PubMed]
106. Chen, Z.; Guo, Y.; Wertz, E.; Shi, J. Merits and Challenges of Ruddlesden–Popper Soft Halide Perovskites in Electro-Optics and Optoelectronics. *Adv. Mater.* **2019**, *31*, 1803514. [CrossRef] [PubMed]
107. Gong, J.; Hao, M.; Zhang, Y.; Liu, M.; Zhou, Y. Layered 2D Halide Perovskites beyond the Ruddlesden–Popper Phase: Tailored Interlayer Chemistries for High-Performance Solar Cells. *Angew. Chem.* **2022**, *134*, e202112022. [CrossRef]
108. Chen, Y.; Sun, Y.; Peng, J.; Tang, J.; Zheng, K.; Liang, Z. 2D Ruddlesden–Popper Perovskites for Optoelectronics. *Adv. Mater.* **2018**, *30*, 1703487. [CrossRef]
109. Li, J.; Yu, Q.; He, Y.; Stoumpos, C.C.; Niu, G.; Trimarchi, G.G.; Guo, H.; Dong, G.; Wang, D.; Wang, L.; et al. $Cs_2PbI_2Cl_2$, All-Inorganic Two-Dimensional Ruddlesden–Popper Mixed Halide Perovskite with Optoelectronic Response. *J. Am. Chem. Soc.* **2018**, *140*, 11085–11090. [CrossRef] [PubMed]
110. Ruddlesden, S.N.; Popper, P. New Compounds of the K_2NIF_4 Type. *Acta Cryst* **1957**, *10*, 538–539. [CrossRef]
111. Kresse, G.; Furthmüller, J. Efficiency of Ab-Initio Total Energy Calculations for Metals and Semiconductors Using a Plane-Wave Basis Set. *Comput. Mater. Sci.* **1996**, *6*, 15–50. [CrossRef]
112. Kresse, G.; Furthmüller, J. Efficient Iterative Schemes for Ab Initio Total-Energy Calculations Using a Plane-Wave Basis Set. *Phys. Rev. B* **1996**, *54*, 11169–11186. [CrossRef]
113. Acharyya, P.; Ghosh, T.; Pal, K.; Kundu, K.; Singh Rana, K.; Pandey, J.; Soni, A.; Waghmare, U.V.; Biswas, K. Intrinsically Ultralow Thermal Conductivity in Ruddlesden–Popper 2D Perovskite $Cs_2PbI_2Cl_2$: Localized Anharmonic Vibrations and Dynamic Octahedral Distortions. *J. Am. Chem. Soc.* **2020**, *142*, 15595–15603. [CrossRef] [PubMed]
114. Delaire, O.; Ma, J.; Marty, K.; May, A.F.; McGuire, M.A.; Du, M.-H.; Singh, D.J.; Podlesnyak, A.; Ehlers, G.; Lumsden, M.D.; et al. Giant Anharmonic Phonon Scattering in PbTe. *Nat. Mater.* **2011**, *10*, 614–619. [CrossRef] [PubMed]
115. Dissanayake, P.D.; Yeom, K.M.; Sarkar, B.; Alessi, D.S.; Hou, D.; Rinklebe, J.; Noh, J.H.; Ok, Y.S. Environmental Impact of Metal Halide Perovskite Solar Cells and Potential Mitigation Strategies: A Critical Review. *Environ. Res.* **2023**, *219*, 115066. [CrossRef]
116. Lyu, M.; Yun, J.; Chen, P.; Hao, M.; Wang, L. Addressing Toxicity of Lead: Progress and Applications of Low-Toxic Metal Halide Perovskites and Their Derivatives. *Adv. Energy Mater.* **2017**, *7*, 1602512. [CrossRef]
117. Xiao, Z.; Song, Z.; Yan, Y. From Lead Halide Perovskites to Lead-Free Metal Halide Perovskites and Perovskite Derivatives. *Adv. Mater.* **2019**, *31*, 1803792. [CrossRef]
118. Hu, H.; Dong, B.; Zhang, W. Low-Toxic Metal Halide Perovskites: Opportunities and Future Challenges. *J. Mater. Chem. A* **2017**, *5*, 11436–11449. [CrossRef]
119. De Souza Carvalho, T.A.; Magalhaes, L.F.; Do Livramento Santos, C.I.; De Freitas, T.A.Z.; Carvalho Vale, B.R.; Vale Da Fonseca, A.F.; Schiavon, M.A. Lead-Free Metal Halide Perovskite Nanocrystals: From Fundamentals to Applications. *Chem. Eur. J.* **2023**, *29*, e202202518. [CrossRef] [PubMed]
120. Zhang, L.; Wang, K.; Zou, B. Bismuth Halide Perovskite-Like Materials: Current Opportunities and Challenges. *ChemSusChem* **2019**, *12*, 1612–1630. [CrossRef]
121. Ayscue, R.L.; Vallet, V.; Bertke, J.A.; Réal, F.; Knope, K.E. Structure–Property Relationships in Photoluminescent Bismuth Halide Organic Hybrid Materials. *Inorg. Chem.* **2021**, *60*, 9727–9744. [CrossRef] [PubMed]
122. Creutz, S.E.; Liu, H.; Kaiser, M.E.; Li, X.; Gamelin, D.R. Structural Diversity in Cesium Bismuth Halide Nanocrystals. *Chem. Mater.* **2019**, *31*, 4685–4697. [CrossRef]
123. Li, X.; Du, X.; Zhang, P.; Hua, Y.; Liu, L.; Niu, G.; Zhang, G.; Tang, J.; Tao, X. Lead-Free Halide Perovskite $Cs_3Bi_2Br_9$ Single Crystals for High-Performance X-Ray Detection. *Sci. China Mater.* **2021**, *64*, 1427–1436. [CrossRef]
124. Leng, M.; Yang, Y.; Zeng, K.; Chen, Z.; Tan, Z.; Li, S.; Li, J.; Xu, B.; Li, D.; Hautzinger, M.P.; et al. All-Inorganic Bismuth-Based Perovskite Quantum Dots with Bright Blue Photoluminescence and Excellent Stability. *Adv. Funct. Mater.* **2018**, *28*, 1704446. [CrossRef]

125. Bass, K.K.; Estergreen, L.; Savory, C.N.; Buckeridge, J.; Scanlon, D.O.; Djurovich, P.I.; Bradforth, S.E.; Thompson, M.E.; Melot, B.C. Vibronic Structure in Room Temperature Photoluminescence of the Halide Perovskite $Cs_3Bi_2Br_9$. *Inorg. Chem.* **2017**, *56*, 42–45. [CrossRef]
126. Li, X.; Zhang, G.; Hua, Y.; Cui, F.; Sun, X.; Liu, J.; Liu, H.; Bi, Y.; Yue, Z.; Zhai, Z.; et al. Dimensional and Optoelectronic Tuning of Lead-free Perovskite $Cs_3Bi_2I_{9-n}Br_n$ Single Crystals for Enhanced Hard X-ray Detection. *Angew. Chem. Int. Ed.* **2023**, *62*, e202315817. [CrossRef] [PubMed]
127. Zhou, Y.; Fei, C.; Uddin, M.A.; Zhao, L.; Ni, Z.; Huang, J. Self-Powered Perovskite Photon-Counting Detectors. *Nature* **2023**, *616*, 712–718. [CrossRef] [PubMed]
128. Zhang, X.; Zhu, T.; Ji, C.; Yao, Y.; Luo, J. In Situ Epitaxial Growth of Centimeter-Sized Lead-Free $(BA)_2CsAgBiBr_7/Cs_2$ $AgBiBr_6$ Heterocrystals for Self-Driven X-Ray Detection. *J. Am. Chem. Soc.* **2021**, *143*, 20802–20810. [CrossRef]
129. McCall, K.M.; Stoumpos, C.C.; Kostina, S.S.; Kanatzidis, M.G.; Wessels, B.W. Strong Electron–Phonon Coupling and Self-Trapped Excitons in the Defect Halide Perovskites $A_3M_2I_9$ (A = Cs, Rb; M = Bi, Sb). *Chem. Mater.* **2017**, *29*, 4129–4145. [CrossRef]
130. McCall, K.M.; Liu, Z.; Trimarchi, G.; Stoumpos, C.C.; Lin, W.; He, Y.; Hadar, I.; Kanatzidis, M.G.; Wessels, B.W. α-Particle Detection and Charge Transport Characteristics in the $A_3M_2I_9$ Defect Perovskites (A = Cs, Rb; M = Bi, Sb). *ACS Photonics* **2018**, *5*, 3748–3762. [CrossRef]
131. Lei, X.-W.; Yue, C.-Y.; Zhao, J.-Q.; Han, Y.-F.; Yang, J.-T.; Meng, R.-R.; Gao, C.-S.; Ding, H.; Wang, C.-Y.; Chen, W.-D. Low-Dimensional Hybrid Cuprous Halides Directed by Transition Metal Complex: Syntheses, Crystal Structures, and Photocatalytic Properties. *Cryst. Growth Des.* **2015**, *15*, 5416–5426. [CrossRef]
132. Cheng, P.; Sun, L.; Feng, L.; Yang, S.; Yang, Y.; Zheng, D.; Zhao, Y.; Sang, Y.; Zhang, R.; Wei, D.; et al. Colloidal Synthesis and Optical Properties of All-Inorganic Low-Dimensional Cesium Copper Halide Nanocrystals. *Angew. Chem.* **2019**, *131*, 16233–16237. [CrossRef]
133. Zhang, P.; Yan, Z.; Li, C.; Du, Y.; Ma, L.; Wang, Z.; Lin, T.; Zhao, L.; Xiao, J. Low-Dimensional Hybrid Copper(I) Halides Single Crystals: Synthesis, Structures, and Tunable Photoluminescence. *Chem. Eng. J.* **2024**, *496*, 154106. [CrossRef]
134. Wang, Q.; Wang, C.; Wang, Z.; Sun, X.; Nikl, M.; OuYang, X.; Wu, Y. Achieving Efficient Neutron and Gamma Discrimination in a Highly Stable^6 Li-Loaded $Cs_3Cu_2I_5$ Perovskite Scintillator. *J. Phys. Chem. Lett.* **2022**, *13*, 9066–9071. [CrossRef] [PubMed]
135. Liu, D.; Wei, Q.; Tong, Y.; Xiang, P.; Cai, P.; Tang, G.; Shi, H.; Qin, L. A Novel Li^+-Doped $CsCu_2I_3$ Single Crystal for Dual Gamma–Neutron Detection. *CrystEngComm* **2023**, *25*, 58–63. [CrossRef]
136. Combes, C.M.; Dorenbos, P.; Van Eijk, C.W.E.; Krämer, K.W.; Güdel, H.U. Optical and Scintillation Properties of Pure and Ce^{3+}-Doped Cs_2LiYCl_6 and Li_3YCl_6:Ce^{3+} Crystals. *J. Lumin.* **1999**, *82*, 299–305. [CrossRef]
137. Glodo, J.; Hawrami, R.; Shah, K.S. Development of Cs_2LiYCl_6 Scintillator. *J. Cryst. Growth* **2013**, *379*, 73–78. [CrossRef]
138. Cheng, S.; Nikl, M.; Beitlerova, A.; Kucerkova, R.; Du, X.; Niu, G.; Jia, Y.; Tang, J.; Ren, G.; Wu, Y. Ultrabright and Highly Efficient All-Inorganic Zero-Dimensional Perovskite Scintillators. *Adv. Opt. Mater.* **2021**, *9*, 2100460. [CrossRef]
139. Wang, Q.; Zhou, Q.; Nikl, M.; Xiao, J.; Kucerkova, R.; Beitlerova, A.; Babin, V.; Prusa, P.; Linhart, V.; Wang, J.; et al. Highly Resolved X-Ray Imaging Enabled by In(I) Doped Perovskite-Like $Cs_3Cu_2I_5$ Single Crystal Scintillator. *Adv. Opt. Mater.* **2022**, *10*, 2200304. [CrossRef]
140. Belousov, M.P.; Gromyko, M.V.; Ignatyev, O.V. Scintillation γ Spectrometers for Use at Nuclear Power Plants (Review). *Instrum. Exp. Tech.* **2017**, *60*, 1–19. [CrossRef]
141. Dorenbos, P. (INVITED) The Quest for High Resolution γ-Ray Scintillators. *Opt. Mater. X* **2019**, *1*, 100021. [CrossRef]
142. Yuan, D. Air-Stable Bulk Halide Single-Crystal Scintillator $Cs_3Cu_2I_5$ by Melt Growth: Intrinsic and Tl Doped with High Light Yield. *ACS Appl. Mater. Interfaces* **2020**, *12*, 38333–38340. [CrossRef]
143. Tsuji, M.; Sasase, M.; Iimura, S.; Kim, J.; Hosono, H. Room-Temperature Solid-State Synthesis of $Cs_3Cu_2I_5$ Thin Films and Formation Mechanism for Its Unique Local Structure. *J. Am. Chem. Soc.* **2023**, *145*, 11650–11658. [CrossRef]
144. Wang, Z.; Wang, L.; Xie, J.; Yang, Y.; Song, Y.; Xiao, G.; Fu, Y.; Zhang, L.; Fang, Y.; Yang, D.; et al. $HCOO^-$ Doping-Induced Multiexciton Emissions in $Cs_3Cu_2I_5$ Crystals for Efficient X-Ray Scintillation. *Small* **2024**, *20*, 2309922. [CrossRef]
145. Chen, H.; Pina, J.M.; Yuan, F.; Johnston, A.; Ma, D.; Chen, B.; Li, Z.; Dumont, A.; Li, X.; Liu, Y.; et al. Multiple Self-Trapped Emissions in the Lead-Free Halide Cs_3 Cu_2 I_5. *J. Phys. Chem. Lett.* **2020**, *11*, 4326–4330. [CrossRef] [PubMed]
146. Lian, L.; Zheng, M.; Zhang, W.; Yin, L.; Du, X.; Zhang, P.; Zhang, X.; Gao, J.; Zhang, D.; Gao, L.; et al. Efficient and Reabsorption-Free Radioluminescence in $Cs_3Cu_2I_5$ Nanocrystals with Self-Trapped Excitons. *Adv. Sci.* **2020**, *7*, 2000195. [CrossRef] [PubMed]
147. Jun, T.; Sim, K.; Iimura, S.; Sasase, M.; Kamioka, H.; Kim, J.; Hosono, H. Lead-Free Highly Efficient Blue-Emitting $Cs_3Cu_2I_5$ with 0D Electronic Structure. *Adv. Mater.* **2018**, *30*, 1804547. [CrossRef]
148. Li, X.; Chen, J.; Yang, D.; Chen, X.; Geng, D.; Jiang, L.; Wu, Y.; Meng, C.; Zeng, H. Mn^{2+} Induced Significant Improvement and Robust Stability of Radioluminescence in $Cs_3Cu_2I_5$ for High-Performance Nuclear Battery. *Nat. Commun.* **2021**, *12*, 3879. [CrossRef] [PubMed]
149. Heo, J.H.; Shin, D.H.; Park, J.K.; Kim, D.H.; Lee, S.J.; Im, S.H. High-Performance Next-Generation Perovskite Nanocrystal Scintillator for Nondestructive X-Ray Imaging. *Adv. Mater.* **2018**, *30*, 1801743. [CrossRef]

150. Van Loef, E.V.; Ciampi, G.; Shirwadkar, U.; Soundara Pandian, L.; Shah, K.S. Crystal Growth and Scintillation Properties of Thallium-Based Halide Scintillators. *J. Cryst. Growth* **2020**, *532*, 125438. [CrossRef]
151. Shirwadkar, U.; Loyd, M.; Du, M.-H.; Van Loef, E.; Ciampi, G.; Soundara Pandian, L.; Stand, L.; Koschan, M.; Zhuravleva, M.; Melcher, C.; et al. Thallium-Based Scintillators for High-Resolution Gamma-Ray Spectroscopy: Ce^{3+}-Doped Tl_2LaCl_5 and Tl_2LaBr_5. *Nucl. Instrum. Methods Phys. Res. Sect. A Accel. Spectrometers Detect. Assoc. Equip.* **2020**, *962*, 163684. [CrossRef]
152. Hawrami, R.; Ariesanti, E.; Buliga, V.; Burger, A. Thallium Strontium Iodide: A High Efficiency Scintillator for Gamma-Ray Detection. *Opt. Mater.* **2020**, *100*, 109624. [CrossRef]
153. Kim, H.; Khan, A.; Daniel, J.; Rooh, G.; Vuong, P.Q. Thallium-Based Heavy Inorganic Scintillators: Recent Developments and Future Perspectives. *CrystEngComm* **2022**, *24*, 450–464. [CrossRef]
154. Pan, L.; Sholom, S.; McKeever, S.W.S.; Jacobsohn, L.G. Magnesium Aluminate Spinel for Optically Stimulated Luminescence Dosimetry. *J. Alloys Compd.* **2021**, *880*, 160503. [CrossRef]
155. Sakai, T.; Koshimizu, M.; Fujimoto, Y.; Nakauchi, D.; Yanagida, T.; Asai, K. Luminescence and Scintillation Properties of Tl- and In-Doped CsCl Crystals. *Jpn. J. Appl. Phys.* **2017**, *56*, 062601. [CrossRef]
156. Zazubovich, S.; Voloshinovskii, A.; Stryganyuk, G. Luminescence of CsCl: Tl Crystal under Synchrotron Excitation. *Phys. Status Solidi (B)* **2002**, *233*, 238–249. [CrossRef]
157. Nagirnyi, V.; Zazubovich, S.; Jaanson, N. Luminescence of CsCl: Tl Single Crystals. *Phys. Status Solidi (B)* **1993**, *175*, 155–164. [CrossRef]
158. Shiran, N.V.; Gektin, A.V.; Boyarintseva, Y.; Vasyukov, S.; Boyarintsev, A.; Pedash, V.; Tkachenko, S.; Zelenskaya, O.; Kosinov, N.; Kisil, O.; et al. Eu Doped and Eu, Tl Co-Doped NaI Scintillators. *IEEE Trans. Nucl. Sci.* **2010**, *57*, 1233–1235. [CrossRef]
159. Lin, W.; He, J.; McCall, K.M.; Stoumpos, C.C.; Liu, Z.; Hadar, I.; Das, S.; Wang, H.; Wang, B.; Chung, D.Y.; et al. Inorganic Halide Perovskitoid $TlPbI_3$ for Ionizing Radiation Detection. *Adv. Funct. Mater.* **2021**, *31*, 2006635. [CrossRef]
160. Rodi, F.; Babel, D. Ternäre Oxide der Übergangsmetalle. IV. Erdalkaliiridium(IV)-oxide: Kristallstruktur von $CaIrO_3$. *Z. Anorg Allge Chem.* **1965**, *336*, 17–23. [CrossRef]
161. Li, X.; He, Y.; Kepenekian, M.; Guo, P.; Ke, W.; Even, J.; Katan, C.; Stoumpos, C.C.; Schaller, R.D.; Kanatzidis, M.G. Three-Dimensional Lead Iodide Perovskitoid Hybrids with High X-Ray Photoresponse. *J. Am. Chem. Soc.* **2020**, *142*, 6625–6637. [CrossRef]
162. Vtyurina, D.N.; Kaurova, I.A.; Kuz'micheva, G.M.; Rybakov, V.B.; Chernyshov, D.Y.; Khramov, E.V.; Firstov, S.V.; Korchak, V.N. Structural Peculiarities, Point Defects and Luminescence in Bi-Doped $CsCdX_3$ (X = Cl, Br) Single Crystals. *J. Alloys Compd.* **2019**, *803*, 912–921. [CrossRef]
163. Vtyurina, D.N.; Eistrikh-Geller, P.A.; Kuz'micheva, G.M.; Rybakov, V.B.; Khramov, E.V.; Kaurova, I.A.; Chernyshov, D.Y.; Korchak, V.N. Influence of Monovalent Bi+ Doping on Real Composition, Point Defects, and Photoluminescence in $TlCdCl_3$ and $TlCdI_3$ Single Crystals. *Sci. China Mater.* **2017**, *60*, 1253–1263. [CrossRef]
164. Yang, G.; Phan, Q.V.; Liu, M.; Hawari, A.; Kim, H. Material Defect Study of Thallium Lead Iodide ($TlPbI_3$) Crystals for Radiation Detector Applications. *Nucl. Instrum. Methods Phys. Res. Sect. A Accel. Spectrometers Detect. Assoc. Equip.* **2020**, *954*, 161516. [CrossRef]
165. Liu, Z.; Zhang, T.; Wang, Y.; Wang, C.; Zhang, P.; Sarvari, H.; Chen, Z.; Li, S. Electronic Properties of a New All-Inorganic Perovskite $TlPbI_3$ Simulated by the First Principles. *Nanoscale Res. Lett.* **2017**, *12*, 232. [CrossRef]
166. Yan, Y.; Liu, F.; Wu, J.; Zhang, Z.; Wu, Z.; Cong, W.-Y.; Guan, C.; Lu, Y.-B. Evaluation of the Electronic Properties of Layered $TlPbI_3$ Perovskites. *Phys. B Condens. Matter* **2023**, *667*, 415190. [CrossRef]
167. Khyzhun, O.Y.; Fochuk, P.M.; Kityk, I.V.; Piasecki, M.; Levkovets, S.I.; Fedorchuk, A.O.; Parasyuk, O.V. Single Crystal Growth and Electronic Structure of $TlPbI_3$. *Mater. Chem. Phys.* **2016**, *172*, 165–172. [CrossRef]
168. Lin, W.; Stoumpos, C.C.; Liu, Z.; Das, S.; Kontsevoi, O.Y.; He, Y.; Malliakas, C.D.; Chen, H.; Wessels, B.W.; Kanatzidis, M.G. $TlSn_2I_5$, a Robust Halide Antiperovskite Semiconductor for γ-Ray Detection at Room Temperature. *ACS Photonics* **2017**, *4*, 1805–1813. [CrossRef]
169. Li, W.; Li, M.; He, Y.; Song, J.; Guo, K.; Pan, W.; Wei, H. Arising 2D Perovskites for Ionizing Radiation Detection. *Adv. Mater.* **2024**, *36*, 2309588. [CrossRef] [PubMed]
170. Ni, Z.; Zhao, J.; Zhang, G.; Chen, Y.; Lian, H.; Lin, J. Synthesis of Pb-free Double-Perovskite Heterojunction (Cs_2SnCl_6-Bi^{3+}&Te^{4+}/Cs_2GeF_6-Mn^{4+}) with White Emission for LED Application. *Adv. Opt. Mater.* **2023**, *11*, 2203029. [CrossRef]
171. Zhou, J.; Luo, J.; Rong, X.; Wei, P.; Molokeev, M.S.; Huang, Y.; Zhao, J.; Liu, Q.; Zhang, X.; Tang, J.; et al. Lead-Free Perovskite Derivative Cs_2 $SnCl_{6-x}Br_x$Single Crystals for Narrowband Photodetectors. *Adv. Opt. Mater.* **2019**, *7*, 1900139. [CrossRef]
172. Sun, J.; Zheng, W.; Huang, P.; Zhang, M.; Zhang, W.; Deng, Z.; Yu, S.; Jin, M.; Chen, X. Efficient Near-Infrared Luminescence in Lanthanide-Doped Vacancy-Ordered Double Perovskite Cs_2ZrCl_6 Phosphors via Te^{4+} Sensitization. *Angew. Chem. Int. Ed.* **2022**, *61*, e202201993. [CrossRef]
173. Wang, A.; Yan, X.; Zhang, M.; Sun, S.; Yang, M.; Shen, W.; Pan, X.; Wang, P.; Deng, Z. Controlled Synthesis of Lead-Free and Stable Perovskite Derivative Cs_2SnI_6 Nanocrystals via a Facile Hot-Injection Process. *Chem. Mater.* **2016**, *28*, 8132–8140. [CrossRef]

174. Scott, S.M.; Zhu, W.; Yao, T.; Vienna, J.D.; Ewing, R.C.; Lian, J. The Thermal Stability and Consolidation of Perovskite Variant Cs_2SnCl_6 Using Spark Plasma Sintering. *J. Am. Ceram. Soc.* **2018**, *101*, 2060–2065. [CrossRef]
175. Chang, T.; Wei, Q.; Zeng, R.; Cao, S.; Zhao, J.; Zou, B. Efficient Energy Transfer in Te^{4+}-Doped Cs_2ZrCl_6 Vacancy-Ordered Perovskites and Ultrahigh Moisture Stability via A-Site Rb-Alloying Strategy. *J. Phys. Chem. Lett.* **2021**, *12*, 1829–1837. [CrossRef] [PubMed]
176. Qing, X.; Wu, C.; Han, X. Efficient Near-Infrared Luminescence Based on Double Perovskite Cs_2SnCl_6. *Molecules* **2023**, *28*, 3593. [CrossRef] [PubMed]
177. Jin, M.; Zheng, W.; Gong, Z.; Huang, P.; Li, R.; Xu, J.; Cheng, X.; Zhang, W.; Chen, X. Unraveling the Triplet Excited-State Dynamics of Bi^{3+} in Vacancy-Ordered Double Perovskite Cs_2SnCl_6 Nanocrystals. *Nano Res.* **2022**, *15*, 6422–6429. [CrossRef]
178. Das Adhikari, S.; Echeverría-Arrondo, C.; Sánchez, R.S.; Chirvony, V.S.; Martínez-Pastor, J.P.; Agouram, S.; Muñoz-Sanjosé, V.; Mora-Seró, I. White Light Emission from Lead-Free Mixed-Cation Doped Cs_2SnCl_6 Nanocrystals. *Nanoscale* **2022**, *14*, 1468–1479. [CrossRef] [PubMed]
179. Jia, X.; Wang, B.; Zhu, M.; Zhang, Y.; Chong, X.; Li, J.; Wang, J. Growth Process and Optical Properties of Cs_2ZrCl_6 Doped with Ce^{3+} and Li^+ Crystals. *Cryst. Growth Des.* **2023**, *23*, 3589–3594. [CrossRef]
180. Buryi, M.; Babin, V.; Ligthart, R.A.M.; Nagorny, S.S.; Mikhailik, V.B.; Vaněček, V.; Prouzová Prochazková, L.; Kandel, R.; Nahorna, V.V.; Wang, P. Correlation of Emission, Scintillation and Charge Trapping Properties in Cs_2HfCl_6 and Cs_2ZrCl_6 Single Crystals. *J. Mater. Chem. C* **2021**, *9*, 2955–2968. [CrossRef]
181. Vanecek, V.; Kral, R.; Paterek, J.; Babin, V.; Jary, V.; Hybler, J.; Kodama, S.; Kurosawa, S.; Yokota, Y.; Yoshikawa, A.; et al. Modified Vertical Bridgman Method: Time and Cost Effective Tool for Preparation of Cs_2HfCl_6 Single Crystals. *J. Cryst. Growth* **2020**, *533*, 125479. [CrossRef]
182. Jia, X.; Jiang, H.; Chong, X.; Ju, W.; Feng, T.; Zhao, Z.; Wei, H.; Li, J.; Cheng, S.; Wang, J. A Novel Co-Substitution Strategy for the Design, Growth, and Optoelectronic Performance Optimization of Cs_2ZrCl_6:Ce,Li Single Crystals. *J. Mater. Chem. C* **2025**, *13*, 2844–2852. [CrossRef]
183. Song, J.; Cui, Q.; Li, J.; Xu, J.; Wang, Y.; Xu, L.; Xue, J.; Dong, Y.; Tian, T.; Sun, H.; et al. Ultralarge All-Inorganic Perovskite Bulk Single Crystal for High-Performance Visible–Infrared Dual-Modal Photodetectors. *Adv. Opt. Mater.* **2017**, *5*, 1700157. [CrossRef]

inorganics

MDPI

Review

A Comprehensive Review of Recent Advances in Perovskite Materials: Electrical, Dielectric, and Magnetic Properties

Faouzia Tayari, Silvia Soreto Teixeira, Manuel Pedro F. Graca and Kais Iben Nassar *

I3N-Aveiro, Department of Physics, University of Aveiro, 3810-193 Aveiro, Portugal; faouzia.tayari@ua.pt (F.T.); silvia.soreto@ua.pt (S.S.T.); mpfg@ua.pt (M.P.F.G.)

* Correspondence: kais.nassar@ua.pt

Abstract: Perovskite materials have emerged as one of the most promising classes of compounds in recent years due to their unique combination of electrical, dielectric, and magnetic properties, which make them ideal candidates for a wide range of advanced technological applications. This comprehensive review explores the latest developments in the electrical, dielectric, and magnetic behavior of perovskites, providing an in-depth analysis of the underlying mechanisms and their potential for improving device performance. The review covers the fundamental aspects of charge transport, polarization, and magnetic interactions in perovskite structures including the impact of crystal symmetry, ion migration, and external stimuli on their properties. Moreover, it highlights the various strategies used to tailor these properties through compositional engineering, doping, and structural modifications, resulting in enhanced efficiency, stability, and multifunctionality in applications such as photovoltaics, capacitors, dielectric resonators, and spintronic devices. Additionally, the paper addresses the challenges associated with the practical implementation of perovskite materials including stability issues under harsh environmental conditions and scalability for industrial applications. The review concludes with an outlook on future directions, emphasizing the need for further research to overcome these challenges and unlock the full potential of perovskite materials in next-generation electronics, energy storage, and magnetic devices.

Keywords: perovskite; sol-gel route; impedance complex; electrical conductivity; dielectric relaxation; magnetic properties

1. Introduction

Perovskite materials, named after the mineral calcium titanate ($CaTiO_3$), have garnered immense attention due to their extraordinary structural versatility and remarkable functional properties. With a general formula of ABX_3, where 'A' and 'B' represent cations of differing sizes and 'X' is an anion, typically oxygen or halide, perovskites exhibit a wide range of tunable properties that are invaluable for modern technological applications. The flexibility of the perovskite structure, stemming from the diverse choices for A, B, and X components, allows for intricate tailoring of their electrical, dielectric, and magnetic properties, making them a cornerstone in materials science research [1–6].

The general formula for perovskites is ABX_3, where 'A' and 'B' are cations of differing sizes and 'X' is an anion that bonds to both. Typically, 'A' is a larger cation such as a rare-earth or alkaline-earth metal, 'B' is a smaller transition metal cation, and 'X' is an anion like oxygen [7–11]. This structure allows for a vast array of elemental substitutions, resulting in a rich variety of physical properties including notable magnetic behaviors [11–15]. The perovskite structure is characterized by a three-dimensional framework of corner-sharing BX_6 octahedra, with the 'A' cations occupying the interstitial sites [14–19]. This unique architecture not only stabilizes the structure, but also enables diverse magnetic interactions, driven by the intricate electronic correlations between the 'A' and 'B' site cations and anions [20–22]. The flexibility in the perovskite composition allows for the fine-tuning of these interactions, leading to phenomena such as ferroelectricity, magnetoresistance, magnetocaloric effects, and spintronics applications.

The structural and chemical aspects of perovskites play a critical role in dictating their physical properties. The ability to accommodate various ions in the A and B sites results in a vast library of perovskite compounds with distinct features. Structural distortions, such as tilting of the BX_6 octahedra, influence key parameters such as bandgap, ferroelectricity, and magnetic interactions. Furthermore, advancements in synthesis techniques have enabled precise control over compositional engineering, defect management, and nanostructuring, opening avenues for enhancing their performance in practical applications [23–25]. Since then, the study of perovskites has evolved significantly, particularly in the latter half of the 20th century with the advent of high-resolution characterization techniques and advanced synthesis methods. These advancements have enabled researchers to unravel the complex magnetic behaviors exhibited by perovskites, spurring a wave of innovations in both fundamental research and practical applications.

In terms of electrical and dielectric properties, perovskites have shown unprecedented potential in applications such as capacitors, energy storage devices, and piezoelectric sensors. Their ability to exhibit high dielectric constants, ferroelectricity, and ion conductivity makes them highly versatile for both fundamental research and industrial applications. The interplay between structural features and electronic behavior, including ion migration and polarization dynamics, has been a central theme in understanding and optimizing these properties. Similarly, perovskites have demonstrated unique magnetic properties, including ferromagnetism, antiferromagnetism, and spin-glass behavior, which are highly dependent on their composition and structural ordering. These properties are critical for spintronic applications, magnetic sensors, and data storage devices. The coupling between magnetic ordering and other functionalities, such as ferroelectricity in multiferroic perovskites, provides exciting opportunities for multifunctional device integration [26–28].

Ferroelectricity: Ferroelectric perovskites, such as $BaTiO_3$, exhibit spontaneous electric polarization that can be reversed by the application of an external electric field. This property is intimately linked with the material's crystal structure, where the displacement of the B-site cation relative to the anions creates a dipole moment [29–34]. The coupling between ferroelectricity and magnetism in certain perovskites has led to the discovery of multiferroic materials, which hold promise for novel electronic devices that exploit both electric and magnetic orderings.

Magnetoresistance: The phenomenon of magnetoresistance, where a material's electrical resistance changes in response to an applied magnetic field, is particularly pronounced in perovskite manganites [35–41]. This colossal magnetoresistance (CMR) effect is a consequence of the interplay between the electronic structure and magnetic ordering, driven by the double exchange mechanism. CMR materials are being explored for their potential in magnetic field sensors and spintronic devices.

Magnetocaloric effect: The magnetocaloric effect (MCE), which refers to the change in temperature of a material upon the application or removal of a magnetic field, is another intriguing property of perovskites. Materials exhibiting a large MCE are of great interest for magnetic refrigeration technologies, which offer an energy-efficient and environmentally friendly alternative to conventional gas-compression refrigeration. Perovskites such as

La(Fe, Si)$_{13}$-based compounds have shown significant potential in this area, prompting ongoing research into optimizing their MCE for practical applications [42–46].

Spintronics and magnetochemistry: Spintronics, or spin-based electronics, represents a burgeoning field where the spin of electrons, in addition to their charge, is utilized for information processing. Perovskites play a crucial role in this domain due to their ability to exhibit spin-polarized currents and giant magnetoresistance. The interplay between the spin degrees of freedom and the lattice structure in perovskites enables the development of spintronic devices with enhanced functionality and performance [47–52]. Magnetochemistry, which explores the chemical aspects of magnetic materials, benefits greatly from the study of perovskites. The tunable nature of the perovskite structure allows for systematic investigations into how chemical substitutions and modifications affect magnetic properties. This understanding is crucial for designing materials with specific magnetic behaviors tailored for applications in catalysis, sensing, and magnetic data storage [53–57].

Applications and future directions: The diverse magnetic properties of perovskites have led to their incorporation into various applications. In addition to the aforementioned uses in sensors, memory devices, and refrigeration, perovskites are being explored for use in energy conversion and storage [58–61] as well as in biomedical applications such as magnetic resonance imaging (MRI) contrast agents and targeted drug delivery systems [59–69]. Despite the significant progress made in understanding and utilizing the magnetic properties of perovskites, several challenges remain. These include improving the stability and reproducibility of perovskite materials, enhancing their performance at room temperature, and developing scalable synthesis methods for industrial applications. Future research efforts are likely to focus on addressing these challenges as well as exploring new compositions and structures to unlock further potential in this versatile class of materials. The study of magnetic properties in perovskite materials is a dynamic and rapidly evolving field, driven by the unique structural characteristics and tunable nature of these compounds. Figure 1 illustrates the trends in publications related to perovskite materials, highlighting significant growth in the fields of electrical, dielectric, and magnetic properties. The data showed a consistent increase in research activity across all categories, underscoring the growing interest and advancements in these key areas of perovskite material applications. In addition to their diverse applications in sensors, memory devices, energy conversion, and biomedical fields, perovskite-based materials also show great promise for next-generation optoelectronic applications, particularly in light-emitting devices, which are expected to drive future innovations in the field [70].

This review aims to provide a comprehensive overview of the recent advances in the study of perovskite materials, with a focus on the structural and chemical aspects that underlie their diverse functional properties. It delves into the structural and chemical fundamentals of perovskites, discussing their crystalline frameworks, chemical tunability, and defect management and highlights their electrical and dielectric properties, emphasizing the mechanisms of charge transport, polarization, and dielectric behavior. It explores the magnetic properties of perovskites, discussing the key factors influencing their magnetic ordering and their implications for technological applications. The review concludes with a discussion of the challenges and future directions in the field, aiming to inspire further research into the design and application of perovskite materials.

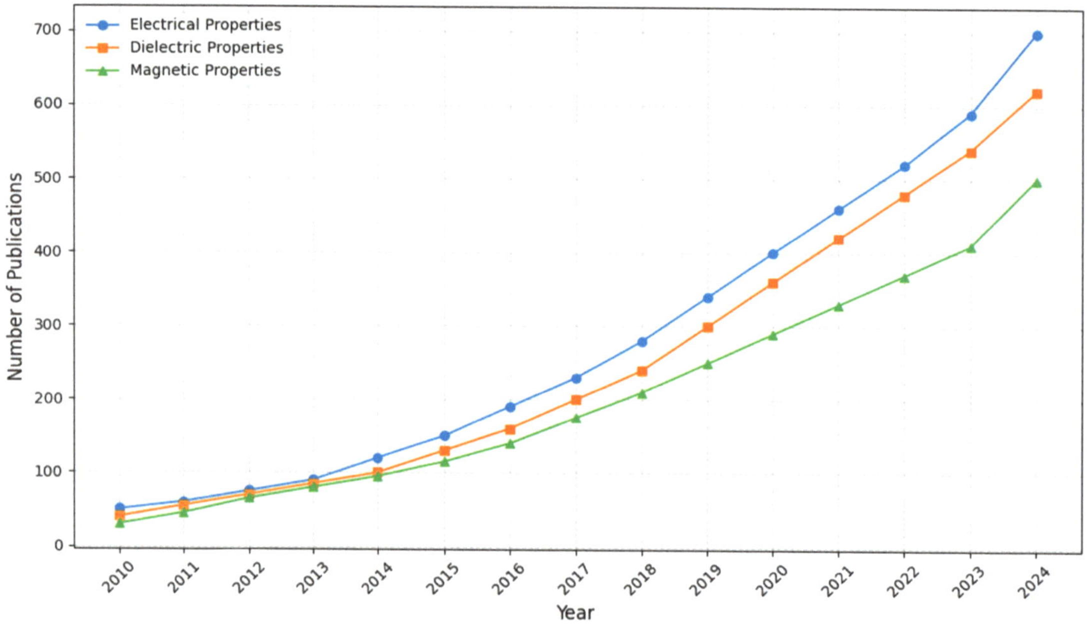

Figure 1. Trends in publications on perovskite materials highlighting electrical, dielectric, and magnetic properties.

2. Structural and Chemical Aspects of Perovskites

Perovskite materials, characterized by their ABX_3 crystal structure, exhibit a remarkable diversity in their chemical compositions and structural variations. Understanding the structural and chemical aspects is fundamental to elucidating their magnetic properties and exploring their potential applications across various fields (Figure 2).

Crystal structure and symmetry: The perovskite crystal structure consists of a three-dimensional network of corner-sharing BX_6 octahedra, where 'B' cations occupy the octahedral centers and 'X' anions reside at the vertices (Figure 3) [71,72]. The 'A' cations occupy the larger cubic sites between the octahedra. This arrangement gives rise to the general formula ABX_3, where 'A' is typically a larger cation (e.g., La, Ba) occupying the A-site, 'B' is a smaller transition metal cation (e.g., Ti, Mn) at the B-site, and 'X' is an anion (e.g., O, F) bonding to both 'A' and 'B'. The cubic perovskite structure (space group $Pm3^{-}$ m) is stable under ambient conditions for many materials but can undergo distortions and phase transitions under different temperatures, pressures, or upon chemical substitution. These structural variations are crucial as they strongly influence the material's magnetic and electrical properties.

A-site and B-site cation substitutions: One of the defining features of perovskites is their ability to substitute both the 'A' and 'B' cations, leading to a vast array of materials with tailored properties. Substituting 'A'-site cations can influence lattice parameters, stability, and sometimes introduce charge carriers through doping. For instance, replacing La with Ca in $LaMnO_3$ can modify the material's magnetic and electronic properties. Similarly, B'-site cation substitutions can alter magnetic interactions and electronic configurations within the octahedral framework. This flexibility allows researchers to fine-tune properties such as magnetism, ferroelectricity, and conductivity. For example, replacing Mn with Fe or Co in perovskite manganites can significantly affect the magnetic ordering and magneto resistive behavior due to changes in the electronic structure and spin configurations. To gain deeper insights into charge generation and transfer within perovskite crystals, ultrafast laser transient absorption spectroscopy (TAS) serves as a valuable figure of merit tool. This technique is crucial for understanding the dynamics of charge carriers, which strongly correlate with the performance of perovskite-based devices [73,74].

Influence of anion (X) variations: The choice of anion in perovskite compounds also plays a critical role in determining their properties. Oxygen (O) is the most common anion in traditional perovskites, but other halides (e.g., F, Cl) or mixed anions can be incorporated to modify properties such as bandgap, conductivity, and magnetic ordering. Mixed-anion perovskites, like $BaZr_{(1-x)}$ $TixO_3$ or $LaMnO_{3-x}F_x$, exhibit intriguing properties that combine the characteristics of their constituent anions, offering new avenues for material design and applications.

Synthesis methods and techniques: The synthesis of perovskite materials is a pivotal aspect in controlling their structural and chemical properties (Figure 4). Various synthesis techniques, including solid-state reactions, sol-gel methods, chemical vapor deposition (CVD), and hydrothermal synthesis, offer different levels of control over particle size, crystallinity, and composition. Characterization techniques such as X-ray diffraction (XRD), scanning electron microscopy (SEM), transmission electron microscopy (TEM), and spectroscopic methods (e.g., X-ray photoelectron spectroscopy, Raman spectroscopy) are essential for elucidating the crystal structure, phase purity, and chemical composition of perovskite materials [75]. These techniques provide insights into how structural and chemical modifications impact the magnetic behaviors and overall performance of perovskite-based devices. The structural and chemical aspects of perovskite materials are pivotal in understanding their magnetic properties and tailoring them for specific applications. By manipulating the composition, doping, and synthesis conditions, researchers can unlock new functionalities and enhance the performance of perovskite materials in fields ranging from electronics and magnetism to energy conversion and beyond. Future research efforts will continue to explore novel compositions and synthesis strategies to further expand the capabilities of perovskite-based materials in emerging technologies.

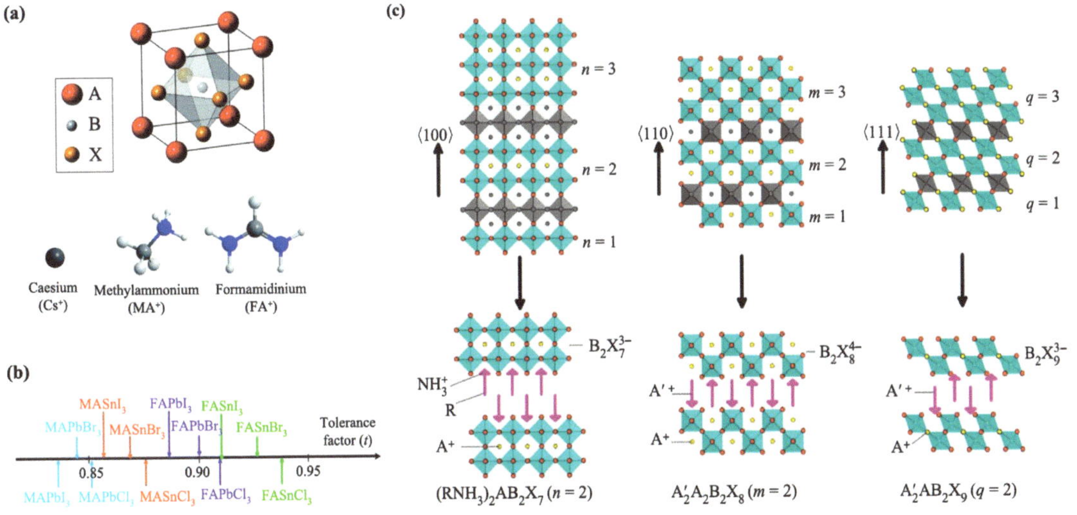

Figure 2. (**a**) Crystal structure of 3D perovskite ABX_3 with Cs^+, MA^+, or FA^+ cations. (**b**) Tolerance factors (t) of 3D halide perovskites. (**c**) Structure derivation of layered 2D perovskite with long cations cutting the 3D perovskite along the (100), (110), and (111) planes. Reproduced with permission from [76]. Copyright © 2019, Springer Nature.

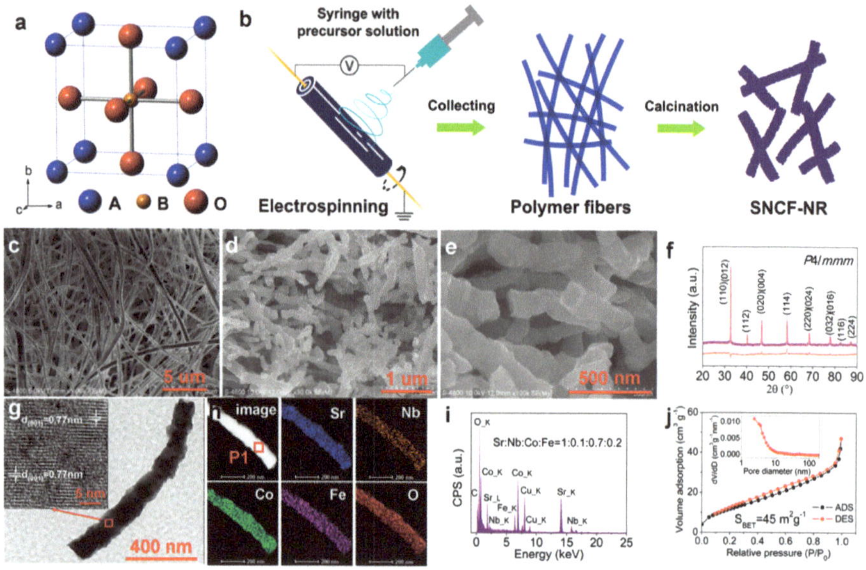

Figure 3. Structural and morphological insights into perovskite materials and solar cells. Reprinted with permission [77]. Copyright © 2017 Wiley-VCH Verlag GmbH & Co. KGaA, Weinheim.

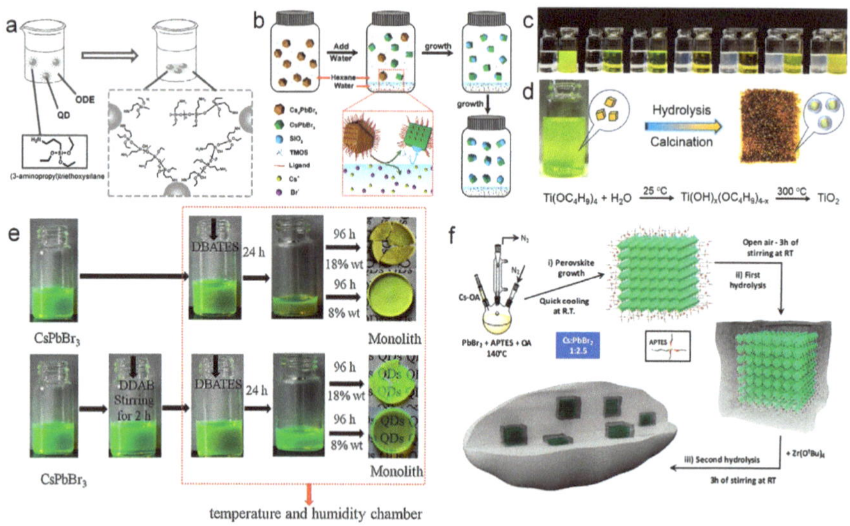

Figure 4. Synthesis (sol-gel) of perovskites. Reprinted with permission from Duan, Y., Wang, D., and Costa, R. D [78]. Copyright © 2021 Wiley-VCH Verlag GmbH & Co. KGaA, Weinheim.

Characterization Techniques

Characterizing perovskite materials is essential for understanding their structural, optical, electrical, and magnetic properties. Key characterization techniques include X-ray diffraction (XRD), which determines the crystal structure and phase purity of perovskite materials. Scanning electron microscopy (SEM) and transmission electron microscopy (TEM) provide detailed morphological and microstructural information at the nanoscale. Spectroscopic techniques including UV-Vis spectroscopy, Fourier-transform infrared spectroscopy (FTIR), and Raman spectroscopy analyze optical and vibrational properties. Electrical and magnetic measurements, such as Hall effect measurements, electrical conductivity, and magnetization measurements, probe electronic and magnetic properties [79–90].

3. Electrical and Dielectric Properties

Perovskite materials exhibit a fascinating array of electrical, dielectric, and magnetic properties, which are central to their widespread application in diverse technological fields. This section explores these properties in detail and discusses the role of impedance spectroscopy in understanding their complex behavior [91–95].

3.1. Electrical Properties

Perovskite materials demonstrate a wide range of electrical behaviors, spanning from insulation to metallic and superconducting states. This variability arises from the ability to finely tune their electronic structure through doping, compositional variations, and defect engineering. Understanding and controlling the electrical conductivity of perovskites are crucial for applications in electronic devices such as transistors, sensors, and energy storage systems. Recent breakthroughs in perovskite solar cells (PSCs) highlight their potential for achieving high efficiency and stability, promising advancements in renewable energy technologies. Achary et al. [96]. conducted a comprehensive study on the structural, electrical, and dielectric properties of two lead-free double perovskites, $BiHoZnZrO_6$ (BHZZR) and $BiHoCuTiO_6$ (BHCTO), synthesized using a cost-effective solid-state reaction technique. Their research revealed that substituting the pair of elements Zn and Zr with Cu and Ti induces a structural transition from a monoclinic to an orthorhombic crystal system (Figure 5). Microscopic imaging and energy-dispersive spectra confirmed the homogeneous distribution of grains and the formation of the desired compounds in both perovskites. Detailed impedance spectroscopy studies highlighted that the dielectric and electrical properties of these materials are significantly influenced by frequency and temperature variations [97–100].

It was observed that the grains, grain boundaries, and electrodes profoundly impact these properties, providing insights into the electrical conduction mechanisms and microstructural behaviors. Equivalent circuit models were developed to_6 extract electrical parameters such as bulk and grain boundary resistance and capacitance, facilitating a detailed comparison of the AC conductivity, electrical modulus, and transport properties of BHZZR and BHCTO. In examining the dielectric properties, the study focused on the dielectric loss (tan δ), which is the ratio of the real and imaginary parts of a capacitor's impedance at a given frequency. A higher tan δ indicates substantial dielectric absorption [101–104]. For capacitors constructed with lossy dielectrics, capacitance (C) changes more significantly with frequency, whereas good dielectric materials exhibit slower changes in C. The dielectric constant (ε_r) was defined as the ratio of a capacitor's capacity with BHZZR as the dielectric medium to that with a vacuum as the dielectric medium. The dielectric constant is temperature and frequency-dependent and was calculated using the formula $\varepsilon_r = Cp/C_0$, where C_p is the parallel capacitance obtained from impedance analysis, and C_0 is determined from the physical dimensions of the sample [105]. For instance, with a pellet diameter of 1.191 cm and thickness of 0.2295 cm, the calculated area was 1.113508×10^{-4} m^2, resulting in a C_0 value of 0.43 pF. Using this value, the dielectric constant was computed from the C_p data derived from the impedance results. The plots of dielectric constant versus frequency for BHZZR and BHCTO showed that the εr was initially high but decreased with increasing frequency for both samples. This behavior is influenced by interfacial, dipolar, ionic, atomic, and electronic polarizations. At lower frequencies, dipoles struggle to align with the alternating electric field, leading to higher dielectric constants. However, at higher frequencies, electronic polarization predominates, causing the dielectric constant to decline. This study elucidates the complex interplay between structural modifications and the resulting electrical and dielectric properties, offering valuable insights for the development and optimization of perovskite-based materials in various technological applications [106].

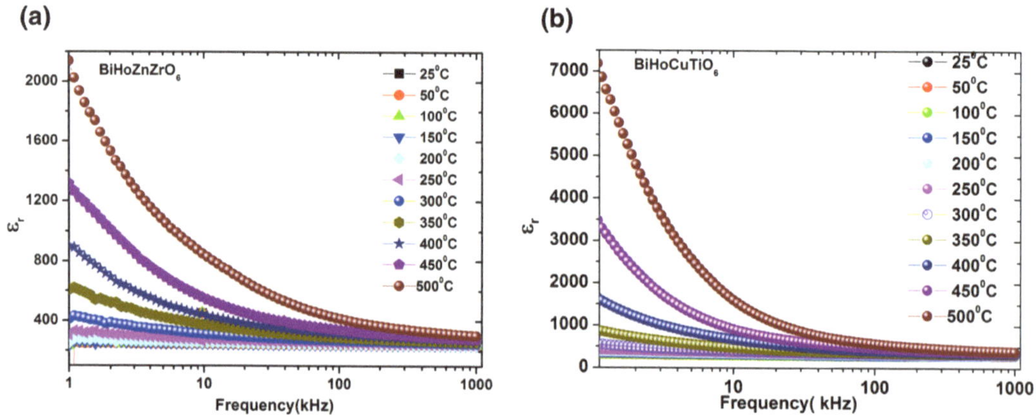

Figure 5. (**a**) The plot of dielectric constant with frequency for BHZZR. (**b**) The plot of dielectric constant with frequency for BHCTO. Reprinted with permission from Achary, P. G. R. et al. (2018) [106]. Copyright © 2018 Springer Science Business Media, LLC, part of Springer Nature.

In addition, Nassar et al. [107]. conducted an in-depth investigation into the AC electrical conductivity and dielectric properties of the double perovskite oxide $BiBaFeZnO_6$ (BiBFZO), synthesized using the sol-gel method. Their structural analysis revealed that BiBFZO crystallizes in a rhombohedral system with the R-3C space group, confirming its single-phase nature. The study employed impedance spectroscopy to examine the dielectric behavior and physical phenomena of the material over a temperature range of 200 to 380 K and frequencies up to 1 MHz. Nyquist diagrams at various temperatures displayed a flat semi-circular arc, modeled using an equivalent circuit comprising an RC element and an RQ element. The frequency-dependent behavior of the imaginary component of impedance (Z'') indicated a relaxation process, while electrical conductivity analysis, based on Jonscher's law, enabled the calculation of activation energy [108–110]. The results validated the non-overlapping small polaron tunneling (NSPT) model as a suitable explanation for the electrical conduction mechanism in BiBFZO. Electrical conductivity exhibited two distinct regions: a low-frequency region where conductivity remained constant, and a high-frequency region where the conductivity increased with frequency. The temperature-dependent variation in conductivity showed a linear behavior, indicative of a thermally activated transport mechanism. The dielectric properties of BiBFZO were also temperature and frequency-dependent (Figure 6). The complex impedance spectra showed a single flat semi-circular arc, which was interpreted using an equivalent circuit model comprising a constant phase element impedance (RQ element) and a parallel resistance-capacitance network (R_1-CPE_1 and R_2-C_1). Activation energy values obtained from both electrical conductivity and complex impedance measurements were found to be very similar, suggesting that the relaxation mechanism is attributable to the same type of charge carriers [111–115]. The temperature dependence of the exponents provided further insight into the conduction mechanism, supporting the NSPT model as the most appropriate for explaining the electrical conduction phenomena in BiBFZO.

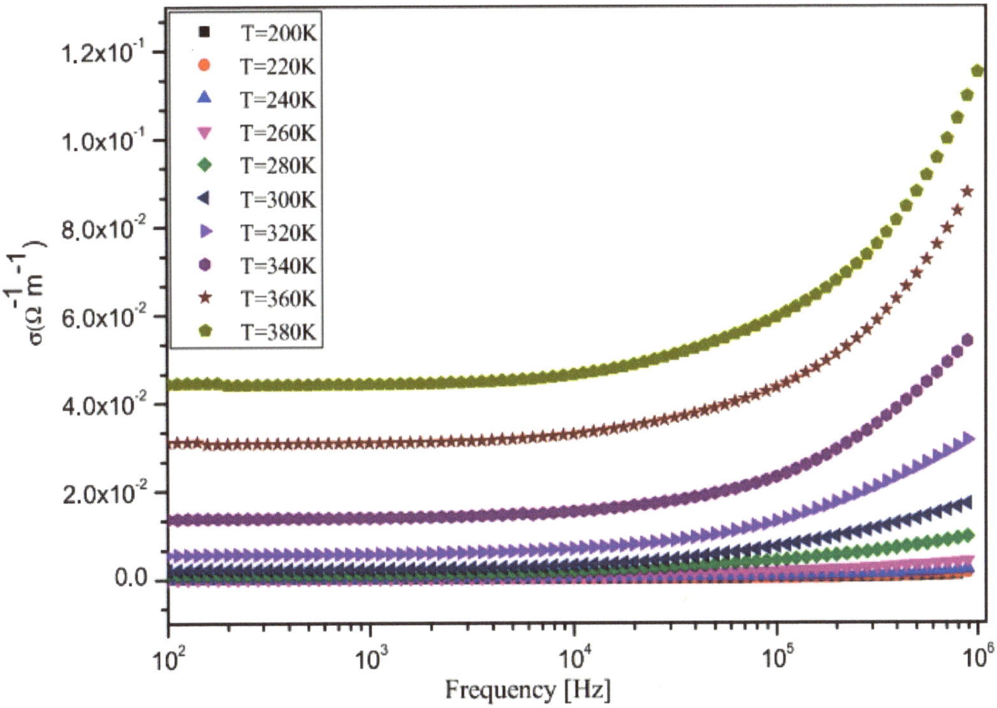

Figure 6. Frequency dependency of conductivity σ_{ac} for $BiBaFeZnO_6$. Reprinted with permission from Nassar, K. I. et al. [107]. Copyright © 2021 The Author(s), under exclusive license to Springer Science Business Media, LLC, part of Springer Nature.

Furthermore, K. Iben Nassar and colleagues [116] conducted an in-depth investigation into the structural and electrical properties of the double perovskite oxide $LaPbFeTiO_6$, synthesized using the sol-gel method. X-ray diffraction analysis revealed that $LaPbFeTiO_6$ crystallizes in the rhombohedral system with the space group R3C. The chemical composition and morphology of the grain size were confirmed through energy-dispersive spectroscopy (EDS) and scanning electron microscopy (SEM), respectively. Electrical conductivity and dielectric characterizations were performed as functions of frequency and temperature (ranging from 200 to 380 K) using impedance spectroscopy. Their results indicated that the dielectric constant of $LaPbFeTiO_6$ increased with temperature. In their study of AC conductivity, the frequency dependence was found to adhere to Jonscher's power law, where $\sigma(\omega) = \sigma_{DC} + A\omega^s$, with ω being the angular frequency, σ_{DC} is the DC conductivity, A is a constant, and s represents the interaction between mobile ions and surrounding lattices. The conductivity curves showed two distinct regions: a low-frequency plateau representing DC conductivity, and a high-frequency region where AC conductivity increased with frequency. This behavior suggests a thermally activated charge carrier hopping process, which intensifies with temperature [4,117–120]. The high-frequency conductivity variation, attributed to the relaxation phenomenon, is due to the mobility of charge carriers, implying their displacement from their original positions to new sites.

The experimental data fitting, performed using ORIGIN 8.0 software, confirmed these observations and indicated that $LaPbFeTiO_6$ behaves as a semiconductive material. Additionally, the temperature dependence of DC conductivity followed Arrhenius behavior, where the activation energy (E_a) was derived from the linear relationship between $Ln(\sigma_{DC})$

and the inverse of temperature (1000/T) (see Figure 7). This analysis provided insights into the thermal activation energies and the conduction mechanisms in this material.

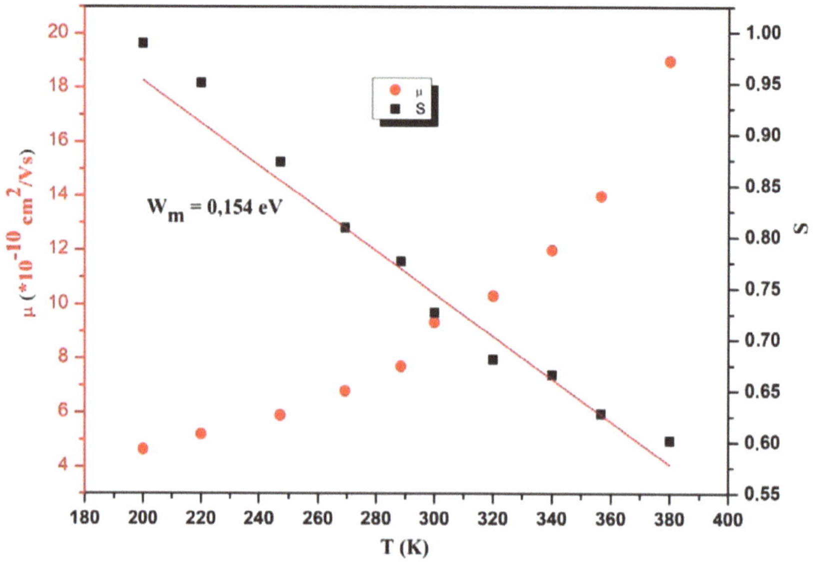

Figure 7. Temperature dependences of the charge carrier mobilities and the exponent s of $LaPbFeTiO_6$. Reprinted with permission from Nassar, K. I. et al. (2021). Applied Physics A: Materials Science & Processing [116]. Copyright © 2021 The Author(s), under exclusive license to Springer-Verlag GmbH, DE, part of Springer Nature.

Indeed, Hasan, Z., Rahman, M.A., Das, D.K. et al. [121] conducted a comprehensive first-principles investigation into the structural, electronic, optical, and mechanical properties of the pure $BaTiO_3$ and Ca-doped $BaTiO_3$ ($Ba_{1-x}Ca_xTiO_3$ with x = 0.125, 0.25, 0.375, 0.500, 0.625) perovskites, aiming to evaluate their potential for optoelectronic, photorefractive, and photovoltaic applications. Utilizing density-functional-theory calculations, the study revealed significant modifications in the properties of $BaTiO_3$ upon Ca doping. The geometrically optimized structures showed a phase transition from cubic to another crystal phase at a doping level of x = 0.25. The electronic band structure analysis indicated a change from an indirect to a direct band gap upon Ca doping, with a shift in the conduction band toward higher energy levels. The bandgap for pure $BaTiO_3$ was found to be 1.729 eV, which increased slightly with Ca doping, reaching 1.749 eV at x = 0.125 and 1.775 eV at x = 0.25, demonstrating a direct band gap at these doping levels. This direct band gap characteristic is crucial for optoelectronic and photovoltaic applications, as it enhances the material's efficiency in light absorption and emission processes [122–125]. The optical properties, including absorption, reflectivity, refractive index, extinction coefficient, conductivity, dielectric function, and loss function, were extensively studied over an energy range from 0 to 30 eV. The results highlighted a prominent absorption peak and significant optical activity in the UV light energy region, suggesting that Ca-doped $BaTiO_3$ is highly suitable for UV photo detectors and other optoelectronic devices. The mechanical properties were also analyzed, showing increased Debye temperatures with higher Ca doping levels, indicating enhanced thermal stability. The elastic constants confirmed the mechanical stability and the presence of covalent bonds in the compounds. Overall, the modification of $BaTiO_3$ by Ca doping significantly improved its structural, electronic, and optical properties, making it a multifunctional material with potential applications in various advanced technologies. The study provides a solid foundation for further experimental and theoretical research to optimize and utilize Ca-doped $BaTiO_3$ in practical applications. Extra, Md. Azizur Rahman, and their team [126] conducted an extensive study on the impact of A-cations (Ba, Sr, and Ca) on the structural, electronic, optical, mechanical, and solar cell performance of novel A_3NCl_3 perovskites using DFT and SCAPS-1D (version 3.3.10) simulation software. This research aimed to investigate lead-free halide perovskites, known for their excellent optical absorption, tunable band gap, and high carrier mobility. The study revealed that these perovskites exhibit a semiconducting nature with a direct band gap and mechanical stability, confirmed by phonon studies. The direct band gap values for Ba_3NCl_3, Sr_3NCl_3, and Ca_3NCl_3 were found to be 0.58(1.20), 1.258(1.75), and 1.683(2.30) eV, respectively, using PBE(HSE) functionals, indicating a decrease in bandgap as the A-cation changed from Ba to Sr to Ca. This trend is attributed to the increase in the cation radius, which weakens electrostatic interactions between the cations and Cl atoms, reducing the energy needed for electron transition from the valence band to the conduction band. The solar cell performance analysis, using the structure Al/FTO/SnS_2/A_3NCl_3/Au, showed that Ba_3NCl_3-based cells achieved the highest power conversion efficiency (PCE) of 28.81%, with a JSC of 38.26 mA/cm^2, FF of 79.91%, and VOC of 0.94 V. In contrast, cells with Sr_3NCl_3 and Ca_3NCl_3 had PCEs of 18.11% and 8.54%, respectively. The electronic band structures indicated a transition from an indirect to a direct bandgap, essential for photovoltaic and optoelectronic applications [126]. The study also examined the partial density of states, revealing hybridization of the Ca, Sr, Ba, and N states with Cl, indicating ionic bonding between these elements. The valence band maximum (VBM) was mainly influenced by the Cl-2p orbital, while the conduction band minimum (CBM) was determined by the N-2p and A (Ca, Sr, Ba)-p orbitals. This comprehensive analysis provides valuable insights into the experimental development of efficient A_3NCl_3-based perovskite solar cells (Figure 8).

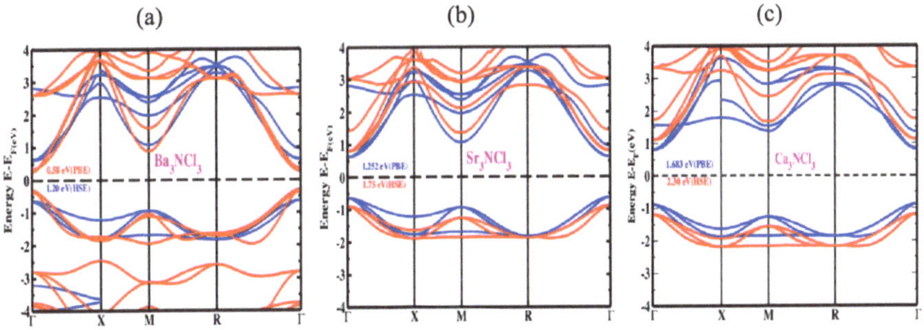

Figure 8. (**a**) Ba_3NCl_3, (**b**) Sr_3NCl_3, and (**c**) Ca_3NCl_3 electronic band structures using the PBE and HSE functions of inorganic novel A_3NCl_3 perovskites. Reprinted with permission from Rahman, M. A. et al. (2024). Energyfuels.3c03175 [126]. Copyright © 2024 American Chemical Society.

Nassar, K.I., Benamara, M., Kechiche, L., and their colleagues [127] studied the structural, microstructural, and electrical properties of bismuth and niobium-doped $LaNiO_3$ perovskite synthesized via the sol-gel technique for potential electronic device applications. The X-ray diffraction analysis with Rietveld refinement indicated a monoclinic phase with a $P_{21/n}$ space group at room temperature. Scanning electron microscopy revealed closely packed grains with high density and minimal porosity. The dielectric relaxation phenomenon analyzed using the Cole-Cole model and the complex impedance data were modeled with an equivalent circuit comprising a resistance and a constant phase element [128]. The frequency-dependent conductivity followed Jonscher's power law, while an Arrhenius-type conduction mechanism with an activation energy of 0.168 eV was identified. Notably, the electrical properties remained stable across different temperatures, highlighting the material's potential for thermally stable capacitor applications. The ac conductivity study showed two plateaus in the spectra, indicating the presence of two distinct processes contributing to the bulk conduction behavior, with one relaxing in the high-frequency region and the other appearing as a plateau at high frequencies (Figure 9).

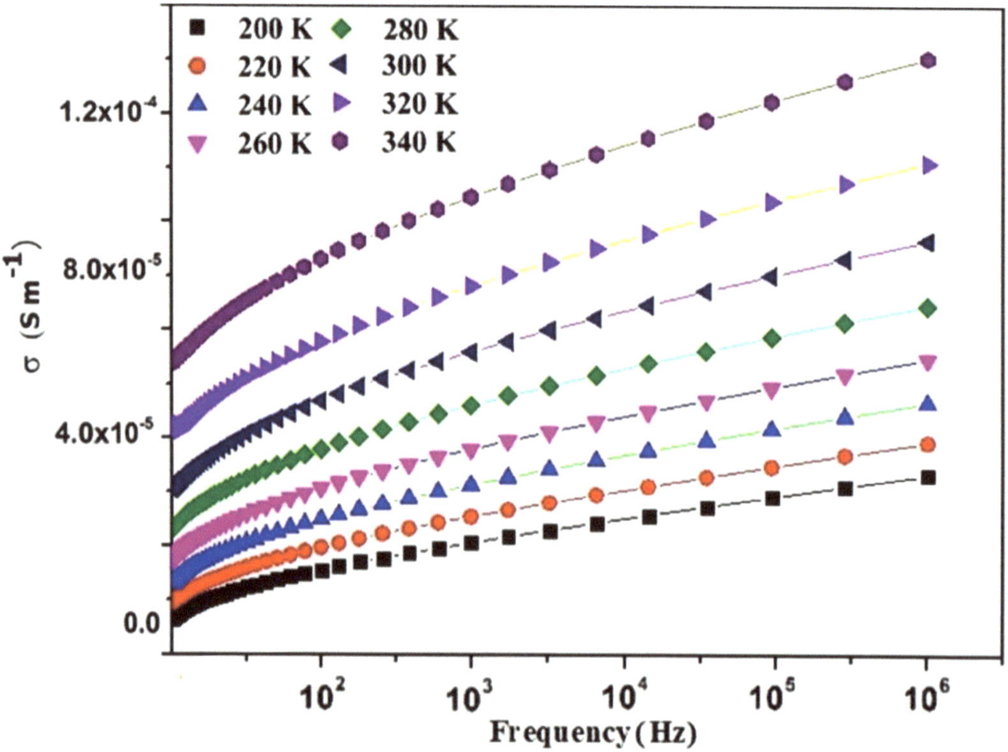

Figure 9. Variation in the conductivity σ_{AC} as a function of temperature and frequency. Reprinted with permission from Nassar, K. I. et al. (2024). Indian Journal of Physics [127]. Copyright © 2024 Indian Association for the Cultivation of Science, published by Springer Nature.

3.2. Dielectric Properties

The dielectric properties of perovskite materials are equally significant, particularly for their applications in capacitors, tunable dielectrics, and memory devices. Many perovskite compounds exhibit high dielectric constants combined with low dielectric losses, making them ideal candidates for advanced electronic components. The frequency-dependent behavior of dielectrics in perovskites, including dielectric relaxation and polarization mechanisms, plays a critical role in optimizing their performance across various applications. In their study, Iben Nassar, K., et al. [129] explored the dielectric properties of the polycrystalline sample $LaBaFe_{1.2}Mn_{0.8}O_6$ synthesized via the sol-gel method. X-ray diffraction confirmed the sample's single-phase formation in a rhombohedral crystal system with the RC space group. Detailed dielectric and impedance spectroscopy analyses over various temperatures revealed a relaxation phenomenon. The real part of the dielectric permittivity (ε') was found to increase with temperature and decrease with frequency. A notable shift in the maximum permittivity curves was observed as the temperature rose from 300 K to 620 K, indicating a dielectric relaxor behavior. The permittivity increased up to a transition temperature (T_d) of 425 K, beyond which it decreased, suggesting a metal-semiconductor transition. Additionally, the analysis of dielectric loss showed almost negligible dissipation at low temperatures, which increased significantly around 400 K and then decreased with higher frequencies at elevated temperatures, consistent with the Debye model (Figure 10). This behavior suggests that charge carriers are unable to follow the applied electric field at high frequencies, leading to reduced dielectric dissipation in this range. These findings underscore the material's potential for applications requiring high permittivity and low dielectric loss such as capacitors.

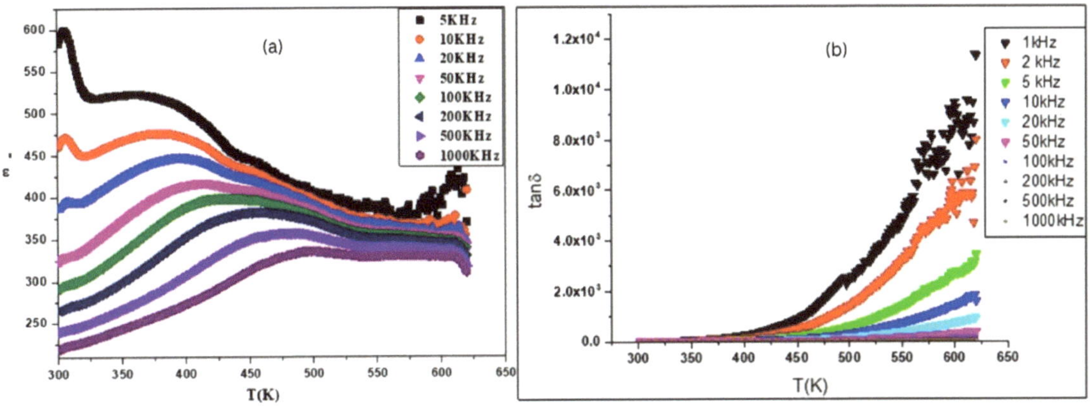

Figure 10. Temperature variation in the real part of permittivity (ε′) at several frequencies (**a**) and the temperature variation in the loss tangent (tanα) at several frequencies of the sample (**b**). Reprinted with permission from Nassar, K. I. et al. [129]. Copyright © 2022 Indian Association for the Cultivation of Science, published by Springer Nature.

The study by Tayari et al. [130] delved into the dielectric and electrical conductivity properties of the $Ba_{0.75}Ni_{0.25}Tc_{0.88}Mn_{0.12}O_3$ (BNTMO) perovskite ceramic, synthesized using the sol-gel method. Structural analysis via X-ray diffraction confirmed a single-phase cubic structure with a Pm-3m space group, while scanning electron microscopy showed a well-defined morphology with an average particle size of 243 nm. Dielectric studies revealed that the real part of the complex permittivity (ε′) increased with temperature and decreased with frequency, suggesting a dielectric polarization mechanism. This behavior, primarily due to interfacial polarization, aligns with observations in similar perovskite and metallic oxide materials. The dielectric loss also decreases with increasing frequency, attributed to relaxation time distribution and charging carrier density. These properties indicate the material's potential for energy storage applications [131]. The consistent activation energy values from various analyses further support the coherence of the charge carrier relaxation process in BNTMO, making it a viable candidate for advanced technological applications (Figure 11).

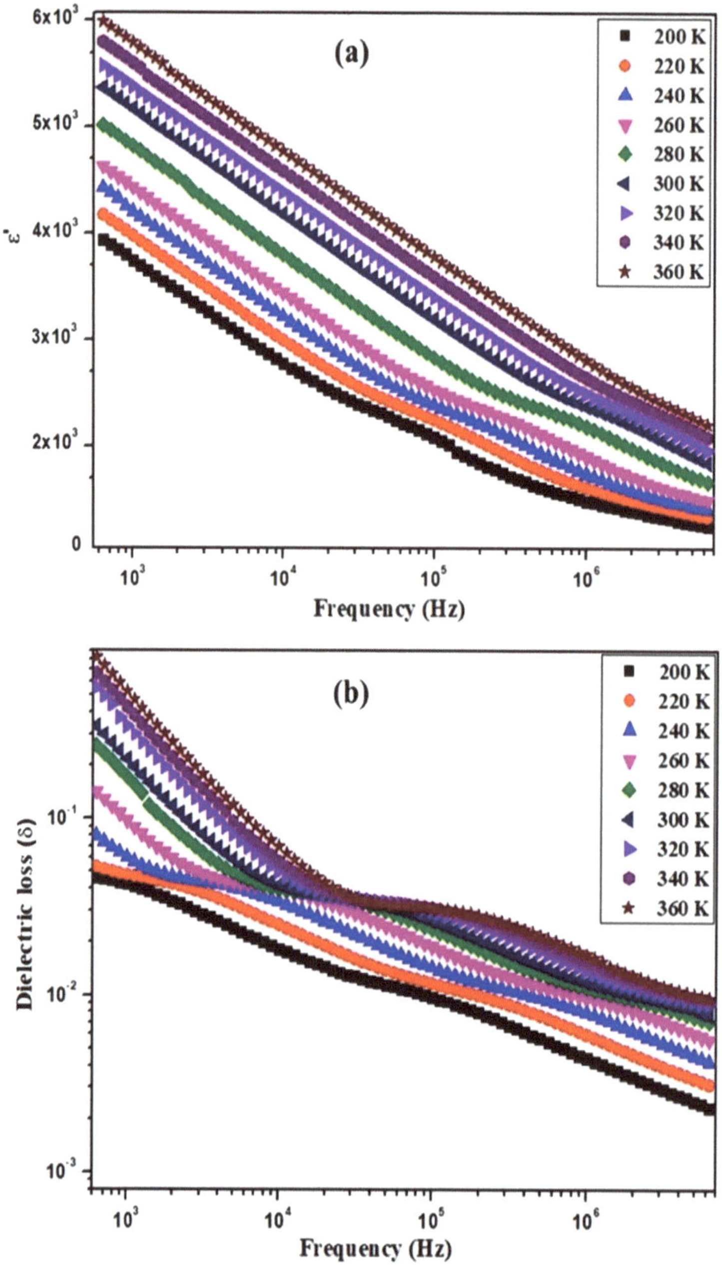

Figure 11. (**a**) Evolution of the real part of the permittivity and (**b**) the variation in dielectric loss as a function of frequency at various temperatures. Reprinted with permission from Tayari et al. (2024). Journal of Sol-Gel Science and Technology [130]. Copyright © 2024 The Author(s), published by Springer Nature.

The study by Kumar, A., Kumar, V., and Kumar, A. et al. [132] examined the dielectric and electrical properties of a $Bi_5Yb_3O_{12}$ (BYO) ceramic material synthesized via a chemical route. X-ray diffraction (XRD) and Le Bail fitting confirmed the single-phase formation of a BYO ceramic with a cubic crystal system and space group Fm-3m. X-ray photoelectron spectroscopy (XPS) verified the oxidation states of the elements, while energy-dispersive X-ray spectroscopy (EDS) determined the elemental composition. The dielectric constant of the BYO ceramic reached a peak value of 337 at 382 K and 1 kHz, and its lowest value was 238 at 523 K and 1 kHz, with a tangent loss (tan δ) of 0.2 at 293 K and 1 kHz, and 0.37 at 293 K and 1 MHz The dielectric constant (ε') and dielectric loss (tan δ) of the BYO ceramic were studied over various temperatures and frequencies, showing that the dielectric constant was higher at low frequencies due to contributions from various polarization types, such as electronic, ionic, dipolar, and interfacial, and decreased with increasing frequency, where electronic polarization dominated. Additionally, temperature influenced the dielectric constant, which initially increased but decreased beyond a certain point, reflecting the material's relaxation behavior. Factors like material thickness and moisture content also affect the dielectric constant. This behavior aligns with the Maxwell-Wagner and Koop models, describing ceramic materials as having highly conducting grains and less conducting grain boundaries. The dielectric loss decreases with increasing frequency, indicating the material's relaxation dynamics and charge carrier density contributions. The study highlights BYO ceramic's potential for technical applications, particularly in capacitors, due to its high permittivity and low dielectric loss, providing insights into its suitability for electronic devices and energy storage applications (Figure 12).

In their study, Suresh et al. [133] explored the structural, optical, and dielectric properties of nanostructured $La_{1-x}Sr_xMnO_3$ perovskites synthesized via the combustion method using citric acid as fuel. X-ray diffraction (XRD) confirmed the formation of the desired compositions, revealing that increasing the Sr doping concentration reduced the crystallite size from 30 nm to 14 nm. Scanning electron microscopy (SEM) displayed a spongy, porous structure in the samples. Optical analysis showed good absorbance in the UV and higher wavelength regions, with the optical bandgap energy increasing. In their study, Jabarov et al. investigated the dielectric and electrical properties of $La_{0.5}Ba_{0.5}MnO_3$ and $La_{0.97}Ba_{0.03}MnO_3$ perovskites over a temperature range of 25–225 °C and a frequency range of 20–10^6 Hz. Their analysis revealed that the electrical conductivity of these compounds increased with temperature and frequency, and a semiconductor-metal phase transition occurred at around 140 °C. The dielectric constant and losses were characterized, showing a stable real part of the dielectric constant under varying temperature and frequency, while the imaginary part of the dielectric constant decreased with frequency. For $La_{0.5}Ba_{0.5}MnO_3$, the imaginary part stabilized after approximately 5×10^3 Hz, and significant dielectric losses were noted at lower frequencies due to the charge carrier dynamics. The temperature-dependent behavior of the imaginary part of the dielectric constant indicated the presence of impurities or radicals, influencing its increase below 90 °C and resonant interactions between 90 and 160 °C. In contrast, $La_{0.97}Ba_{0.03}MnO_3$ showed a clear increase in dielectric constant up to 140 °C, followed by a decrease, indicative of the phase transition. The findings underscore the significance of these perovskites in applications requiring controlled dielectric and conductive properties, highlighting their potential in electronic and catalytic processes. From 1.4 eV to 2.5 eV, as the Sr doping concentration increased, the dielectric analysis highlighted that $La_{0.3}Sr_{0.7}MnO_3$ exhibited a high dielectric constant at ambient temperature and low frequencies, attributed to interfacial dislocations, charge defects, oxygen vacancies, and grain boundary effects as explained by the Maxwell-Wagner model (Figure 13). The dielectric constant decreased with increasing frequency but increased with higher Sr doping concentrations, making these materials suitable for supercapacitor applications. The study also noted that ac conductivity followed Johnscher's universal power law at high frequencies, with electrode polarization effects evident from the log σ log f plot. The dielectric loss tangent (tan δ) decreased with increasing frequency, indicating higher resistivity at the grain boundaries compared with the grains. The behavior of tan δ, which decreased with increased doping, suggests its suitability for high-frequency device applications. Moreover, in their study, Jabarov et al. [134] investigated the dielectric and electrical properties of $La_{0.5}Ba_{0.5}MnO_3$ and $La_{0.97}Ba_{0.03}MnO_3$ perovskites over a temperature range of 25–225 °C and a frequency range of 20–106 Hz. Their analysis revealed that the electrical conductivity of these compounds increased with temperature and frequency, and a semiconductor-metal phase transition occurred at around 140 °C. The dielectric constant and losses were characterized, showing a stable real part of the dielectric constant under varying temperature and frequency, while the imaginary part of the dielectric constant decreased with frequency. For $La_{0.5}Ba_{0.5}MnO_3$, the imaginary part stabilized after approximately 5×10^3 Hz, and significant dielectric losses were noted at lower frequencies due to the charge carrier dynamics. The temperature-dependent behavior of the imaginary part of the dielectric constant indicated the presence of impurities or radicals, influencing its increase below 90 °C and resonant interactions between 90 and 160 °C. In contrast, $La_{0.97}Ba_{0.03}MnO_3$ showed a clear increase in dielectric constant up to 140 °C, followed by a decrease, indicative of the phase transition. The findings underscore the significance of these perovskites in applications requiring controlled dielectric and conductive properties, highlighting their potential in electronic and catalytic processes.

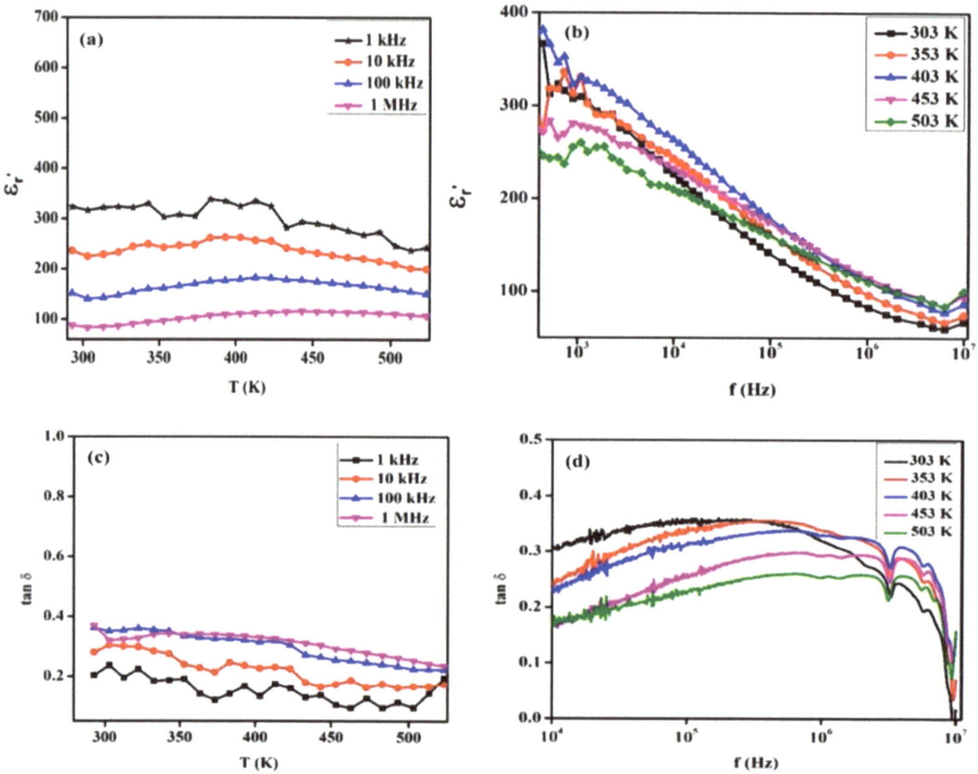

Figure 12. (**a**,**b**) Temperature and frequency dependent variation in dielectric constant (ε_r) and (**c**,**d**) temperature and frequency-dependent variation in tangent loss (tan δ) of BYO sintered at 1123 K for a duration of 8 h. Reprinted with permission from Kumar, A. et al. (2024). Journal of Materials Science: Materials in Electronics [132]. Copyright © 2024 The Author(s), under exclusive license to Springer Science Business Media, LLC, part of Springer Nature.

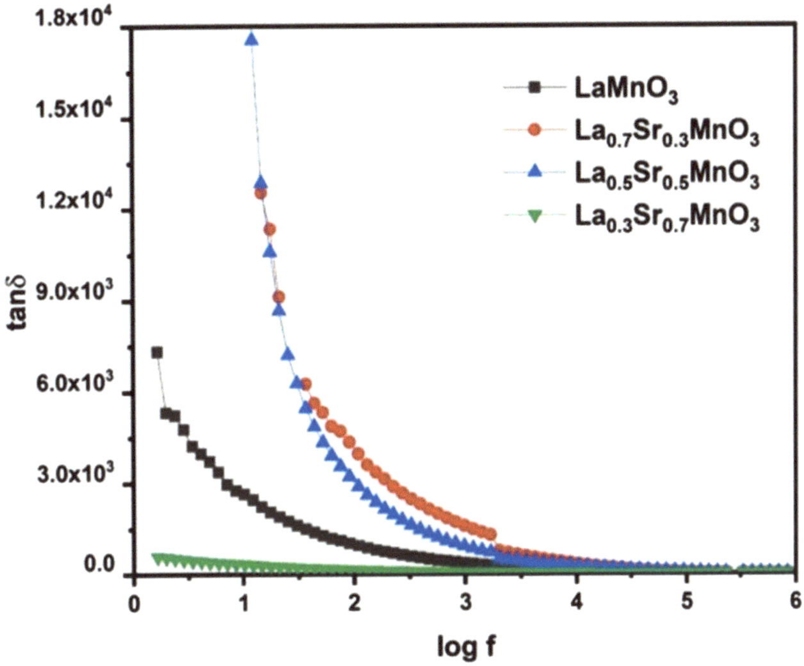

Figure 13. Variation in loss factor tan δ of $LaMnO_3$, $La_{0.7}Sr_{0.3}MnO_3$, $La_{0.5}Sr_{0.5}MnO_3$, and $La_{0.3}Sr_{0.7}MnO_3$ nanoparticles with frequency. Reprinted with permission from Suresh, S. et al. (2023). Materials Today Communications [133]. Copyright © 2023 Elsevier, published by Materials Today Communications.

Similarly, the study by Algahtani et al. [135] explored the structural, optical, and microwave dielectric properties of $Sr_2Fe_{1+x}Nb_{1-x}O_6$ ceramics synthesized via a solid-state route. Their investigation employed X-ray powder diffraction (XRD), scanning electron microscopy (SEM), Fourier transform infrared (FTIR) spectroscopy, impedance analysis, and photoluminescence (PL) spectroscopy. XRD analysis confirmed a single-phase tetragonal structure with space group I4/m across the $Sr_2Fe_{1+x}Nb_{1-x}O_6$ composition range of $0.015 \leq x \leq 0.030$. The findings indicated enhanced structure stability with increasing x. Impedance analysis revealed that increasing frequency enhanced carrier hopping and mobility, impacting conductivity. The dielectric constant showed variations with the frequency and relative densities, indicating a rise with increasing density but was affected by the impurity phases at microscopic levels. Dielectric loss, decreased with increasing frequency, was attributed to the reduced power loss from current leakage in the capacitor configuration. PL analysis highlighted an excitation energy suitable for green light lasers. Overall, the research underscores the potential of $Sr_2Fe_{1+x}Nb_{1-x}O_6$ ceramics in microwave applications, emphasizing their structural stability and favorable dielectric properties across varying compositions and densities.

4. Magnetic Properties

Perovskite materials display diverse magnetic properties, ranging from ferromagnetism and antiferromagnetism to complex magnetic ordering phenomena. These properties are essential for applications in spintronics, magnetic storage, and multifunctional materials that integrate magnetic and electronic functionalities [136]. The interplay between the structural, electronic, and magnetic characteristics of perovskites offers fertile ground for exploring new physical phenomena and developing innovative technologies. The magnetic properties of RE_2NiCrO_6 (RE = Ce, Pr, and Nd) double perovskite oxides were thoroughly investigated using experimental and theoretical approaches in the study by Irfan et al. [137]. The compounds were synthesized via the hydrothermal method, and their structural, electronic, and magnetic properties were characterized. X-ray diffraction confirmed the orthorhombic structure, while SEM images displayed well-formed particles, and EDS mapping verified the elemental composition of Ce, Pr, Nd, Ni, Cr, and O in appropriate proportions. Density functional theory calculations using the FP-LAPW method revealed the half-metallic nature of these compounds, with the spin-polarized electronic structures and density of states indicating ferromagnetic coupling between the Ni and Cr ions. The magnetic properties were further elucidated through magnetization measurements at room temperature, as depicted in Figure 14a–c. The presence of hysteresis loops with residual magnetization and coercivity confirmed the ferromagnetic behavior of the RE_2NiCrO_6 compounds. According to the Goodenough-Kanamori rule, this ferromagnetic order arises from the virtual electron transfer between transition metal ions. The experimentally calculated values of magnetic saturation (Ms), coercivity (Hc), and retentivity (Mr) had Ms values of 5.96 emu/g, 7.93 emu/g, and 9.95 emu/g for the Ce, Pr, and Nd compounds, respectively. These values correlated well with the theoretical predictions and highlight the potential of RE_2NiCrO_6 double perovskite oxides for applications in spintronics due to their robust ferromagnetic properties and half-metallic character [137].

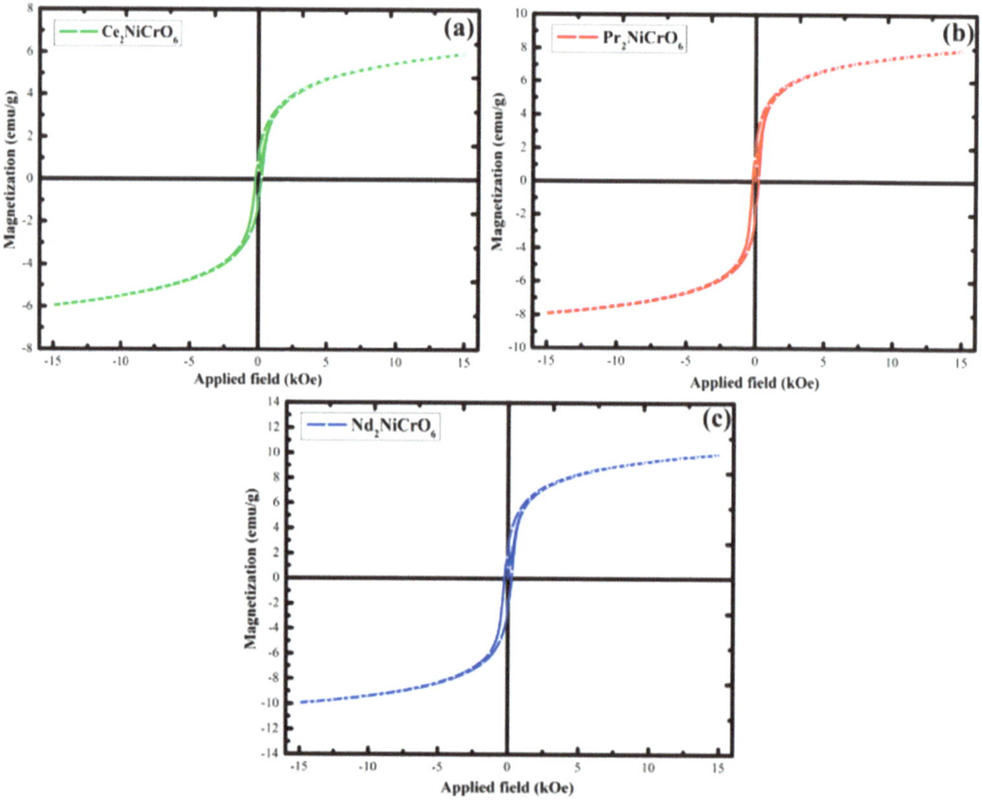

Figure 14. (**a–c**) Magnetization curves of (**a**) Ce_2NiCrO_6, (**b**) Pr_2NiCrO_6, and (**c**) Nd_2NiCrO_6 double perovskite oxides. Reprinted with permission from the author(s) (2023). Physica E: Low-dimensional Systems & Nanostructures [137]. Copyright © Elsevier, published by Elsevier BV.

The investigations conducted by Marouane et al. [138] looked at the impact of potassium (K) substitution at the A-site on the structural, magnetic, and magnetocaloric properties of brownmillerite oxides $La_{1-x}K_xSrMn_2O_{5+\delta}$ ($0.1 \leq x \leq 0.3$). Synthesized via solid-state reaction at high temperatures, the compounds were found to crystallize in the orthorhombic system with a Pnma space group, as revealed by structural analysis. Magnetic measurements, including Arrott's analysis, identified a second-order ferromagnetic phase transition in all samples, where the Curie temperature (T_C) decreased with increasing potassium content. Specifically, T_C values were observed at 357 K, 354 K, and 350 K for x = 0.1, 0.2, and 0.3, respectively. This reduction in T_C with higher K substitution can be attributed to changes in the Mn^{3+} and K^{3+} concentrations, influencing super exchange interactions. The magnetocaloric effect, quantified by the isothermal magnetic entropy change $|\Delta SM|$, indicated values of 1.2 J kg^{-1} K^{-1}, 1.08 J kg^{-1} K^{-1}, and 0.8 J kg^{-1} K^{-1} for x = 0.1, 0.2, and 0.3, respectively. Overall, these findings underscore the tunability of magnetic and magnetocaloric properties in $La_{1-x}K_xSrMn_2O_{5+\delta}$ through controlled potassium substitution, highlighting their potential for magnetic refrigeration applications (Figure 15). Similarly, S. Belhamra et al. [84] studied the structural, electronic, magnetic properties, and magnetocaloric effects of $LaRO_3$ (R = Mn, Cr, Fe) perovskite oxides using first-principles calculations and Monte Carlo simulations. Their research revealed that these materials exhibited intrinsic weak ferromagnetism at room temperature, attributed to the unique structural arrangements of transition metal ions and oxygen octahedra. The temperature-dependent magnetization curves indicated broad peaks in magnetic entropy change (ΔS) around room temperature: 137 K for $LaCrO_3$, 350 K for $LaMnO_3$, and 785 K for LaFeO3. Furthermore, the comparative analysis of magnetic hysteresis loops showed distinctive coercive fields (hC) and remanent magnetizations (Mr) among the compounds, with $LaCrO_3$ displaying superparamagnetic behavior and $LaMnO_3$ and $LaFeO_3$ exhibiting typical ferromagnetic characteristics. These findings underscore the potential of $LaRO_3$ perovskites as efficient materials for magnetic refrigeration applications due to their competitive relative cooling powers.

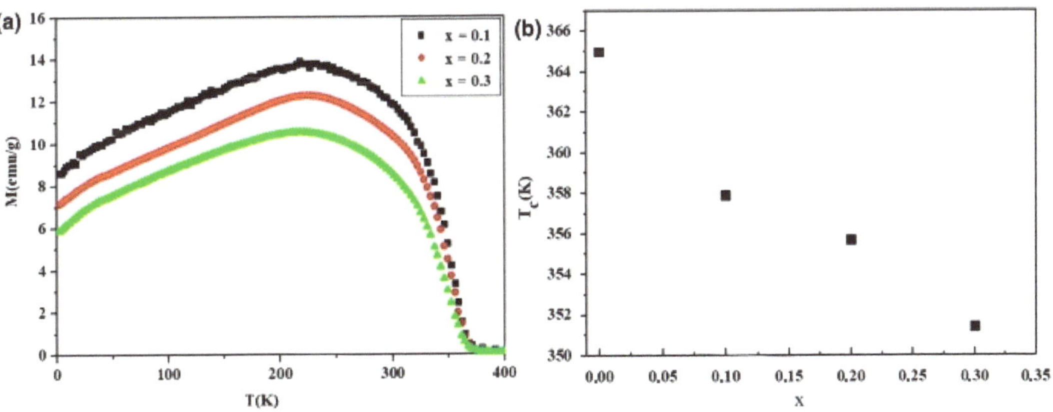

Figure 15. (**a**) Temperature dependence of the magnetization at H = 0.005, TH = 0.005 T for $La_{1-x}K_xSrMn_2O_{5+}\delta$ and (**b**) Curie temperature dependence as a function of xx for $La_{1-x}K_xSrMn_2O_{5+}\delta$ with x = 0.1, 0.2, 0.3, x = 0.1, 0.2, 0.3. Reprinted with permission from Mohamed, M. et al. (2022). Indian Journal of Physics [138]. Copyright © 2022 Indian Association for the Cultivation of Science, published by Springer Nature.

The study by Quanli Hu et al. [139] investigated the electrochemical and magnetic properties of perovskite type rare earth manganite nanofibers, specifically $LaMnO_3$,

$NdMnO_3$, and $SmMnO_3$, fabricated via electrospinning. The structural characterization revealed a rhombohedral structure for $LaMnO_3$ and orthorhombic structures for $NdMnO_3$, $SmMnO_3$, and $EuMnO_3$ NFs. The $RMnO_3$ were examined for their electrochemical behavior as supercapacitor electrodes, demonstrating promising specific capacitances and cycle stability, despite some needing improvement. Regarding magnetic properties, the NFs exhibited paramagnetic behavior at room temperature. At low temperatures, $LaMnO_3$ and $NdMnO_3$ NFs displayed a transition to ferromagnetic states, while $SmMnO_3$ and $EuMnO_3$ NFs exhibited transitions to antiferromagnetic states. Magnetic hysteresis loops at 10 K indicated ferromagnetic behavior for $LaMnO_3$ and $NdMnO_3$ NFs, with saturation magnetizations of 44.58 and 69.81 emu/g, respectively. In contrast, $SmMnO_3$ and $EuMnO_3$ NFs showed weak ferromagnetic components within antiferromagnetic phases. The coercivity of $EuMnO_3$ NFs was notably high at 3915 Oe. These findings underscore the multifaceted applications of $RMnO_3$ NFs in both the electrochemical and magnetic fields, warranting further exploration into enhancing their performance for advanced energy storage and magnetic device applications. In the same context, in their study, Idrissi et al. [140] explored the structural, electronic, and magnetic properties of the rare earth-based solar perovskites $GdAlO_3$, $DyAlO_3$, and $HoAlO_3$ by using density functional theory implemented in Quantum Espresso. The research focused on optimizing the unit cells of these materials to assess their stability and ferromagnetic behavior. They found that cubic $HoAlO_3$ exhibited the highest stability among the studied compounds, with $DyAlO_3$ being more stable than $GdAlO_3$ within this structure. Electronic structure analysis via band structure and density of states revealed that $GdAlO_3$ behaved as a semiconductor, while $DyAlO_3$ and $HoAlO_3$ displayed a half-metallic ferromagnetic character. The spin-polarized magnetic moments calculated for these materials indicate a clear ferromagnetic nature, with magnetic moments of 7.02 μ_B for $GdAlO_3$, 5.00 μ_B for $DyAlO_3$, and 4.00 μ_B for $HoAlO_3$. These findings suggest that these perovskite materials hold promise for applications in spintronics, optoelectronics, and potentially as photovoltaic materials. The study underscores their potential in advancing next-generation electronic and energy technologies. Extra, Yaiza Asensio et al. [141] discussed the magnetic properties of layered hybrid organic-inorganic metal-halide perovskites (HOIPs), specifically focusing on how different compositions and perovskite phases influenced these properties. It highlighted the role of transition metals in the crystal structure as crucial for inducing magnetic behavior. The study compared various systems with Cu^{2+} ions, detailing their temperature-dependent magnetization (M(T)) curves and magnetic hysteresis loops (M(H)s). For example, in PEA_2CuCl_4 RP crystals, long-range ferromagnetic (FM) order is observed with a Curie temperature (T_C) around 12 K, attributed to intralayer Cu–Cu spin interactions and weak interlayer interactions. Conversely, EA_2CuCl_4 RP crystals exhibited FM behavior at $T_C \approx 11$ K, with stronger interlayer antiferromagnetic (AFM) contributions due to a shorter interlayer distance. The text further explored DJ phase HOIPs like $PEAACuCl_4$ and $EDACuCl_4$, noting shifts in TC and magnetization saturation fields (HS) due to changes in crystal structure and interlayer distances. In Mn^{2+} HOIPs, such as PEA_2MnCl_4 RP crystals, the magnetic behavior was dominated by intralayer AFM interactions with a weak FM component arising from spin-canting effects due to crystal structure distortions. Overall, the discussion underscores how structural variations in HOIPs influence their magnetic properties, impacting their potential applications in spintronics and related fields. Furthermore, Yaiza Asensio et al. [142] investigated the impact of La doping on the structural and magnetic properties of $Ca_{1-x}La_xFe_{0.5}Mn_{0.5}O_{3-\delta}$ perovskites synthesized via the solid-state method.

Using X-ray diffraction and magnetic measurements, they explored these materials across a doping range ($0.05 \leq x \leq 0.15$). Their structural analysis revealed that all phases crystallized in the cubic system with a Pm-3m space group. Electron density studies highlighted the formation of partial covalent and ionic bonds involving Mn/Fe-O and Ca/La-O units, alongside an ordered arrangement of oxygen vacancies. Magnetic investigations indicated a paramagnetic to antiferromagnetic transition as the temperature decreased, with a decrease in Neel temperature (T_N) attributed to enhanced ferromagnetic interactions due to La doping. Furthermore, they observed weak ferromagnetism at room temperature, as evidenced by non-null magnetization and magnetic hysteresis loops. Below T_N, complex magnetic responses including antiferromagnetic, ferromagnetic, and ferrimagnetic ordering were observed, influenced by exchange interactions and antiphase boundaries. This study underscores the potential technological relevance of these materials, particularly their observed exchange bias effect and ambient temperature ferromagnetism (Figure 16) [143].

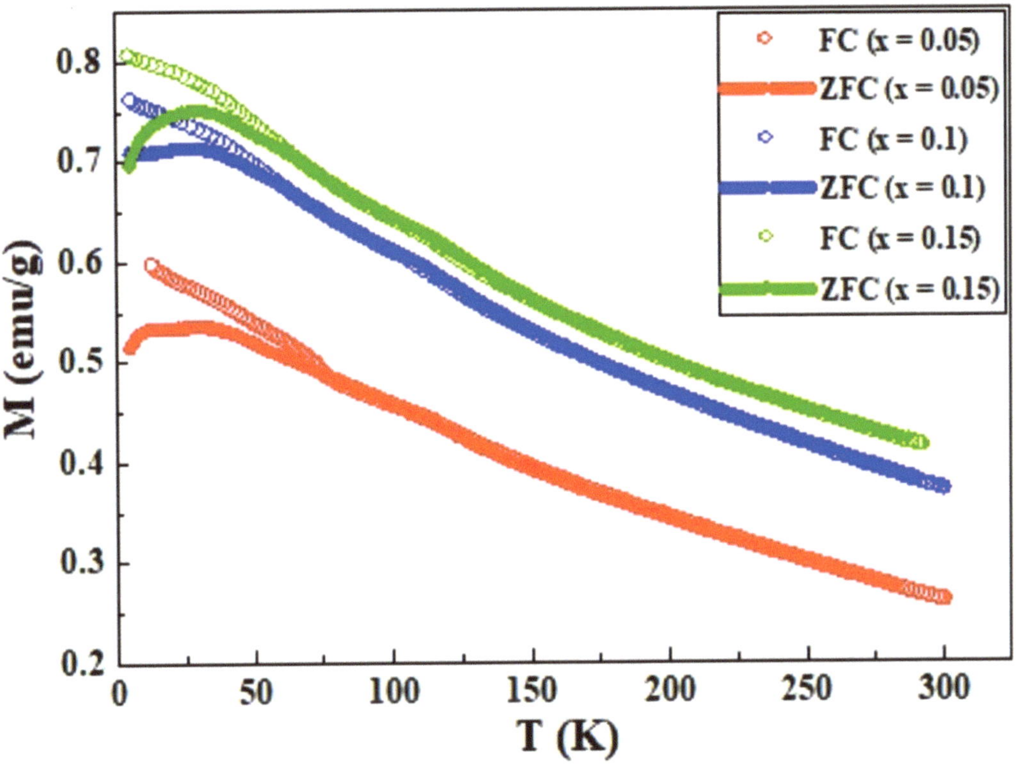

Figure 16. Thermal variation in the ZFC and FC magnetization of $Ca_{1-x}La_xFe_{0.5}Mn_{0.5}O_{3-\delta}$ ($0.05 \leq x \leq 0.15$) ceramics in an applied magnetic field of 0.2 T. Reprinted with permission from Selmi, R. et al. [143]. Copyright © 2022 The Author(s), under exclusive license to Springer-Verlag GmbH, DE, part of Springer.

In their study, Zheng et al. [144] investigated the effects of titanium (Ti) substitution on the magnetic properties and magnetocaloric effect (MCE) of Gd_2CoMnO_6 double perovskites. These materials are of interest for magnetic refrigeration due to their potential in achieving enhanced MCE at cryogenic temperatures. The researchers synthesized $Gd_2CoMn_{1-x}Ti_xO_6$ samples (x = 0–1.0) using a solid-state reaction method and characterized them for their crystal structure, surface morphology, magnetic behavior, and magnetocaloric properties.

The researchers proposed a mechanism attributing this enhanced cryogenic magnetocaloric effect to the magnetic phase transition behavior influenced by Ti substitution.

They conducted detailed magnetic field dependence studies (M vs. H) at 5 K and 50 K to analyze the magnetization behavior. At 5 K, the M vs. H curves displayed antiferromagnetic characteristics with negligible hysteresis, indicating the potential for excellent MCE performance at low temperatures [145–147]. In contrast, at 50 K, hysteresis behaviors were observed in the M vs. H curves, reflecting weakened ferromagnetic interactions with increasing Ti content. Overall, the comprehensive investigation by Zheng et al. underscores the effectiveness of Ti substitution in enhancing the magnetocaloric properties of Gd_2CoMnO_6 perovskites, making them promising candidates for advanced magnetic refrigeration technologies.

Furthermore, in their study on $La_{1-x}R_xFeO_3$ (R = Co, Al, Nd, Sm) [148] perovskite nanomaterials synthesized via the sol-gel method, Fang Yang et al. investigated the structural and magnetic properties across different doping compositions. They employed X-ray diffraction, thermogravimetric analysis-differential thermal analysis (TG-DTA), and a vibrating sample magnetometer (VSM) to characterize these materials. For $La_{1-x}CoFeO_3$, increasing the Co^{2+} content led to a shift in diffraction peaks toward larger angles, accompanied by an increase in grain size (50.7 nm to 133.5 nm) and enhanced magnetic loop area and magnetization. Conversely, doping with Al^{3+} in $La_{1-x}Al_xFeO_3$ resulted in a reduction in grain size (down to 17.9 nm) and an increase in magnetization, although the material exhibited weakened ferromagnetic properties possibly due to altered sintering conditions. Nd^{3+} and Sm^{3+} doping improved magnetization and particle refinement, with $La_{0.85}Co_{0.15}FeO_3$ showing increased magnetization with longer calcination times. The magnetic hysteresis loops confirmed saturation magnetization at 8000 Oe for the $La_{1-x}Co_xFeO_3$ samples, reinforcing the influence of doping on magnetic behavior. These findings underscore the role of controlled doping in tailoring structural and magnetic properties of perovskite nanomaterials, crucial for optimizing their performance in various applications. Significant advancements have been made in the study of magnetic properties, leading to novel insights and potential applications. Veliu et al. [149] elucidated the intricate dynamics of antiferromagnetically coupled sublattices, revealing the occurrence of two-step spin-crossover transitions.

These findings underscore the complex equilibrium and nonequilibrium behaviors inherent in such systems, thereby enhancing our understanding of magnetic phase transitions and their applications in advanced magnetic materials. Meanwhile, Thach et al. [150] conducted a comparative analysis of domain wall motion driven by spin transfer torque (STT) and spin orbit torque (SOT) in ferrimagnetic GdFeCo wires. Their research highlights the distinct mechanisms and efficiencies of STT and SOT in manipulating magnetic domains, which is crucial for the development of next-generation spintronic devices.

In their study, Su et al. [151] investigated the electrochemical and magnetic properties of $PrFeO_3$ and $EuFeO_3$ nanofibers synthesized using a straightforward fabrication method. Both types of nanofibers exhibited uniform, continuous morphologies with orthorhombic perovskite structures and average diameters of 233 nm for $PrFeO_3$ and 176 nm for $EuFeO_3$.

The electrochemical performance, characterized by specific capacitances of 123 F g^{-1} and 143 F g^{-1} at 1 A g^{-1} for $PrFeO_3$ and $EuFeO_3$, respectively, was attributed to the redox reactions involving Fe^{3+}/Fe^{4+} and $Fe(CN)_{63}/Fe(CN)_{64}$. Magnetic measurements revealed distinct behaviors: $PrFeO_3$ nanofibers demonstrated stable antiferromagnetic properties without spin-glass behavior or magnetic phase transitions, whereas $EuFeO_3$ nanofibers exhibited inhomogeneous magnetic properties with a transition from paramagnetic to ferromagnetic states at low temperatures, confirmed by a bifurcation in the ZFC and FC magnetization curves below 10 K. Both materials showed strong local short-range ferromagnetic interactions at 5 K, transitioning to paramagnetic states at 300 K. These findings suggest potential applications of these nanofibers in supercapacitors and condensed matter physics, leveraging their unique electrochemical and magnetic properties. Hossain et al. [152] synthesized $Nd_2Ni_{1-x}Co_xMnO_{6-\delta}$ ($0.2 \leq x \leq 0.5$) perovskite oxides using the citric-nitrate method and investigated their structural, optical, and magnetic properties. X-ray powder diffraction (XRD) and Rietveld analysis revealed that the unit cell parameters and volume increased with cobalt doping. Optical studies using UV-Vis spectroscopy showed an absorption peak around 330 nm, with a slight influence on the energy band gap due to Co doping. Magnetically, the partial substitution of Ni with Co at the B-site led to a decrease in Curie temperature (T_C) and increased the irreversibility between zero-field cooled (ZFC) and field-cooled (F_C) states below T_C. Magnetization studies as a function of external magnetic field at various temperatures confirmed the presence of ferromagnetic and paramagnetic phases, with saturation magnetization derived from the law of approach to saturation. At 2 K, the remanence saturation ratio increased with Co doping but decreased at higher temperatures (165 K and 200 K), and the coercivity increased monotonically with Co concentration due to magnetic domain wall pinning effects. The study suggested that Co doping increased the concentration of mixed Mn^{3+}/Mn^{4+} and Ni^{3+}/Co^{3+} ions, and the observed magnetic moments were lower than expected due to antiferromagnetic interactions and diamagnetic low-spin Co^{3+} ions. This led to magnetic frustration, aligning with the temperature-dependent ZFC and FC data, highlighting the combined magnetic effects of NdNi (Co)$O_{3-\delta}$ and $NdMnO_{3-\delta}$.

In addition, Dzara et al. [153] investigated the influence of the rare earth cation on the magnetic properties of 12R-$Ba_4M^{4+}Mn_3O_{12}$ (M = Ce, Pr) perovskites, focusing on cluster magnetism in the Mn_3O_{12} trimers and the role of the bridging MO_6 octahedron. The study confirmed the frustrated antiferromagnetic ground state of 12R-$Ba_4PrMn_3O_{12}$ below TN $\approx$ 7.75 K. The Pr^{4+} cation caused more complex magnetic properties than Ce^{4+}, with 12R-$Ba_4PrMn_3O_{12}$ exhibiting a sharp antiferromagnetic transition at $T_2 \approx 12.15$ K and an additional transition at $T_1 \approx 200$ K, likely related to canted antiferromagnetic order. In contrast, 12R-$Ba_4CeMn_3O_{12}$ showed no hysteresis or net moment at low temperatures. The study demonstrated that the magnetic properties of these materials could be tuned by varying the rare earth cation, highlighting the role of Pr^{4+} in introducing complex magnetic interactions and potential for material design within the hexagonal, layered perovskite family (Figure 17).

Magnetic properties play a pivotal role in perovskite materials, influencing their applications across various technological domains such as spintronics, data storage, and magnetic refrigeration [154]. Perovskite oxides, with their versatile crystal structures and tunable magnetic behaviors, have attracted significant research interest due to their potential for high-performance magnetic and electronic devices. Perovskites exhibit a rich spectrum of magnetic phenomena including ferromagnetism, antiferromagnetism, ferrimagnetism, and even complex magnetic orders such as spin glass and frustrated magnetism. For instance, the compound $LaMnO_3$ demonstrates colossal magnetoresistance due to its strong electron-lattice interactions, making it promising for next-generation memory devices. Moreover, $LaMnO_3$ is a fascinating compound within the perovskite family, known for its intricate magnetic properties and diverse applications. It crystallizes in the orthorhombic perovskite structure (space group Pbnm), where La and Mn occupy the A and B sites, respectively, and O fills the X site. The magnetic behavior of $LaMnO_3$ is primarily governed by the electronic configuration and the coupling between the magnetic moments of Mn ions. At ambient conditions, it typically exhibits antiferromagnetic ordering, where neighboring Mn spins align antiparallelly within the crystal lattice. This antiferromagnetic phase arises from the super exchange interactions mediated by oxygen ions between adjacent Mn ions, leading to a Néel temperature around 140 K.

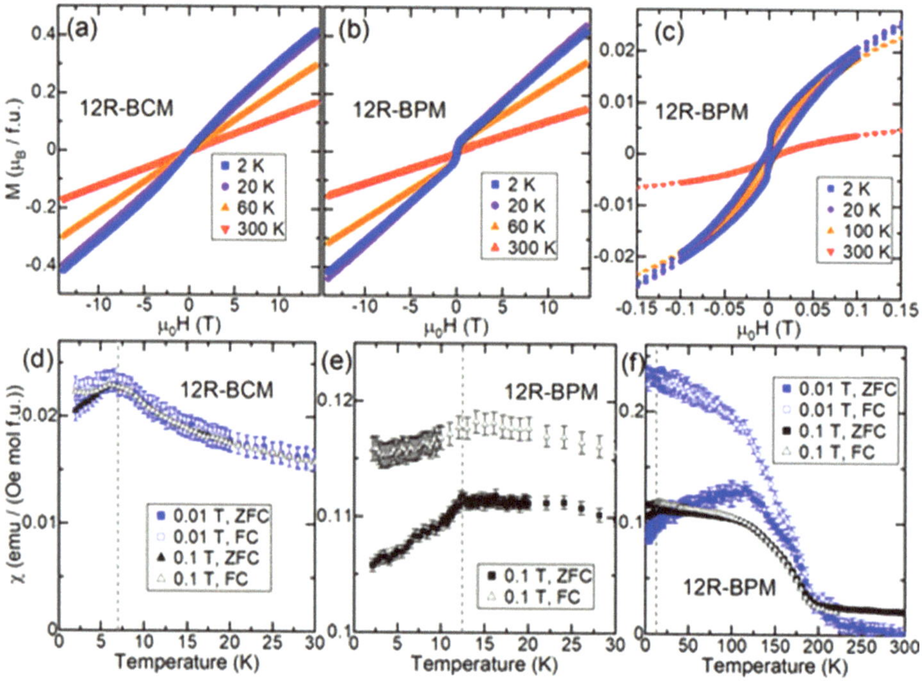

Figure 17. (**a**,**b**) Magnetization (M) as a function of the applied field for (**a**) 12R-BCM and (**b**) 12R-BPM. (**c**) Magnified low-field view of M vs. applied field for 12R-BPM. (**d**) DC susceptibility (χ) of 12R-BCM in low applied fields. (**e**,**f**) χ of 12R-BPM in low applied fields. The dashed lines denote the approximate position of each material's low-temperature transition. ZFC signifies zero-field-cooled, and FC signifies field-cooled. Reprinted with permission from Dzara, M. J. et al. Chemistry of Materials, 36(6) [153]. Copyright © 2024 American Chemical Society.

Below TN, the spins order in a manner that minimizes the magnetic energy, resulting in a unique magnetic structure that influences its physical properties. It is also known to display complex magnetic behaviors under different conditions such as the application of external fields or doping with other ions. For instance, upon doping with divalent ions like Ca^{2+}, the material can exhibit ferromagnetic or ferrimagnetic behavior, depending on the doping concentration and the nature of the substituting ions [155]. This tunability makes $LaMnO_3$ a versatile platform for studying magnetic phase transitions and for potential applications in spintronics and magnetic memory devices. On the other hand, materials like $SrFeO_{3-\delta}$ exhibit high-temperature antiferromagnetism, essential for stable magnetic data storage applications. $SrFeO_{3-\delta}$, another intriguing perovskite compound, exhibits complex magnetic properties that have made it a subject of intensive research for various technological applications. This material crystallizes in the perovskite structure, where Sr and Fe occupy the A and B sites, respectively, and O fills the X site. The magnetic behavior of $SrFeO_{3-\delta}$ is influenced by its oxygen deficiency, which can vary and affect its electronic and magnetic structure.

Perovskite-based magnetic refrigerants offer an alternative to conventional refrigerants like fluorocarbons, addressing environmental concerns while improving energy efficiency. Magnetochemistry in perovskites explores the interplay between magnetic ions, crystal structure distortions, and electronic states, which collectively dictate their magnetic properties. For example, doping $LaFeO_3$ with various transition metals alters its magnetic structure, affecting its magnetic and electrical transport properties. Understanding these relationships is crucial for tailoring perovskite materials with the desired magnetic functionalities for specific applications. Furthermore, the discovery of multiferroic perovskites, exemplified by $BiFeO_3$, where magnetic and ferroelectric orders coexist and couple, opens new avenues for multifunctional device integration [156]. Such materials exhibit promising characteristics for spin-electronics and novel sensor technologies, leveraging the coupling between magnetic and electrical properties. In the realm of entropy and magnetic perovskites, the study of magnetic entropy changes provides insights into their potential for magnetocaloric applications. Materials exhibiting large magnetic entropy changes, such as $La_{0.67}Ca_{0.33}MnO_3$, are studied for their efficiency in magnetic refrigeration cycles. This research area bridges fundamental thermodynamics with material science, aiming to optimize cooling efficiency and minimize environmental impact. Overall, the magnetic properties of perovskite materials represent a dynamic field of research with profound implications for both fundamental science and technological innovation. Continued advancements in magnetochemistry, entropy effects, and magnetic refrigeration will further expand the scope of perovskite applications in emerging technologies. The comprehensive understanding of these properties enables the design and development of advanced materials tailored for next-generation magnetic devices and sustainable cooling solutions.

5. Conclusions and Perspectives

Perovskite materials stand at the forefront of advanced material research due to their extraordinary electrical, dielectric, and magnetic properties, coupled with their structural flexibility and compositional tunability. This review has outlined the critical advancements in understanding and optimizing these properties for diverse technological applications including energy storage, spintronics, and magnetic refrigeration. Despite the immense progress achieved, several challenges remain. These include the need for improved stability under environmental stress, enhanced reproducibility in synthesis, and scalable manufacturing processes for industrial applications. Addressing these issues will be pivotal in transitioning perovskite technologies from research to commercial viability.

Future research should focus on developing novel compositions and hybrid systems that combine perovskites with other material classes to achieve multifunctionality and enhanced performance. Additionally, integrating machine learning and advanced characterization techniques could accelerate the discovery of new perovskite materials and provide deeper insights into their behavior. The field of perovskite materials research continues to expand rapidly, driven by the growing demand for high-performance, sustainable, and multifunctional materials. By addressing existing challenges and leveraging new technologies, perovskites have the potential to revolutionize a wide range of applications, paving the way for innovative solutions in energy, electronics, and beyond.

6. Challenges and Future Directions

Despite the remarkable advancements in perovskite materials, several challenges hinder their full-scale application. Stability under environmental conditions such as moisture, temperature, and radiation remain a critical issue, particularly for photovoltaics and sensors. Additionally, scalable and cost-effective synthesis methods that maintain high quality, coupled with reproducibility in material properties, are essential for industrial adoption. Many perovskite materials demonstrate optimal performance under specific conditions, such as low temperatures, limiting their practical use at ambient environments. Furthermore, the integration of perovskites into complex device architectures without compromising their functionality poses significant challenges. Addressing these issues requires innovative approaches including advanced compositional engineering to enhance stability and multifunctionality as well as the use of machine learning to accelerate material discovery and optimization. This is summarized in Table 1, which shows the challenges and strategies in perovskite device implementation. Hybrid systems that combine perovskites with polymers or nanocomposites offer potential solutions to overcome these limitations. Advanced characterization techniques are critical for understanding and fine-tuning the structural, electrical, and magnetic properties of perovskites. Moreover, focusing on eco-friendly synthesis methods and recyclable device designs aligns with the global sustainability goals. Overcoming these challenges will unlock the full potential of perovskites in transformative applications, such as renewable energy, advanced electronics, and next-generation magnetic devices, paving the way for a sustainable and technologically advanced future.

Table 1. Challenges and strategies in perovskite device implementation.

Challenge	Current Strategies	Future Directions
Stability under harsh conditions	- Protective coatings (e.g., encapsulation with polymers or inorganic layers) - Optimized synthesis methods (e.g., controlled temperature and humidity during deposition) - Use of stabilizers and additives to enhance material stability in air and moisture	- Development of more robust materials (e.g., stable perovskite variants) - Advanced encapsulation techniques (e.g., hybrid materials for better protection) - Engineering new perovskite formulations for better resistance to environmental factors
Scalability for industrial use	- Solution-processing methods (e.g., spin coating, spray coating)	- Improved scalability of synthesis methods (e.g., continuous processing methods)
	- Roll-to-roll printing for large-scale perovskite film production	- Automation in production to reduce cost and improve efficiency
	- Large-area deposition techniques for uniform films	- Optimization of material deposition techniques for seamless large-area integration
Efficiency limitations	- Doping (e.g., metal ion doping for conductivity improvement)	- Development of new perovskite compositions with improved intrinsic properties
	- Compositional engineering (e.g., tuning the A-site or B-site cation for desired properties)	- Hybrid perovskite materials combining organic and inorganic components for better performance
	- Light management techniques (e.g., photonic structures, light trapping)	- Integration with other energy conversion technologies for improved performance (e.g., tandem solar cells)
Material defects and grain boundaries	- Grain boundary engineering (e.g., post-deposition annealing, additive treatments)	- Creation of defect-free perovskite materials via novel synthesis methods
	- Use of surfactants to reduce defects and improve crystallinity	- Development of self-healing perovskite materials to address defects in the crystal structure
Interfacial issues with electrodes	- Use of electron and hole transport layers (e.g., TiO_2, Spiro-OMeTAD)	- Development of new interface engineering methods to improve charge transport across the interface
	- Optimized electrode materials to reduce charge recombination	- Use of transparent conductive materials for better electrode performance in devices

Table 1. *Cont.*

Challenge	Current Strategies	Future Directions
Cost of materials	- Use of low-cost precursors and solution-based processing methods	- Research into more abundant and less expensive perovskite materials
	- Use of cheaper substrates and simplified fabrication techniques	- Development of scalable fabrication techniques to lower production costs
Environmental concerns of lead-based perovskites	- Development of lead-free perovskites (e.g., tin-based perovskites, halide perovskites)	- Continued research into the safe disposal or recycling of lead-based materials
	- Encapsulation techniques to reduce the risk of lead leakage	- Development of environmentally friendly materials and processes for perovskite devices
Long-term stability and operational lifetime	- Use of multi-layer structures to protect perovskites from degradation over time	- Research on perovskite longevity, focusing on reducing degradation during operation
	- Improvement of moisture, temperature, and UV stability through composite films	- Integration with advanced weather-resistant coatings or coatings that repair themselves over time
Device architecture design	- Development of optimized cell architecture (e.g., monolithic, tandem solar cells)	- New perovskite-based device architectures for higher efficiency and performance
	- Advanced interlayer design to facilitate charge transport	- Hybrid device configurations combining perovskites with other materials for energy applications (e.g., perovskite-silicon tandems)

Author Contributions: F.T. and K.I.N. conceptualized and designed the review framework. F.T. conducted the primary literature search and wrote the initial manuscript draft. S.S.T. and M.P.F.G. provided critical revisions and contributed to the refinement of the manuscript. K.I.N. supervised the project, provided expert guidance on the structure and content, and coordinated the contributions. All authors have read and agreed to the published version of the manuscript.

Funding: The authors acknowledge the support of FCT-Fundação para a Ciência e a Tecnologia, I.P., within the scope of the project's LA/P/0037/2020, UIDP/50025/2020, and UIDB/50025/2020 of the Associate Laboratory Institute of Nanostructures, Nanomodeling, and Nanofabrication-i3N.

Data Availability Statement: Data sharing is not applicable to this article as no datasets were generated or analyzed during the current study.

Acknowledgments: The authors acknowledge the support of FCT-Fundação para a Ciência e a Tecnologia, I.P., within the scope of the project's LA/P/0037/2020, UIDP/50025/2020, and UIDB/50025/2020 of the Associate Laboratory Institute of Nanostructures, Nanomodeling, and Nanofabrication-i3N.

Conflicts of Interest: The authors declare no conflicts of interest.

References

1. Labhasetwar, N.; Saravanan, G.; Megarajan, S.K.; Manwar, N.; Khobragade, R.; Doggali, P.; Grasset, F. Perovskite-type catalytic materials for environmental applications. *Sci. Technol. Adv. Mater.* **2015**, *16*, 036002. [CrossRef] [PubMed]
2. Dong, H.; Zhang, C.; Liu, X.; Yao, J.; Zhao, Y.S. Materials chemistry and engineering in metal halide perovskite lasers. *Chem. Soc. Rev.* **2020**, *49*, 951–982. [CrossRef] [PubMed]

3. Liang, J.; Xu, W.; Wang, T.; Yang, Q.; Zhang, S. Implementing NAT Hole Punching with QUIC. In Proceedings of the 2024 IEEE 100th Vehicular Technology Conference (VTC2024-Fall), Washington, DC, USA, 7–10 October 2024; IEEE: Piscataway, NJ, USA, 2024.
4. Shi, X.; Ishtiaq, U.; Din, M.; Akram, M. Fractals of interpolative Kannan mappings. *Fractal Fract.* **2024**, *8*, 493. [CrossRef]
5. Zhang, X.; Chen, A.L.; Piao, X.; Yu, M.; Zhang, Y.; Zhang, L. Is AI chatbot recommendation convincing customer? An analytical response based on the elaboration likelihood model. *Acta Psychol.* **2024**, *250*, 104501. [CrossRef]
6. Shi, X.; Zhang, Y.; Pujahari, A.; Mishra, S.K. When latent features meet side information: A preference relation-based graph neural network for collaborative filtering. *Expert Syst. Appl.* **2025**, *260*, 125423. [CrossRef]
7. Xu, C.; Zhu, M.; Wang, Q.; Cui, J.; Huang, Y.; Huang, X.; Huang, J.; Gai, J.; Li, G.; Qiao, P.; et al. TROP2-directed nanobody-drug conjugate elicited potent antitumor effect in pancreatic cancer. *J. Nanobiotechnol.* **2023**, *21*, 410. [CrossRef] [PubMed]
8. Zhang, H.H.; Chao, J.B.; Wang, Y.W.; Liu, Y.; Yao, H.M.; Zhao, Z.P.; Niu, K. 5G Base Station Antenna Array with Heatsink Radome. *IEEE Trans. Antennas Propag.* **2024**, *72*, 2270–2278. [CrossRef]
9. Wang, Z.; Liao, Z.; Zhou, B.; Yu, G.; Luo, W. SwinURNet: Hybrid transformer-cnn architecture for real-time unstructured road segmentation. *IEEE Trans. Instrum. Meas.* **2024**, *73*. [CrossRef]
10. Luo, Y.X.; Dong, Y.L. Strain Measurement at up to 3000 °C Based on Ultraviolet-Digital Image Correlation. *NDT E Int.* **2024**, *146*, 103155. [CrossRef]
11. Pi, X.; Yan, D.; Xu, Y.; Pan, M.; Wang, Z.; Chang, M.; Qi, Z. TLRs signaling pathway regulation, antibacterial and apoptotic activity of largemouth bass ECSIT during Edwardsiella piscicida infection. *Aquaculture* **2025**, *595*, 741615. [CrossRef]
12. Lv, H.; Zeng, J.; Zhu, Z.; Dong, S.; Li, W. Study on prestress distribution and structural performance of heptagonal six-five-strut alternated cable dome with inner hole. In *Structures*; Elsevier: Amsterdam, The Netherlands, 2024; Volume 65.
13. Cui, M.; Han, D.; Liu, H.; Li, K.-C.; Tang, M.; Chang, C.-C.; Ayaz, F.; Sheng, Z.; Guan, Y.L. Secure Data Sharing for Consortium BlockchainEnabled Vehicular Social Networks. *IEEE Trans. Veh. Technol.* **2024**, *73*, 19682–19695. [CrossRef]
14. Shi, J.; Han, D.; Chen, C.; Shen, X. KTMN: Knowledge-driven Two-stage Modulation Network for visual question answering. *Multimed. Syst.* **2024**, *30*, 350. [CrossRef]
15. Nakajima, T.; Fujio, Y.; Sugahara, T.; Tsuchiya, T. Flexible ceramic film sensors for free-form devices. *Sensors* **2022**, *22*, 1996. [CrossRef] [PubMed]
16. Song, X.; Han, D.; Chen, C.; Shen, X.; Wu, H. Vman: Visual-modified attention network for multimodal paradigms. *Vis. Comput.* **2024**, 1–18. [CrossRef]
17. Li, J.; Han, D.; Weng, T.-H.; Wu, H.; Li, K.-C.; Castiglione, A. A secure data storage and sharing scheme for port supply chain based on blockchain and dynamic searchable encryption. *Comput. Stand. Interfaces* **2025**, *91*, 103887. [CrossRef]
18. Han, D.; Shi, J.; Zhao, J.; Wu, H.; Zhou, Y.; Li, L.-H.; Khan, M.K.; Li, K.-C. LRCN: Layer-residual Co-Attention Networks for visual question answering. *Expert Syst. Appl.* **2025**, *263*, 125658. [CrossRef]
19. Han, D.; Pan, N.; Li, K.-C. A traceable and revocable ciphertext-policy attribute-based encryption scheme based on privacy protection. *IEEE Trans. Dependable Secur. Comput.* **2020**, *19*, 316–327. [CrossRef]
20. Han, D.; Zhu, Y.; Li, D.; Liang, W.; Souri, A.; Li, K.-C. A blockchain-based auditable access control system for private data in service-centric IoT environments. *IEEE Trans. Ind. Inform.* **2021**, *18*, 3530–3540. [CrossRef]
21. Sun, B.-L.; Shi, X.-L.; Hu, Q.; Du, M.; Gong, H.; Wang, J.; Fan, Y.; Zhang, J.; Li, Z. Ultrahigh-strength textile fiber-supported schiff base copper complexes for photocatalytic degradation of methyl orange. *ACS Appl. Polym. Mater.* **2024**, *6*, 13077–13088. [CrossRef]
22. Zhang, T.; Shi, X.-L.; Hu, Q.; Gong, H.; Shi, K.; Li, Z. Ultrahigh-Performance Fiber-Supported Iron-Based Ionic Liquid for Synthesizing 3,4-Dihydropyrimidin-2-(1*H*)-ones in a Cleaner Manner. *Langmuir* **2024**, *40*, 9579–9591. [CrossRef]
23. Shen, J.; Bao, X.; Li, J.; Chen, X.; Cui, H. Study on the mechanism of EPWP dissipation at the joints of shield tunnel in liquefiable strata during seismic events. *Soil Dyn. Earthq. Eng.* **2025**, *188*, 109089. [CrossRef]
24. Chen, X.; Dai, A. Quantifying contributions of external forcing and internal variability to Arctic warming during 1900–2021. *Earth's Future* **2024**, *12*, e2023EF003734. [CrossRef]
25. Guan, D.; Xing, Y.; Huang, J.; Xiao, A.; El Saddik, A.; Lu, S. S2Match: Self-paced sampling for data-limited semi-supervised learning. *Pattern Recognit.* **2025**, *159*, 111121. [CrossRef]
26. Xie, J.; Wen, M.; Ding, P.; Tu, Y.; Wu, D.; Liu, K.; Tang, K.; Chen, M. One-dimensional consolidation analysis of layered unsaturated soils: An improved model integrating interfacial flow and air contact resistance effects. *Comput. Geotech.* **2024**, *176*, 106791. [CrossRef]
27. Li, W.; Xie, Z.; Zhao, J.; Wong, P.K.; Li, P. Fuzzy finite-frequency output feedback control for nonlinear active suspension systems with time delay and output constraints. *Mech. Syst. Signal Process.* **2019**, *132*, 315–334. [CrossRef]
28. Li, W.; Xie, Z.; Wong, P.K.; Cao, Y.; Hua, X.; Zhao, J. Robust nonfragile H∞ optimum control for active suspension systems with time-varying actuator delay. *J. Vib. Control* **2019**, *25*, 2435–2452. [CrossRef]

29. Li, W.; Xie, Z.; Wong, P.K.; Ma, X.; Cao, Y.; Zhao, J. Nonfragile H∞ control of delayed active suspension systems in finite frequency under nonstationary running. *J. Dyn. Syst. Meas. Control* **2019**, *141*, 061001. [CrossRef]
30. Zhao, J.; Wong, P.K.; Xie, Z.; Ma, X.; Hua, X. Design and control of an automotive variable hydraulic damper using cuckoo search optimized PID method. *Int. J. Automot. Technol.* **2019**, *20*, 51–63. [CrossRef]
31. Li, W.; Xie, Z.; Zhao, J.; Wong, P.K. Velocity-based robust fault tolerant automatic steering control of autonomous ground vehicles via adaptive event triggered network communication. *Mech. Syst. Signal Process.* **2020**, *143*, 106798. [CrossRef]
32. Xu, Y.; Xie, Z.; Zhao, J.; Li, W.; Li, P.; Wong, P.K. Robust non-fragile finite frequency H∞ control for uncertain active suspension systems with time-delay using TS fuzzy approach. *J. Frankl. Inst.* **2021**, *358*, 4209–4238. [CrossRef]
33. Li, W.; Xie, Z.; Cao, Y.; Wong, P.K.; Zhao, J. Sampled-data asynchronous fuzzy output feedback control for active suspension systems in restricted frequency domain. *IEEE/CAA J. Autom. Sin.* **2020**, *8*, 1052–1066. [CrossRef]
34. Nassar, K.I.; Slimi, M.; Rammeh, N.; Bouhamed, A.; Njeh, A.; Kanoun, O. Investigation of AC electrical conductivity and dielectric properties of $BiBaFeZnO_6$ double perovskite oxides. *J. Mater. Sci. Mater. Electron.* **2021**, *32*, 24050–24057. [CrossRef]
35. Iben Nassar, K.; Slimi, M.; Rammeh, N.; Soreto Teixeira, S.; Graça, M.P.F. Structural and electrical properties of double perovskite oxide $LaPbFeTiO_6$ synthesized by a sol-gel process. *Appl. Phys. A* **2021**, *127*, 940. [CrossRef]
36. Li, W.; Xie, Z.; Wong, P.K.; Mei, X.; Zhao, J. Adaptive-event-trigger-based fuzzy nonlinear lateral dynamic control for autonomous electric vehicles under insecure communication networks. *IEEE Trans. Ind. Electron.* **2020**, *68*, 2447–2459. [CrossRef]
37. Li, W.; Xie, Z.; Wong, P.K.; Hu, Y.; Guo, G.; Zhao, J. Event-triggered asynchronous fuzzy filtering for vehicle sideslip angle estimation with data quantization and dropouts. *IEEE Trans. Fuzzy Syst.* **2021**, *30*, 2822–2836. [CrossRef]
38. Nassar, K.I.; Benamara, M.; Kechiche, L.; Teixeira, S.S.; Graça, M.P.F. Structural, electrical properties of bismuth and niobium-doped $LaNiO_3$ perovskite obtained by sol-gel route for future electronic device applications. *Indian J. Phys.* **2024**, *98*, 2745–2753. [CrossRef]
39. Iben Nassar, K.; Rammeh, N. Study of structural, dielectric, and electrical properties of the $LaBaFe_{1.2}Mn_{0.8}O_6$ double perovskite. *Indian J. Phys.* **2023**, *97*, 1749–1757. [CrossRef]
40. Tayari, F.; Nassar, K.I.; Benamara, M.; ben Moussa, S.; Alzahrani, A.Y.A.; Teixeira, S.S.; Graça, M.P.F. Insights into dielectric and electrical conductivity dynamics in sol-gel synthesized $Ba_{0.75}Ni_{0.25}Tc_{0.88}Mn_{0.12}O_3$ perovskite ceramic. *J. Sol-Gel Sci. Technol.* **2024**, *111*, 132–143. [CrossRef]
41. Gupta, V.; Ncho, C.M.; Goel, A.; Jeong, C.M.; Choi, Y.H. Effects of in ovo injection of α-ketoglutaric acid on hatchability, growth, plasma metabolites, and antioxidant status of broilers. *Antioxidants* **2022**, *11*, 2102. [CrossRef] [PubMed]
42. Liu, J.; Xie, Z.; Zhao, J.; Wong, P.K. Probabilistic Adaptive Dynamic Programming for Optimal Reliability-Critical Control with Fault Interruption Estimation. *IEEE Trans. Ind. Inform.* **2024**, *20*, 8524–8535. [CrossRef]
43. Liu, J.; Xie, Z.; Gao, J.; Hu, Y.; Zhao, J. Failure characteristics of the active-passive damping in the functionally graded piezoelectric layers-magnetorheological elastomer sandwich structure. *Int. J. Mech. Sci.* **2022**, *215*, 106944. [CrossRef]
44. Li, W.; Xie, Z.; Zhao, J.; Gao, J.; Hu, Y.; Wong, P.K. Human-machine shared steering control for vehicle lane keeping systems via a fuzzy observer-based event-triggered method. *IEEE Trans. Intell. Transp. Syst.* **2021**, *23*, 13731–13744. [CrossRef]
45. Chu, S.; Xie, Z.; Wong, P.K.; Li, P.; Li, W.; Zhao, J. Observer-based gain scheduling path following control for autonomous electric vehicles subject to time delay. *Veh. Syst. Dyn.* **2022**, *60*, 1602–1626. [CrossRef]
46. Xie, Z.; Li, S.; Wong, P.K.; Li, W.; Zhao, J. An improved gain-scheduling robust MPC for path following of autonomous independent-drive electric vehicles with time-varying and uncertainties. *Veh. Syst. Dyn.* **2024**, 1–27. [CrossRef]
47. Xie, Z.; You, W.; Wong, P.K.; Li, W.; Zhao, J. Fuzzy robust non-fragile control for nonlinear active suspension systems with time varying actuator delay. *Proc. Inst. Mech. Eng. Part D J. Automob. Eng.* **2024**, *238*, 46–62. [CrossRef]
48. Cao, Y.; Xie, Z.; Li, W.; Wang, X.; Wong, P.K.; Zhao, J. Combined Path Following and Direct Yaw-Moment Control for Unmanned Electric Vehicles Based on Event-Triggered T–S Fuzzy Method. *Int. J. Fuzzy Syst.* **2024**, *26*, 2433–2448. [CrossRef]
49. Hu, Y.; Sugiyama, Y. Well-posedness of the initial-boundary value problem for 1D degenerate quasilinear wave equations. *Adv. Differ. Equ.* **2025**, *30*, 177–206. [CrossRef]
50. Du, J.; Liu, H.; Wang, F.; Bao, W.; Li, H. Solidification Microstructure Reconstruction and Its Effects on Phase Transformation, Grain Boundary Transformation Mechanism, and Mechanical Properties of TC4 Alloy Welded Joint. *Metall. Mater. Trans. A* **2024**, *55*, 1193–1206. [CrossRef]
51. Lv, S.; Liu, H.; Wang, F.; Liu, X.; Peng, M.; Wei, Y.; Li, C. Effect of axial misalignment on the microstructure, mechanical, and corrosion properties of magnetically impelled arc butt welding joint. *Mater. Today Commun.* **2024**, *40*, 109866. [CrossRef]
52. Bao, W.; Liu, H.; Wang, F.; Du, J.; Wang, Y.; Li, H.; Ye, X. Keyhole critical failure criteria and variation rule under different thicknesses and multiple materials in K-TIG welding. *J. Manuf. Process.* **2024**, *126*, 48–59. [CrossRef]
53. Zhou, H.; Guo, J.; Zhu, G.; Xie, F.; Tang, X.; Luo, X. Highly efficient preparation of crystalline yttrium carbonate in sodium carbonate system: Formation and growth mechanism. *J. Rare Earths*, 2024, *in press*. [CrossRef]
54. Zhou, H.; Guo, J.; Zhu, G.; Xu, H.; Tang, X.; Luo, X. Flotation behavior and mechanism of smithsonite under the system of bidentate ligand sulfide sodium thiocyanate. *Sep. Purif. Technol.* **2024**, *334*, 126086. [CrossRef]

55. Luan, S.; Ji, Y.; Liu, Y.; Zhu, L.; Zhao, H.; Zhou, H.; Li, K.; Zhu, W.; Zhu, B. AI-powered ultrasonic thermometry for HIFU therapy in deep organ. *Ultrason. Sonochemistry* **2024**, *111*, 107154. [CrossRef]
56. Zhang, M.; Xu, Y.-C.; Wang, J.-F.; Peng, K.-R.; Dai, S.; Wu, J.-M.; Zhu, L. Cold-sprayed Cu matrix composite coatings with core-shell structured Co@ WC reinforcements on Q235 steel. *Surf. Interfaces* **2025**, *56*, 105577. [CrossRef]
57. Feng, C.; Feng, Z.; Mao, R.; Li, G.; Zhong, Y.; Ling, K. Prediction of vitrinite reflectance of shale oil reservoirs using nuclear magnetic resonance and conventional log data. *Fuel* **2023**, *339*, 127422. [CrossRef]
58. Wehrenfennig, C.; Eperon, G.E.; Johnston, M.B.; Snaith, H.J.; Herz, L.M. High charge carrier mobilities and lifetimes in organolead trihalide perovskites. *Adv. Mater.* **2013**, *26*, 1584–1589. [CrossRef] [PubMed]
59. Wehrenfennig, C.; Eperon, G.E.; Stranks, S.D.; Snaith, H.J.; Herz, L.M. Charge carrier recombination channels in the low-temperature phase of organic-inorganic lead halide perovskite thin films. *APL Mater.* **2014**, *2*, 081513. [CrossRef]
60. Serpetzoglou, E.; Kymakis, E.; Stratakis, E.; Koutselas, I. In situ monitoring of the charge carrier dynamics of $CH_3NH_3PbI_3$ perovskite crystallization process. *J. Mater. Chem. C* **2019**, *7*, 12170–12179. [CrossRef]
61. Serpetzoglou, E.; Kymakis, E.; Stratakis, E.; Koutselas, I. Charge carrier dynamics in different crystal phases of $CH_3NH_3PbI_3$ perovskite. *Opto-Electron. Sci.* **2022**, *1*, 210005. [CrossRef]
62. Yu, Z.; Zhao, J.; Markov, V.; Han, D. Effects of hydrogen addition on ignition characteristics and engine performance of ammonia-hydrogen blended fuel: A kinetic analysis. *Int. J. Hydrogen Energy* **2024**, *87*, 722–735. [CrossRef]
63. Feng, G.; Liu, T.; Jiang, F.; Guo, Z.; Xiao, L.; Wu, Q.; Zhang, X.; Hu, Q.; Liu, J.; Liang, J. Novel (Ni, Mn) co-doping $CuFe_5O_8$ black ceramic pigment with pinning strengthen effect in high-temperature black zirconia ceramic application. *Ceram. Int.* 2024, *in press*. [CrossRef]
64. Yue, T.; Zhang, W.; Pei, H.; Danzeng, D.; He, J.; Yang, J.; Luo, Y.; Zhang, Z.; Xiong, S.; Yang, X.; et al. Monascus pigment-protected bone marrow-derived stem cells for heart failure treatment. *Bioact. Mater.* **2024**, *42*, 270–283. [CrossRef]
65. Wang, X.; Zhang, Y.; Wen, M.; Mang, H.A. A simple hybrid linear and nonlinear interpolation finite element for the adaptive Cracking Elements Method. *Finite Elem. Anal. Des.* **2025**, *244*, 104295. [CrossRef]
66. Wang, Y.; Xia, S.; Chen, K.; Zhang, J.; Tan, H.; Yu, C.; Cui, J.; Zeng, J.; Wu, J.; Wang, P.; et al. Atomic-Scale Tailoring C–N Coupling Sites for Efficient Acetamide Electrosynthesis over Cu-Anchored Boron Nitride Nanosheets. *ACS Nano* **2024**, *18*, 34403–34414. [CrossRef] [PubMed]
67. Li, Z.; Li, W.; Li, X.; Zhang, Y. Guardians of the Ledger: Protecting Decentralized Exchanges From State Derailment Defects. *IEEE Trans. Reliab.* **2024**, 1–13. [CrossRef]
68. Thakur, P.; Sharma, N.; Pathak, D.; Sharma, P.; Kishore, K.; Dhar, S.; Lal, M. State-of-art review on smart perovskites materials: Properties and applications. *Emergent Mater.* **2024**, *7*, 667–694. [CrossRef]
69. Zhang, F.; Miao, X.; van Dijk, N.; Brück, E.; Ren, Y. Advanced Magnetocaloric Materials for Energy Conversion: Recent Progress, Opportunities, and Perspective. *Adv. Energy Mater.* **2024**, *14*, 2400369. [CrossRef]
70. Konidakis, I.; Karagiannaki, A.; Stratakis, E. Advanced composite glasses with metallic, perovskite, and two-dimensional nanocrystals for optoelectronic and photonic applications. *Nanoscale* **2022**, *14*, 2966–2989. [CrossRef]
71. Zhu, Y.; Zhou, W.; Zhong, Y.; Bu, Y.; Chen, X.; Zhong, Q.; Liu, M.; Shao, Z. A Perovskite Nanorod as Bifunctional Electrocatalyst for Overall Water Splitting. *Adv. Energy Mater.* **2017**, *7*, 1602122. [CrossRef]
72. Tayari, F.; Teixeira, S.S.; Graça, M.P.F.; Essid, M.; Nassar, K.I. Investigating structural, dielectric, and electrical characteristics of sol-gel synthesized perovskite ceramic $Bi_{0.7}Ba_{0.3}(FeTi)_{0.5}O_3$. *J. Sol-Gel Sci. Technol.* **2024**, *112*, 601–613. [CrossRef]
73. Manser, J.S.; Kamat, P.V. Band filling with free charge carriers in organometal halide perovskites. *Nat. Photonics* **2014**, *8*, 737–743. [CrossRef]
74. Konidakis, I.; Stratakis, E.; Koutselas, I. Improved charge carrier dynamics of $CH_3NH_3PbI_3$ perovskite films synthesized by means of laser-assisted crystallization. *ACS Appl. Energy Mater.* **2018**, *1*, 5101–5111. [CrossRef]
75. Dhahri, R.; Albargi, H.B.; Jbeli, A.; Elkenany, E.B.; Althumairi, N.A.; Al-Syadi, A.M.; Sharma, N.; Lal, M.; Nassar, K.I. Enhanced electrical and magnetic properties of barium manganese titanium oxide perovskite ceramics synthesized by solid-state reaction. *J. Mater. Sci. Mater. Electron.* **2025**, *36*, 30. [CrossRef]
76. Luo, S.-Q.; Wang, J.-F.; Yang, B.; Yuan, Y.-B. Recent advances in controlling the crystallization of two-dimensional perovskites for optoelectronic devices. *Front. Phys.* **2019**, *14*, 53401. [CrossRef]
77. Meng, F.; Pang, A.; Dong, X.; Han, C.; Sha, X. H∞ optimal performance design of an unstable plant under bode integral constraint. *Complexity* **2018**, *2018*, 4942906. [CrossRef]
78. Duan, Y.; Wang, D.; Costa, R.D. Recent Progress on Synthesis, Characterization, and Applications of Metal Halide Perovskites@Metal Oxide. *Adv. Funct. Mater.* **2021**, *31*, 2104634. [CrossRef]
79. Guo, X.; Zhou, J.; Wang, T. Methane recovery from anaerobic digestion of urea-pretreated wheat straw. *Renew. Energy* **2017**, *113*, 668–675.
80. Li, J.; Wang, X.; Zhang, Y. Role of soil in improving process performance and methane yield of anaerobic digestion with corn straw as substrate. *Energy* **2018**, *151*, 869–878.

81. Chen, H.; Liu, X.; Xu, F. Simultaneous ammonia stripping and anaerobic digestion for efficient thermophilic conversion of dairy manure at high solids concentration. *Energy* **2017**, *138*, 78–86.
82. Luo, T.; Wang, Q.; Zhang, W. Using the properties of soil to speed up the start-up process, enhance process stability, and improve the methane content and yield of solid-state anaerobic digestion of alkaline-pretreated poplar processing residues. *Biotechnol. Biofuels* **2014**, *7*, 160.
83. Zhou, S.; Li, W.; Liu, G. Anaerobic digestion of poplar processing residues for methane production after alkaline treatment. *Bioresour. Technol.* **2013**, *140*, 226–232.
84. Chen, Y.; Cheng, J.; He, Z. Importance of "weak-base" poplar wastes to process performance and methane yield in solid-state anaerobic digestion. *J. Environ. Manag.* **2017**, *192*, 151–159.
85. Liu, X.; Wang, W.; Zhang, Y. A novel and simple approach to the good process performance of methane recovery from lignocellulosic biomass alone. *Biotechnol. Biofuels* **2016**, *9*, 253.
86. Zheng, Y.; Zhao, J.; Xu, F.; Li, Y. Leading pretreatments for enhancing the degradability of lignocellulosic wastes and the final products. *Bioengineered* **2016**, *7*, 356–365.
87. Dhahri, R.; Tayari, F.; Haouas, A.; Alathlawi, H.J.; Albargi, H.B.; Elkenany, E.B.; Al-Syadi, A.M.; Sharma, N.; Lal, M.; Nassar, K.I. Crystal structural characteristics and optical and electrical properties of Bi-doped $(Ba_{0.8}Sr_{0.2})(Ti_{0.85}Zr_{0.15})O_3$ perovskite ceramics. *J. Mater. Sci. Mater. Electron.* **2025**, *36*, 331. [CrossRef]
88. Chen, J.; Wang, Y.; Zhang, Y.; Lu, Y.; Shu, Q.; Hu, Y. Extrinsic-and-Intrinsic Reward-Based Multi-Agent Reinforcement Learning for Multi-UAV Cooperative Target Encirclement. *IEEE Trans. Intell. Transp. Syst.* **2025**, 1–13. [CrossRef]
89. Chen, J.; Wang, Y.; Zhang, Y.; Lu, Y.; Li, Q.; Shu, Q. Non-Preemptive Scheduling of Periodic Tasks with Data Dependencies in Heterogeneous Multiprocessor Embedded Systems. *ACM Trans. Des. Autom. Electron. Syst.* **2025**, *30*, 1–25. [CrossRef]
90. Jing, H.; Lin, Q.; Liu, M.; Liu, H. Electromechanical braking systems and control technology: A survey and practice. *Proc. Inst. Mech. Eng. Part D J. Automob. Eng.* **2024**. [CrossRef]
91. Li, S.; Li, J.; Tian, G.; Shang, H. Stiffness adjustment for a single-link robot arm driven by series elastic actuator in muscle training. *IEEE Access* **2019**, *7*, 65029–65039. [CrossRef]
92. Li, S.; Li, J.; Tian, G.; Shang, H. Variable stiffness control for SEAs in rehabilitation training. *Adv. Robot.* **2019**, *33*, 424–438. [CrossRef]
93. Li, X.; Cai, H.; Tian, G.; Hu, B. Life Field Theory: An Objective Presentation of the Dynamic Evolution of Human Life. *IEEE Trans. Comput. Soc. Syst.* **2023**, *10*, 1423–1427. [CrossRef]
94. Li, Y.S.; Zhang, J.; Tian, G.H.; Shang, H.C.; Tang, H.B. Kirenol, darutoside and hesperidin contribute to the anti-inflammatory and analgesic activities of Siegesbeckia pubescens makino by inhibiting COX-2 expression and inflammatory cell infiltration. *J. Ethnopharmacol.* **2021**, *268*, 113547. [CrossRef]
95. Wang, X.; Tian, G.; Li, Z.; Zheng, L. The crosstalk between miRNA and mammalian circadian clock. *Curr. Med. Chem.* **2015**, *22*, 1582–1588. [CrossRef]
96. Yuan, X.C.; Song, J.L.; Tian, G.H.; Shi, S.H.; Li, Z.G. Proteome analysis on the mechanism of electroacupuncture in relieving acute spinal cord injury at different time courses in rats. *Zhen Ci Yan Jiu = Acupunct. Res.* **2009**, *34*, 75–82.
97. Li, X.; Dai, Q.; Shi, Z.; Chen, H.; Hu, Y.; Wang, X.; Zhang, X.; Tian, G. Clinical efficacy and safety of electroacupuncture in migraine treatment: A systematic review and network meta-analysis. *Am. J. Chin. Med.* **2019**, *47*, 1755–1780. [CrossRef] [PubMed]
98. Li, X.; Lin, S.; Lin, Y.; Su, Y.; Wang, C.; Huang, L.; Zhao, J.; Tian, G. The analgesic mechanism of Xi Shao Formula research on neuropathic pain based on metabolomics. *J. Tradit. Chin. Med. Sci.* **2023**, *10*, 448–460. [CrossRef]
99. Yang, Z.; Niu, X.; Hou, K.; Liang, W.; Guo, Y. Prediction model on maximum potential pollution range of debris flows generated in tailings dam break. *Electron. J. Geotech. Eng.* **2015**, *20*, 4363–4369.
100. Ye, F.; Zhang, S.; Warby, J.; Wu, J.; Gutierrez-Partida, E.; Lang, F.; Shah, S.; Saglamkaya, E.; Sun, B.; Zu, F.; et al. Overcoming C60-induced interfacial recombination in inverted perovskite solar cells by electron-transporting carborane. *Nat. Commun.* **2022**, *13*, 1–12. [CrossRef] [PubMed]
101. Yang, Z.; Niu, X.; Hou, K.; Liang, W.; Guo, Y. Relationships between water-cement ratio and rheological characteristics of power-law cement grouts. *Electron. J. Geotech. Eng.* **2015**, *20*, 5825–5831.
102. Yang, Z.Q.; Zhu, Y.Y.; Zou, D.S. Formation conditions and risk evaluation of glacial debris flow disasters along International Karakorum Highway (KKH). *Ital. J. Eng. Geol. Environ.* **2011**, 1031–1037. [CrossRef]
103. Yang, Z.; Qian, S.; Hou, K. Time-dependent Behavior Characteristics of Power-law Cement Grouts Applied in Geotechnical Engineering. *Electron. J. Geotech. Eng.* **2016**, *21*, 1017–1024.
104. Yang, Z.; Zhu, Y.; Qian, S.; Hou, K. Types and space distribution characteristics of debris flow disasters along China-Pakistan Highway. *Electron. J. Geotech. Eng* **2016**, *21*, 191–200.
105. Liu, C.; Li, Z.; Zhang, H.; Yu, H.; Yan, J.; Wei, D.; Song, Z.; Cao, J.; Sun, Y. Visualization of the elevated levels of hypochlorous acid in Alzheimer's disease with a ruthenium (II) complex-based luminescence probe. *Anal. Chim. Acta* **2023**, *1279*, 341779. [CrossRef]

106. Achary, P.G.R.; Dehury, S.K.; Choudhary, R.N.P. Structural, electrical and dielectric properties of double perovskites: $BiHoZnZrO_6$ and $BiHoCuTiO_6$. *J. Mater. Sci. Mater. Electron.* **2018**, *29*, 6805–6816. [CrossRef]
107. Thakur, P.; Kishore, K.; Kumar, R.; Pathak, D.; Nassar, K.I.; Lal, M. Structural, morphological, dielectric, and magnetic properties of CoFe2O4 ceramics at different sintering temperatures. *J. Mater. Sci. Mater. Electron.* **2024**, *35*, 1932. [CrossRef]
108. Smith, J.; Brown, K.; Davis, R. Data-driven pressure estimation and optimal sensor selection for noisy turbine flow with blocked clustering strategy. *J. Appl. Phys.* **2024**, *145*, 075901.
109. Zhou, M.; Liu, H.; Wang, X. Bagasse-based porous flower-like MoS_2/carbon composites for efficient microwave absorption. *Nano-Micro Lett.* **2024**, *16*, 105.
110. Gao, Y.; Sun, L.; Chen, R. Effects of pristine and photoaged tire wear particles and their leachable additives on key nitrogen removal processes and nitrous oxide accumulation in estuarine sediments. *J. Hazard. Mater.* **2025**, *487*, 137136.
111. Zhang, T.; Wu, Q.; Yang, S. Investigating the structural behavior of double-skin steel tubes filled with basalt fibre reinforced recycled aggregate concrete under high temperature. *J. Build. Eng.* **2025**, *100*, 111782.
112. Sun, P.; Liu, J. A review of seismic resilience of shield tunnels. *Tunn. Undergr. Space Technol.* **2023**, *137*, 105075.
113. Guo, H.; Liu, X. Limosilactobacillus regulating microbial communities to overcome the hydrolysis bottleneck with efficient one-step co-production of H_2 and CH_4. *Adv. Sci.* **2024**, *11*, 2406119.
114. Xiao, Y.; Wang, Z.; Zhang, F. IndVisSGG: VLM-based scene graph generation for industrial spatial intelligence. *Adv. Eng. Inform.* **2024**, *65*, 103107.
115. Tang, X.; Wu, J. An AUV-enabled dockable platform for long-term dynamic and static monitoring of marine pastures. *IEEE J. Ocean. Eng.* **2024**, *50*, 276–293.
116. Kim, Y.H.; Park, J.; Kim, S.; Kim, J.S.; Xu, H.; Jeong, S.H.; Hu, B.; Lee, T.W. Exploiting the full advantages of colloidal perovskite nanocrystals for large-area efficient light-emitting diodes. *Nat. Nanotechnol.* **2022**, *17*, 590–597. [CrossRef] [PubMed]
117. Guo, B.; Lai, R.; Jiang, S.; Zhou, L.; Ren, Z.; Lian, Y.; Li, P.; Cao, X.; Xing, S.; Wang, Y.; et al. Ultrastable near-infrared perovskite light-emitting diodes. *Nat. Photonics* **2022**, *16*, 637–643. [CrossRef]
118. Brinkmann, K.O.; Becker, T.; Zimmermann, F.; Kreusel, C.; Gahlmann, T.; Theisen, M.; Haeger, T.; Olthof, S.; Tückmantel, C.; Günster, M.; et al. Perovskite–organic tandem solar cells with indium oxide interconnect. *Nature* **2022**, *604*, 280–286. [CrossRef] [PubMed]
119. Wang, S.; Xu, J.; Zhu, Y. Facilitated methanogenesis involved in anaerobic digestion of dairy manure by soil. *J. Clean. Prod.* **2019**, *232*, 117640.
120. Zhang, C.; Su, H.; Baeyens, J. Water-free anaerobic co-digestion of vegetable processing waste with cattle slurry for methane production at high total solid content. *Energy* **2014**, *74*, 258–267.
121. Sharma, R.; Hooda, N.; Hooda, A.; Khasa, S. Improved magnetic properties of Fe substituted La_2CoMnO_6 for spintronic applications. *J. Magn. Magn. Mater.* **2023**, *587*, 171347. [CrossRef]
122. Gao, Y.; Fan, Y.; Zhang, C. Process modeling and optimization of anaerobic co-digestion of peanut hulls and swine manure using response surface methodology. *Energy Fuels* **2019**, *33*, 10168–10175.
123. Li, J.; Wang, Y.; Zhang, H. Damage evolution characteristics of siliceous slate with varying initial water content during freeze-thaw cycles. *Sci. Total Environ.* **2024**, *950*, 175200.
124. Wang, L.; Zhang, Y.; Li, P. (±)-Hypandrone A, a pair of polycyclic polyprenylated acylphloroglucinol enantiomers with a caged 7/6/5/6/6 pentacyclic skeleton from Hypericum androsaemum. *Org. Chem. Front.* **2024**, *11*, 3459–3464.
125. Jiang, Q.; Tong, J.; Scheidt, R.A.; Wang, X.; Louks, A.E.; Xian, Y.; Tirawat, R.; Palmstrom, A.F.; Hautzinger, M.P.; Harvey, S.P.; et al. Compositional texture engineering for highly stable wide-bandgap perovskite solar cells. *Science* **2022**, *378*, 1295–1300. [CrossRef] [PubMed]
126. Rahman, M.A.; Rahman, M.F.; Marasamy, L.; Harun-Or-Rashid, M.; Ghosh, A.; Chaudhry, A.R.; Irfan, A. Impact of A-Cations Modified on the Structural, Electronic, Optical, Mechanical, and Solar Cell Performance of Inorganic Novel A_3NCl_3 (A = Ba, Sr, and Ca) Perovskites. *Energy Fuels* **2024**, *38*, 8199–8217. [CrossRef]
127. Lin, R.; Wang, Y.; Lu, Q.; Tang, B.; Li, J.; Gao, H.; Gao, Y.; Li, H.; Ding, C.; Wen, J.; et al. All-perovskite tandem solar cells with 3D/3D bilayer perovskite heterojunction. *Nature* **2023**, *620*, 994–1000. [CrossRef] [PubMed]
128. Krichene, A.; Boujelben, W.; Rathod, K.N.; Gadani, K.; Chen, C.; Kandasami, A.; Shah, N.A.; Solanki, P.S. Electronic structure and room temperature colossal magnetodielectric effect in $La_{0.4}Dy_{0.1}Ca_{0.5}MnO_3$ manganite. *J. Alloys Compd.* **2024**, *999*, 175022. [CrossRef]
129. You, S.; Eickemeyer, F.T.; Gao, J.; Yum, J.-H.; Zheng, X.; Ren, D.; Xia, M.; Guo, R.; Rong, Y.; Zakeeruddin, S.M.; et al. Bifunctional hole-shuttle molecule for improved interfacial energy level alignment and defect passivation in perovskite solar cells. *Nat. Energy* **2023**, *8*, 515–525. [CrossRef]
130. Zhou, S.; Fu, S.; Wang, C.; Meng, W.; Zhou, J.; Zou, Y.; Lin, Q.; Huang, L.; Zhang, W.; Zeng, G.; et al. Aspartate all-in-one doping strategy enables efficient all-perovskite tandems. *Nature* **2023**, *624*, 69–73. [CrossRef] [PubMed]

131. Algahtani, A.; Ali, S.; Hussain, T.; Ali, A.; Quraishi, A.M.; Tirth, V.; Abdullaeva, B.S.; Kamran, M.; Aslam, M.; Zaman, A. Studies on structural, optical and microwave dielectric properties of double perovskite $Sr_2Fe_{1+x}Nb_{1-x}O_6$, ceramics synthesized by solid state route. *Opt. Mater.* **2024**, *148*, 114822. [CrossRef]
132. Kumar, A.; Kumar, V.; Kumar, A.; Tiwary, D.; Mandal, K.D. Studies on electrical & dielectric properties of rare-earth-based complex perovskite oxide $Bi_5Yb_3O_{12}$ ceramic material synthesized via chemical route. *J. Mater. Sci. Mater. Electron.* **2024**, *35*, 664.
133. Suresh, S.; Vindhya, P.S.; Devika, S.; Kavitha, V.T. Structural, optical and dielectric properties of nanostructured $La_{1-x}Sr_xMnO_3$ perovskites. *Mater. Today Commun.* **2023**, *36*, 106657. [CrossRef]
134. Cao, J.; Liu, Y.; Wang, Z.; Liu, L. Arylgermylation of alkenes by a cooperative photoactivation and hydrogen atom transfer strategy. *Org. Chem. Front.* **2024**, *11*, 7098–7106. [CrossRef]
135. Liu, J.; Wen, Z.; Cui, Y.; Zhao, Y.; Chen, Y. The Spin Glass State and Exchange Bias Effect in $SrFe_{1-x}Cu_xO_{3-\delta}$. *J. Supercond. Nov. Magn.* **2022**, *35*, 3403–3409. [CrossRef]
136. Hu, S.; Otsuka, K.; Murdey, R.; Nakamura, T.; Truong, M.A.; Yamada, T.; Handa, T.; Matsuda, K.; Nakano, K.; Sato, A.; et al. Optimized carrier extraction at interfaces for 23.6% efficient tin–lead perovskite solar cells. *Energy Environ. Sci.* **2022**, *15*, 2096–2107. [CrossRef]
137. Irfan, M.; Murtaza, G.; Muhammad, N.; Tahir, S.; Raza, H.H.; Sabir, B.; Iftikhar, M.; Sharif, S. Experimental and theoretical studies of structural, electronic and magnetic properties of RE_2NiCrO_6 (RE = ce, Pr and Nd) double perovskites. *Phys. E Low-Dimens. Syst. Nanostructures* **2023**, *148*, 115635. [CrossRef]
138. Mohamed, M.; Mohamed, M.; Iben Nassar, K.; Rammeh, N. Effect of the partial substitution of K on the structural, magnetic and magnetocaloric properties of brownmillerite oxides $La_{1-x}K_xSrMn_2O_{5+\delta}$ ($0.1 \leq x \leq 0.3$). *Indian J. Phys.* **2023**, *97*, 1717–1725. [CrossRef]
139. Li, H.; Li, Y.; Zhang, G.; Liu, Y.; Han, Z.; Zhang, H.; Xu, Q.; Zhao, J.; Jin, M.; Song, D.; et al. Directly printing high-resolution, high-performance 3D curved electronics based on locally polarized electric-field-driven vertical jetting. *Addit. Manuf.* **2024**, *96*, 104579. [CrossRef]
140. Thach, P.V.; Sumi, S.; Tanabe, K.; Awano, H. Comparison of Current Induced Domain Wall Motion Driven by Spin Transfer Torque and by Spin Orbit Torque in Ferrimagnetic GdFeCo Wires. *Magnetochemistry* **2024**, *10*, 36. [CrossRef]
141. Sharma, S.; Khan, S. Tailoring the structural, optical, electrical and multiferroic properties of $Sm_{1-x}R_xFeO_3$ (x = 0.0 and 0.5; R = Pr, Nd, and Gd) and their synergistic photocatalytic activity. *Ceram. Int.* **2024**, *50*, 9185–9205. [CrossRef]
142. Yang, F.; Yang, X.; Su, K.; Lin, J.; He, Y.; Lin, Q. Structural and Magnetic Properties of Perovskite Functional Nanomaterials $La_{1-x}R_xFeO_3$ (R = Co, Al, Nd, Sm) Obtained Using Sol-Gel. *Molecules* **2023**, *28*, 5745. [CrossRef] [PubMed]
143. Selmi, R.; Cherif, W.; Khammassi, F.; Sarabando, A.R.; Ferreira, N.M. Structural and magnetic properties of $Ca_{1-x}La_xFe_{0.5}Mn_{0.5}O_{3-\delta}$ ($0.05 \leq x \leq 0.15$) perovskites. *Appl. Phys. A* **2022**, *128*, 847. [CrossRef]
144. Su, W.; Hu, Q.; Yue, B.; Wang, Y.; Liu, J. Electrochemical and magnetic properties of perovskite-type $PrFeO_3$ and $EuFeO_3$ nanofibers. *Ionics* **2023**, *29*, 5525–5534. [CrossRef]
145. Zhu, A.; Li, D.; Zhu, T.; Zhu, X. Tailored $SrFeO_{3-\delta}$ for chemical looping dry reforming of methane. *Green Chem. Eng.* **2024**, *6*, 102–115. [CrossRef]
146. Shen, X.; Sun, Y.; Wu, Y.; Wang, J.; Jiang, E.; Xu, X.; Su, J.; Jia, Z. The coupling of CH_4 partial oxidation and CO_2 splitting for syngas production via double perovskite-type oxides $LaFexCo_{1-x}O_3$. *Fuel* **2020**, *268*, 117381. [CrossRef]
147. Zeeshan, M.; Yaqub, M.A.; Khan, S.; Ramay, S.M.; Atiq, S. Fast polarization switching capability mediated by oxygen deficiency in Co-substituted $SrFeO_{3-\delta}$ perovskites. *J. Phys. Chem. Solids* **2023**, *181*, 111557. [CrossRef]
148. Rosa, G.B.; da Silveira Lacerda, L.H.; de Lazaro, S.R. Influence of the oxygen vacancy and Ag-doping on the magnetic and electronic properties of the $SrFeO_3$ material. *J. Magn. Magn. Mater.* **2024**, *591*, 171715. [CrossRef]
149. Amin, S. Synthesis and Multiferroic Properties of $BaFe_2O_4/BaFeO_3$ Ceramics. Ph.D. Thesis, Quaid I Azam University, Islamabad, Pakistan, 2022.
150. El Boukili, A.; Mounkachi, O.; Hamedoun, M.; Lachkar, P.; Hlil, E.K.; Benyoussef, A.; Balli, M.; Ez-Zahraouy, H. A study of structural, magnetic and magnetocaloric properties of $(1-x)La_{0.6}Ca_{0.4}MnO_3/xMn_2O_3$ composite materials. *J. Alloys Compd.* **2021**, *859*, 158392. [CrossRef]
151. Ho, T.; Phan, T.; Putri, W.; Yu, S.; Thang, P. Large Magnetocaloric Effect in Cu-Doped $La_{0.7}Ca_{0.3}MnO_3$ Compounds. *Mater. Trans.* **2023**, *64*, 1991–1999. [CrossRef]
152. Abdel-Latif, I.A.; Hegab, A.F.; Azab, A.A.; Roumaih, K.M. Study on crystal structure, electrical and magnetic properties of $LaFe_xCo_{1-x}O_3$ prepared by sol-gel method. *J. Solid-State Chem.* **2023**, *325*, 124144. [CrossRef]
153. Dzara, M.J.; Campello, A.C.; Breidenbach, A.T.; Strange, N.A.; Park, J.E.; Ambrosini, A.; Coker, E.N.; Ginley, D.S.; Lee, Y.S.; Bell, R.T.; et al. Influence of the Rare Earth Cation on the Magnetic Properties of Layered $12R\text{-}Ba_4M^{4+}Mn_3O_{12}$ (M = Ce, Pr) Perovskites. *Chem. Mater.* **2024**, *36*, 2810–2818. [CrossRef]

154. Nadig, P.R.; Murari, M.S.; Daivajna, M.D. Influence of heat sintering on the physical properties of bulk $La_{0.67}Ca_{0.33}MnO_3$ perovskite manganite: Role of oxygen in tuning the magnetocaloric response. *Phys. Chem. Chem. Phys.* **2024**, *26*, 5237–5252. [CrossRef]
155. Liao, H.; Ma, H.; Meng, H.; Kang, N.; Wang, L. Ropinirole suppresses LPS-induced periodontal inflammation by inhibiting the NAT10 in an ac4C-dependent manner. *BMC Oral Health* **2024**, *24*, 510. [CrossRef] [PubMed]
156. Wang, L.; Hu, H.; Ko, C.-C. Osteoclast-Driven Polydopamine-to-Dopamine Release: An Upgrade Patch for Polydopamine-Functionalized Tissue Engineering Scaffolds. *J. Funct. Biomater.* **2024**, *15*, 211. [CrossRef] [PubMed]

MDPI AG
Grosspeteranlage 5
4052 Basel
Switzerland
Tel.: +41 61 683 77 34

Inorganics Editorial Office
E-mail: inorganics@mdpi.com
www.mdpi.com/journal/inorganics

www.ingramcontent.com/pod-product-compliance
Lightning Source LLC
Chambersburg PA
CBHW041403210926
51850CB00013B/3165
* 9 7 8 3 7 2 5 8 4 2 9 8 8 *